FEROZSONS (PRIVATE) LTD.
LAHORE

WILD FLOWERS OF PAKISTAN

CONTENTS

page

LIST OF PLATES

PLATE *page*

LIST OF FIGURES

LIST OF ABBREVIATIONS AND SYMBOLS USED

auct.	of authors
c.	about or approximate.
cm	centimetre
f.	forma or a form. But when used after a person's name, filius-meaning the son of. As Linn.f.– the son of C. Linnaeus.
fl.	flowers
m	metre
N.W.F.P.	North West Frontier Province
sp.	species
spp.	more than 1 species
subsp.	subspecies
var.	variety

FOREWORD

Having been closely associated for more than twenty years with the preparation and compilation of a detailed flora of Pakistan, I greatly welcome the production of this small volume which is the first attempt to introduce to the lay reader and non-specialist, an account of some of the varied and fascinating flowering plants of our country.

The scientific flora has, so far, covered 193 different plant families, describing those members from each family which occur in Pakistan, and is published in the same number of separate fascicles. I have often been asked by both colleagues and visitors to the University how they can identify the commoner wild flower that they encounter without having the botanical expertise and the necessary resources to be able to consult this published flora series. This introductory book is intended to fill such a need, and with its copious selection of colour photographs will, I hope, stimulate a wider appreciation of our flora and help to foster an interest in the plant life of each province, amongst those who enjoy the countryside.

The rapid industrialization of our country, coupled with the growing needs to feed and clothe our increasing human population, has placed a great strain upon our natural environment. Many of our smaller rivers and streams have become polluted by industrial effluents, more and more of our so-called waste places have been cleared of jungle and brought under cultivation, and the need to harvest more and more fuel wood has destroyed centuries of plant growth in semi-desert areas. As a result, many of our native plants are disappearing and a few are even threatened with extinction unless we take measures to preserve the natural diversity of our plant wealth. To this end we must try to educate our populace about our native flora and try to create a greater appreciation of the aesthetic as well as the economic value of many of our flowering plants. I am sure that this book will make a major contribution to this long-felt need.

(S. I. Ali)

Vice Chancellor
University of Karachi
1 March 1993.

PREFACE

Whilst there has been a great effort to collect, catalogue and describe the vast array of flowering plants which make up Pakistan's vascular plant flora, most of this information is published in highly condensed and technical format, which is difficult for the non-specialist either to obtain or to understand.

It is an unfortunate fact that in the 47 years since the creation of Pakistan as an independent country, no one has produced any identification guide in popular form, aimed at the average citizen who would like to know more about the wild flowering plants which are found in different parts of the country.

This has been the objective of the authors in compiling this book, which is no more than a modest start to fill this important gap. The first step has already been taken in the recent publication of an illustrated guide to the flowers of the capital area (E. & Y. Nasir & R. Akhter, 1987, *Wild Flowers of Rawalpindi and Islamabad).* Pakistan has a great diversity of both climate and terrain, from scorching deserts to relatively moist temperate hill forests, resulting in an estimated 4,940 native plant species presently known. It would be an impossible task to produce a book such as this, aimed at a popular readership and at the same time to attempt a comprehensive coverage of all the plant families represented. Instead, a carefully selected list of just over 650 plants have been illustrated with photographs taken, where they grow in their natural state, accompanied by brief descriptive accounts indicating their relationship within the plant kingdom, their occurrence and most importantly a key by which the non-specialist should be able to identify them. The complete list of species included in this book, is given by scientific, English and local names, in an index at the end.

The main criteria upon which these plants have been selected are as follows:

1. Preference has been given to plants of widespread occurrence and which are common.
2. Plants which are conspicuous because of their striking foliage or attractive flowering parts, including those which are less common.
3. An attempt has been made to give special coverage to plants which are typical, or representational of the four main provinces of Pakistan including all the northern areas.

Because of these rather arbitrary criteria, a few plants have been included which are not truly wild, but are introduced 'exotics' that have become naturalized and are now conspicuous by their ubiquity. For example, the turk's turban *(Holmskiolda sanguinea),* paper mulberry *(Broussonetia papyrifera),* prickly Pear *(Opuntia monacantha)* and the Indian laburnum *(Cassia fistula).* In the same way, plant families characterized by very small or inconspicuous flowers are under represented such as grass family *(Poaceae),* spinach family *(Amaranthaceae)* and goosefoot family *(Chenopodiaceae).*

There is an introductory section describing the main floristic or phytogeographical regions which are found within Pakistan's borders. This will give the reader some insight into their associated climatic and edaphic characteristics which influence the most prevalent type of plants, their growth and distribution within each floristic zone.

The main part of the book then comprises a brief account of each species to accompany the illustration. By way of information, the total number of species found in Pakistan is mentioned just after each generic name, followed by local name and then the common English name where available (e.g. **Ranunculus** 26 spp. *Buttercup*). A simplified botanical key is provided which should help the reader to identify the plant when actually encountered out in the field. This key identifies the various plant genera and the species within a genus, using a coupled system of contrasting characters which are numbered and placed in adjacent lines of the key, thus:-

EUPHORBIACEAE *Castor Oil Family*

1. + Inflorescence a cyathium. Stamen 1. **1. Euphorbia**

– Inflorescence various, but not as above.
Stamens more than 1. 2

2. + Trees 2.6 m tall. **4. Mallotus**

– Generally annual or perennial herbs or shrubs
up to 1.2 m tall (trees in Phyllanthus also). 3

The reader's attention is also drawn to the Glossary at the end of this book in which various technical botanical terms are explained, accompanied by illustrative diagrams where possible. Plants are generally recognized and classified by their flowering and fruiting parts which can only be described accurately yet briefly by the use of these technical terms. It is hoped that the reader will consult this Glossary and thereby be able to use the keys to the genera and species. Some abbreviations and symbols used in the text have also been explained on page xix.

Yasin J. Nasir
Rubina A. Rafiq
T. J. Roberts

ACKNOWLEDGEMENTS

We are extremely grateful to all those who have helped in the production of this book, not least those who generously gave photographs of flowering plants, which though not always selected for reproduction within these pages, nevertheless added to our scanty knowledge of their distribution and frequency of occurrence and who allowed all their photographs to be deposited with the National Herbarium reference collection of colour transparencies.

We specially wish to thank our colleagues Dr. Muhammad Qaiser, Professor of Botany at Karachi University and Dr. Abdul Ghaffar, Editor of the Pakistan Journal of Botany, for kindly reviewing various preliminary drafts of the manuscript and for their helpful criticisms.

In the actual production process we wish to thank Mr. A. Mirza Jamil, Chairman, Elite Publishers Ltd., Karachi, for his care and patience which enabled such a high standard of colour reproduction from slides of variable quality. We wish also to thank Mr. Nazir A. Rakha for help in typing the manuscript and the National Printing and Packaging Company of Rawalpindi for producing the first three preliminary printed texts.

The photographs are acknowledged briefly in the list of plates and since this forms the major contribution to the book, we would like again to thank these people who in most cases gave us more of their slides than we were able to use. The senior author Mr. Yasin J. Nasir contributed 272 of the slides illustrated in this book, whilst Mr. S. A. Sultan contributed 146 and the junior author Mrs. Rubina Akhter Rafiq contributed 108 slides. The Editor, Tom Roberts, included 56 of his pictures and Herr Fritz Berger, as a result of the many expeditions into Swat and northern areas, 51 slides. Mark Mallalieu of the British Overseas Development Agency contributed 6. Chris Chadwell, an expert on the flora of the Indian Himalaya, contributed 4 slides, whilst Major General M. Shoaib Quraishi 3 and Khan Muhammed Khan of the Punjab Wildlife Department contributed 5 slides. We would also like to thank the parents of the late Timothy Hurrell, who was a keen photographer as well as mountaineer, for presenting many of his alpine flower pictures to the National Herbarium, one of which is included in this book.

ACKNOWLEDGEMENTS

[illegible]

INTRODUCTION

Few countries of comparable size have the diversity of habitat and climate that is found in Pakistan. The country lies between 23° 45' to 36° 50' N and 60° 55' to 75° 30' E with a total area of 803943 sq km (excluding Jammu and Kashmir whose sovereignty is disputed with India). More than half of the land in the north and west is mountainous or highland and includes some of the highest peaks in the world (K2 8611 m, Nanga Parbat 8126 m, Broad Peak 8047 m, Rakaposhi 7788 m) and the largest glaciers outside the Arctic such as the Siachen and Baltoro, each about 48 km in length. The land gradually tapers to sea level in the south.

Practically the whole area lies in the watershed of the river Indus and supports one of the greatest networks of irrigation canals in the world. The total irrigated area of farmland in Pakistan is about 14.7 million hectares, of which 12.9 million hectares is canal diversion; the remainder is ground water extraction (Biswas, 1987). This undoubtedly has had a great impact not only on the well-being of the human population but also on modifying the landscape. The monsoons (July to mid-September) are also a major factor in influencing the growth and distribution of plants, especially in the context of so much of the country being semi-arid.

Plants are at once distinguished, by the layman, from animals by their spatial immobility during most of their life cycle. They achieve the new space necessary to feed, grow and reproduce, by dispersal of their fruits and seeds and to a lesser extent by vegetative means such as rhizomes, bulbs, tubers etc. The many fascinating strategies adapted by plants for seed dispersal lie outside the scope of this book, but it is relevant to consider that when compared with animals, plants are more influenced by and subject to vagaries of their immediate environment. Factors of climate, elevation, underlying soil geology are paramount in determining their distribution. In the desert, they cannot burrow underground to escape the diurnal heat as do animals, nor can they migrate in winter to warmer southern latitudes as do a host of palearctic birds which visit Pakistan in the plains each winter.

It is therefore, all the more fascinating and surprising to discover how widely adaptable some plants are to variations in temperature and to extremes of moisture in the atmosphere and the soil.

Animals and flowering plants are not erratically distributed on this planet. Their occurrence shows an understandable pattern within a particular region. It becomes apparent to the botanist that there is a certain cohesion and similarity of types or plant forms. This is because the present day distribution of plants is partly a contribution of geological and evolutionary factors. It becomes apparent that certain plant families are represented by a great diversity of species within a particular region. For example, in the dry steppic conditions prevailing over Balochistan, there are comparatively more *Compositae* (Sunflower family) and *Labiatae* (Mint family) as compared to other plants families, whilst in the moister alpine and forest zones of the North West Frontier Province, plant families like the *Ranunculaceae* (Buttercup family), *Saxifragaceae* (Saxifrage family), and *Gentianaceae* (Gentian family) predominate.

The Phytogeographic Regions

Pakistan has a rich and varied flora. There are about 4940 native species of flowering plants (about 5738 species if cultivated or nautralized taxa are included) which are found in a variety of habitats from seashore and deserts to high mountainous areas to the north. These include about 372 species which are endemic, mostly found in the north and western mountainous regions of Pakistan.

Whilst there are only two major zoogeographical regions represented in Pakistan, the palearctic, mainly to the west of the Indus and to the north and the oriental region in the east and south, this division is not adequate in explaining the distribution of flowering plants, as Pakistan includes within its boundaries four recognized phytogeographical regions which help to explain the richness and diversity of its flora.

Table 1
Phytogeographical Regions in Pakistan
(Adapted after Ali & Qaiser, 1986)

Regions	*Major areas in Pakistan*
1. **Saharo-Sindian**	Sindh, central and southern Punjab, southern Balochistan and plains of North West Frontier Province (N.W.F.P.)
2. **Irano-Turanian**	
(a) Western subregion	Waziristan and North Balochistan
(b) Eastern subregion or central Asiatic	Upper portions of Gilgit (Hunza, Shimshal etc.) and Chitral (partly) Wakhan and adjoining area south-east
3. **Sino-Japanese**	Kashmir, N.W.F.P. (Koh Safed, parts of Hazara, Swat, Dir, Chitral (partly), Astor, Naltar, Bagrot Valleys
4. **Indian**	Sub-Himalayan tracts, east and west of river Jhelum

It must, however, be emphasized that amongst plant geographers there is as yet no unanimous agreement on the limits and subdivisions of the main phytogeographical regions; as such the present text is only a general guide to enable the reader to appreciate that there are these areas each with a different assemblage of plants.

Since growth and distribution of plants is influenced by various factors (i.e. climatic, topographic and biotic), there is no clear-cut boundary between each region and intermingling of floras occur. Islamabad and Rawalpindi floristically are largely Saharo-Sindian, but having adequate rainfall (mean annual 900-1000 mm) a Sino-Japanese flora is also found with a Himalayan one.

1. SAHARO-SINDIAN

Within Pakistan this is the largest area which, according to Boissier (1867) and Eig (1931), stretches from the Atlantic coast of north Africa through Sahara, Sinai Peninsula and includes major parts of the Arabian Peninsula, Palestine, parts of Syria, south Iraq and south Iran.

In Pakistan this region comprises mainly the flat alluvial plain of the Indus and its tributaries covering the major portion of Sindh and Punjab, also foothills of Balochistan and N.W.F.P. and coastal regions of Pakistan. The hot dry deserts of Cholistan, Thar and Thal are found in Sindh and Punjab. There is also the desert region of Chagai in Balochistan along the western border with Afghanistan, where due to inland drainage, dry salt lakes occur. Along Sindh-Balochistan border lies a hilly belt—the Kirthar range. Low detached dry calcareous hills are found near Karachi which extend from Cape Monze to Manghopir.

Climatically, the area is characterized by hot dry summers and mild winters. The mean daily maximum temperature in summer ranges from 41° to 46°C, while temperature of 50°C or above are not uncommon in Sindh and lower Punjab. The highest recorded temperature in June at Jacobabad is 53°C, Sibi 52°C, Multan 51°C. Frost can occur in hilly areas in winter. The mean annual rainfall is low (Karachi 205 mm; Jacobabad 140 mm; Peshawar 347 mm). Parts of Punjab such as Islamabad and Rawalpindi enjoy a more equable climate due to better rainfall.

The most distinctive plant community of coastal Sindh and Balochistan is the mangrove vegetation. These comprise stunted trees and shrubs of *Avicennia, Rhizophora, Aegiceras* and *Ceriops,* the first named being predominant.

The sand dunes and low hills of the desert have an ephemeral flora or include plants that can stand the heat and drought. These include trees and shrubs like *Acacia nilotica, Prosopis cineraria, Tamarix aphylla, Lycium shawii, Salvadora oleoides, Zizyphus sp., Calligonum polygonoides and Ledptadenia pyrotechnica.*

On calcareous hills are found herbs and shrubs like *Inula grantioides, Vernonia cinerascens, Blepharis scindica, Commiphora wightii, Grewia tenax, Salvadora oleoides,* and *Euphorbia caducifolia.*

The soils of the region exhibit considerable variety. The arid tracts like Thar and Thal tend to be poor and sandy and have a good deal of natural salinity. In the Indus riverine tracts, where periodic flooding occurs due to monsoon, more fertile alluvium spreads and produces two to three crops a year with perennial irrigation system. With the spread of irrigation canals and consequent rise in water table and capillary ascent of salts to the surface, the problem of man-induced water logging and salinity is increasing in the upper Sindh and Punjab plains. The loess soils of northern Punjab (Potohar and Salt Range) are characterized by susceptibility to erosion due to poor structure and low fertility.

In the distant past, much of the plains area was covered with sub-tropical scrub vegetation. Since the land has been modified for agriculture and urbanization, the natural vegetation has disappeared. Occasionally, in natural reserves (Rakhs) or graveyards, one can get a glimpse of the former vegetation. These include trees such as *Prosopis cineraria, Salvadora oleoides* and *Capparis decidua.* Where conditions are moist trees like *Acacia modesta* and *A. nilotica* are found.

Floristically the area is poor, comprising only 9.1% of the total flora (Ali & Qaiser, 1986).

2. IRANO-TURANIAN

According to Zohary (1973), this large florsistic region is divided into 2 subregions, the western and eastern Irano-Turanian.

a. The Western Irano-Turanian

The subregion includes parts of south and east Palestine, stretches of the Syrian desert, most of Iran and Afghanistan.

In Pakistan, this subregion lies north-west and approximately between 29º to 33º north, and includes north Balochistan and Waziristan.

The whole region is dry and rugged. The central Brahui range forms the main spur of several ranges of mountains which average 3000 m, the highest being Mount Zarghun (about 3575 m). The Sulaiman ranges in Waziristan are a continuation of the central Brahui range.

Climatically the area is harsh like Afghanistan and neighbouring Iran, characterized by extreme ranges in temperature both diurnal and annual. At elevations above 1500 m, winters are very cold and light snowfall is a normal feature; temperature can dip to -15ºC or lower. In summer the temperature can soar to 40ºC. There is no permanent snowline or glaciers, as the mountains are not high enough. Monsoons do not affect the area. The major precipitation as in adjoining Afghanistan falls during the months of February-March. The average rainfall varies between 200–250 mm per year. The mountains are mostly of limestone, so the drainage is mostly underground, but water springs are occasional.

Floristically, the area is distinct and interesting. Juniper tracts, perhaps the most extensive in the whole world, are found in all the high ranges from Harboi up to Zhob between 2000-3000 m and comprise open forests of juniper *(Juniperus excelsa)*. Other trees are wild pistachio *(Pistacia cabulica)*, ash *(Fraxinus xanthoxyloides)* and hackberry *(Celtis caucasica)*. At higher elevations in the juniper forest zone, spiny cushion shaped plants of *Acantholimon lycopodioides, Onobrychis cornuta, Gypsophila* and *Astragalus* spp. cover the ground.

During late March to April, the lower hills (1700 m and below) and cultivated areas are covered with many pretty bulbous plants like wild onion, tulips, gageas, wild poppies, muscari and irises. *Gentianodes olivieri* is the only wild gentian in the area. On hillsides, shrubs such as *Sophora griffithii, Amygdalus brahuica, Stocksia brahuica,* with red bladdery inflated fruits and the tall biennial *Ferula oopoda* are conspicuous. Along stream-beds are found several species of tamarisk. *Dionysia lacei,* a shrubby rare cliff plant with yellow flowers is endemic to Balochistan.

b. The Eastern Irano-Turanian or Central Asiatic

This subregion includes mountains of Central Asia, Tibet, Mongolia and Gobi desert. In Pakistan, this area lies approximately between 35º and 36º 5' north. Some authors such as Leonard (1989) consider the central Asiatic to be a distinct region.

Unlike north Balochistan and Waziristan (western Irano-Turanian), in this region the mountains are much higher. The main range is the Karakoram with 19 peaks over 7600 m and four, including mount K2, over 8000 m. There are extensive glaciers, which in size are next to those found in Arctic region. Except in valleys, where there is some cultivation and the soil can be irrigated, the whole area is a barren desert with sand dunes and rugged mountains. It is drained by the river Indus and its tributaries, the Shyok, Shigar and Suru rivers.

Climatically the area is harsh and cold with temperatures in winter dipping to as low as -40°C. In summer the heat is intense during the day and at night the temperature dips to zero, so that rocks crack and crumble as water freezes and thaws. The disintegration of rocks in this region is a dramatic on-going process.

The mean rainfall varies from 125 to 500 mm, and sub-humid condition prevail where there is relatively more moisture. Unlike adjoining Ladak there are no high dry plains in Gilgit.

There are two main elements in the flora, namely the alpine and the desert.

Alpine plants are found at elevations above 3100 m, along melting snow and streams. One area of much interest is the Deosai plateau in Baltistan, which lies about 38 km west of Skardu and averages 4000 m in altitude. In July and August, the whole area is covered with many alpines, some Himalayan, other Euro-Siberian or central Asiatic such as *Oxyria digyna, Thymus linearis, Draba nemorosa, Geranium collinum, Primula macrophylla, P. rosea, P. warschenewskiana, Euphrasia himalaica, Rhodiola himalayenses, R. wallichiana, Pedicularis oederi, P. pyramidata, P. bicornuta, Persicaria affine, Aconitum rotundifolium, Biebersteinia odora, Potentilla salesoviana, Anemone pulsatilla, Hippuris vulgaris, Androsace septentrionalis, A. mucronifolia, A. akbaitalensis* etc.

The desertic flora includes plants that grow on cliffs, sand and along streams or springs. In sandy and rocky places there are Saharo-Sindian elements like *Capparis spinosa, Peganum harmala, Aristida adscensionis, Stipagrostis plumosa* etc. There are also others such as *Ephedra regeliana, Perowskia abrotanoides, Chesneya cuneata, Trachomitum venetum,* which are typically central Asiatic or eastern Irano-Turanian. Plants occurring near streams and springs are shrubs like *Rosa webbiana, Tamaricaria elegans, Hippophae rhamnoides* and herbs such as *Cirsium wallichii, Epilobium tibeticum, Inula acuminata* and *Artemisia spp*.

The plants that are naturalized and found as escapes in and around ruderal areas and cultivated fields include many weeds like *Silene conoidea, Lycopsis arvensis, Chenopodium album, C. botrys, Hyoscyamus niger, H. pusillus, Capsella bursa-pastoris, Datura stramonium, Daucus carota, Filago germanica* and grasses like *Bromus japonicus, Avena fatua, Setaria viridis* and *Eragrostis poaoides*.

A sizeable portion of the flora in and around villages is introduced and cultivated. These include the oriental plane *(Platanus orientalis)*, Russian olive *(Elaeagnus angustifolia*,) walnut *(Juglans regia)*, plum *(Prunus armeniaca)*, apple *(Malus pumila)*, mulberry *(Morus alba)* and several species of poplars and willows.

Forests are few. The hardy juniper, *Juniperus excelsa,* is found between 2700–3400 m. In Gilgit (in Bagrot, upper Chaprot and Naltar valleys), there are some mixed forests of spruce *(Picea smithiana)*, blue pine *(Pinus wallichiana)*, fir *(Abies pindrow)* and the birch *(Betula utilis)*. The mountain ash *(Sorbus tianshanica)* and several species of willows *(Salix excelsa, S. illiensis and S. capusii)* are found. The deodar *(Cedrus deodara)* is found only in the Astor valley in Gilgit.

Floristically, the Irano-Turanian area comprises about 45.6% of the total flora (Qaiser & Ali, 1986), including many endemics. Due to overgrazing and deforestation, the habitat of plants is being rapidly altered and degraded. As elsewhere, non-palatable plant species have tended to spread and flourish at the expense of the palatable ones.

3. SINO-JAPANESE REGION

This region is a narrow belt extending from Japan, Korea, south west China westward throughout upper Assam, Taiwan, Bhutan, west Nepal and Himachal Pradesh. In Pakistan, this region is approximately between 32° and 35° 5' north and includes Kashmir and a major portion of the North West Frontier Province. An extension of this area is the Safed Koh (Kurram, N.W.F.P.) and Nuristan (east Afghanistan).

Physically the land is mountainous and comprises the outer ranges of the Himalaya, the Safed Koh mountains and Nuristan in east Afghanistan. The famous hill resorts of Nathia Gali, Murree and Kaghan valley fall in this region. The altitude varies from 600 m in the foothill Himalaya to mountains exceeding 4800 m.

Climatically the area benefits from the summer monsoon rain except in places like upper Kaghan and Swat Kohistan valley (N.W.F.P.) where the monsoon rain is blocked by the high mountain ranges and the area comes under the rain shadow effect.

Floristically the area is very rich and comprises evergreen coniferous forests of blue pine, cedar, spruce and yew. Deciduous trees include oaks, horse chestnut, maples, poplars, Prunus sp. etc. and subtropical forests of *Pinus roxburghii*.

There are many Sino-Japanese elements found in this Himalayan zone. Examples are trees such as yew *(Taxus wallichiana)*, dogwood *(Cornus macrophylla)*, maple *(Acer japonicum)*, amlok *(Diospyros lotus)*, *Viburnum cylindricum*, *Leycesteria formosa*, *Euonymus japonicus*, smoke tree *(Rhus succedanea)*, *R. japonica*. Shrubs include juniper *(Juniperus squamata)*, *Sageretia theezans*, sanatha *(Dodonaea viscosa)*, *Solanum verbascifolium*, *Lyonia ovalifolia*, *Daphne cannabina* and *Lonicera quinquelocularis*. Perennial herbs include *Geranium nepalense*, *Boenninghausenia albiflora*, *Oxalis acetosella*, *Mazus pumilus*, *Lysimachia japonica*, and *Androsace umbellata*.

Due to population pressure, deforestation and grazing, this forested area is fast dwindling. It is estimated that our forest areas are disappearing at the rate of one percent each year (*Environment Profile of Pakistan*, EUAD, GOP, 1987). Some authors regard the Pakistan Himalaya as an extension of the Sino-Japanese region. According to Ali & Qaiser (1986), about 10.6% of the total flora of Pakistan is represented here.

4. INDIAN

This region includes the Indian Peninsula, Sri Lanka, the Gangetic Plain and foothill Himalaya.

In Pakistan, this region lies mainly in the north from 29°5' -32° north and about 73° east, comprising the north and north east Punjab and includes the foothills of sub-Himalayan tracts below 1500 m on either side of Jhelum river and adjoining Pabbi hills (near Kharian), Bhimber (near Gujrat) and Sialkot. Similar floristic areas extend along river systems in parts of the North West Frontier Province.

An extension of this area lies in south west Sindh, between 24°2' to 28° north, known as Nagar Parkar, which includes a hilly belt of the Karunjhar hills.

Physically the land to the north consists of low hills and ravines, where soil erosion due to grazing and deforestation is active. In the adjoining plains, the flat ground is utilized for cultivation. The area benefits largely from the summer monsoon rain. Westerly disturbances in winter contribute to the rainfall as well.

The mean annual rainfall varies from 700-900 mm. Climatically there are extremes in temperature. The maximum temperature in June is 40°-45°C and minimum for January is sub-zero.

The vegetation is of tropical or subtropical type and there is much variety in plant life. During monsoon (July-September), there is a lush green dominance of trees, shrubs and climbers, many of which are of eastern origin.

Nagar Parkar, situated near the Rann of Kutch, is largely Saharo-Sindian but because of higher rainfall than in other parts of Sindh has a few Indian elements like *Delonix alata, Moringa concanensis, Tamarix indica and Pluchea tomentosa*.

According to Ali & Qaiser (1986), the Indian element is represented by about 4.5% of the total flora of Pakistan.

KEY TO PLANT GENERA AND SPECIES FOLLOWED BY BRIEF DESCRIPTIONS OF PLANTS ILLUSTRATED

RANUNCULACEAE *Buttercup Family*

Herbs or shrubs. Leaves radical, alternate or opposite, divided or entire. Flowers regular or irregular of free sepals and petals. The sepals sometimes petaloid, spurred or hood-like. Petals spurred or not, often reduced, nectariferous. Stamens many, free. Fruit a collection of achenes or follicles.

1.	+	Flowers regular.	4
	–	Flowers irregular.	2
2.	+	Flowers spurless.	**1. Aconitum**
	–	Flowers spurred.	3
3.	+	Annuals with solitary fruit.	**2. Consolida**
	–	Perennials with 2 or more fruit.	**3. Delphinium**
4.	+	Sepals or/and petals 5.	5
	–	Sepals or/and petals 4 or many.	9
5.	+	Aquatic herbs with submerged thread-like leaf segments.	**9. Batrachium**
	–	Terrestrial herbs but sometimes found on wet ground or moist places.	6
6.	+	Flowers yellow.	**8. Ranunculus**
	–	Flowers purplish-blue or white, sometimes lemon.	7
7.	+	Petals absent. Sepals petal-like.	**7. Anemone**
	–	Sepals and petals present.	8
8.	+	Sepals and petals with a distinct spur.	**4. Aquilegia**

– Sepals and petals without spur. **5. Paraquilegia**

9. + Sepals and petals many. **11. Trollius**

– Sepals petaliod, 4-10 in number. Petals absent. 10

10. + Low growing plants of marshy or wet habitats with broad rounded or oblong leaves. Sepals 5-10 in number. **10. Caltha**

– Climbers or stragglers. Sepals 4. **6. Clematis**

1. Aconitum 7 spp. *Monks Hood*

Generally tall perennial herbs with flowers dispersed in racemes. Sepals 5, petal-like, the upper one hood-like. Petals 2 or 4, enclosed within the calyx. Fruit of several follicles.

+ Hood c. 2 times as high as broad. 1. **A. laeve**

– Hood about as high as broad. 2. **A. chasmanthum**
3. **A. violaceum**

1. **A. laeve** Royle Pl. 1

Perennial 1-1.5 m tall. Leaves round to kidney-shaped, 20-30 cm broad, deeply divided into 5-9 lobes; lobes coarsely toothed. Flowers pale purple or white, borne on a long spike-like inflorescence. Hood prominent. Follicles 3.

Distribution: Pakistan eastward to west Nepal.

Common in forests from 2700-4000 m in Chitral, Gilgit, Kaghan valley and Kashmir. July-August.

2. **A. chasmanthum** Stapf ex Holmes Pl. 1

Robust perennial up to 90 cm tall. Leaves rounded, 6-9 cm broad, leaf segments broader. Inflorescence up to 30 cm long. Flowers blue or blue-purple. Follicles generally 5.

Distribution: Pakistan, Kashmir.

Fairly common and gregarious in alpine meadows from 3000-4300 m in Chitral, Kaghan, Gilgit and Kashmir. August.

3. **A. violaceum** Jacq. ex Stapf Pl. 1

Similar to *A. chasmanthum* but slender (up to 30 cm tall), with violet flowers and deeply dissected leaves.

Distribution: Pakistan, Kashmir to central Nepal.

Gregarious in humid alpine meadows from 3000-4000 m in Deosai plateau and adjoining areas. July-September.

2. Consolida 7-8 spp.

A genus closely related to *Delphinium,* but differing in its annual habit and single follicles.

4. **C. stocksiana** (Boiss.) Nevski Pl. 1

Branched perennial 20-60 cm tall. Basal leaves long petioled, divided into narrow segments. Flowers creamy-white, in racemes. Sepals 6-10 mm long, with purplish-red median streaks; spur up to 20 mm or longer.

Distribution: Turkmenistan, Pamir-Alai, Iran, Afghanistan, Pakistan.

Found in dry areas in fields and waste places in Balochistan up to 1700 m. April-May.

3. Delphinium 16 spp. *Larkspur, Delphinium*

Erect herbs with deeply lobed leaves and irregular racemed flowers. Sepals 5, petal-like, the upper one forming a spur. Petals 4.

1.	+	Roots tuberous. Flowers generally small, and with short pedicels.	2
	–	Roots cylindrical. Flowers usually larger, and with long pedicels.	4
2.	+	Pedicels with more or less erect hairs. Spur of sepal hooked.	5. **D. uncinatum**
	–	Pedicel with appressed hairs. Spur of sepal not hooked.	3
3.	+	Spur 6-10 mm long. Inflorescence up to 12 - flowered.	6. **D. kohatense**
	–	Spur about 13.5 mm long. Inflorescence many-flowered.	7. **D. roylei**
4.	+	Plants stout, 60-100 cm or more tall. Flowers not saccate. Upper sepal 15 x 7mm, spur 12-15 mm.	8. **D. pyramidale**

– Plants 15-30 cm tall. Flowers more or less saccate.
Upper sepal 22 x 17 mm, spur 10 mm. 9. **D. nordhagenii**

5. **D. uncinatum** Hook. f. & Thoms. Pl. 1

Plants up to 90 cm tall, pubescent or glabrous. Radical leaves 15-50 (-55) cm broad, deeply 3-5 – lobed. Inflorescence 10-30 cm long, with spreading hairs. Flowers pink-mauve. Pedicel with more or less erect hairs. Spur of upper sepal hooked. Follicles 3, glabrous or hairy.

Distribution: Afghanistan, Pakistan, Kashmir.

Variable as to hairiness of plant parts. Glabrous to subglabrous forms are known. Found from 500-2300 m in Swat, Hazara, Murree hills and Ziarat (Balochistan). April.

6. **D. kohatense** (Bruehl) Munz Pl. 1

Plants up to 45 cm tall. Stems few, branched. Petals of basal leaves 20-60 cm long; leaves 20-40 mm, 3-partite into cuneate lobes which are further shallowly lobed. Racemes up to 12-flowered. Pedicels with appressed hairs. Sepals pale blue, upper sepal including spur 11-15 mm long. Follicles 3, tip pubescent.

Distribution: Afghanistan, Pakistan.

Found in drier low hilly areas in Punjab and N.W.F.P. from 350-1200 m. April.

7. **D. roylei** Munz Pl. 2
Syn.: *D. incanum* Royle

Similar to the above species, but with longer upper sepals (including spur 20-25 mm long), and racemes that are many-flowered. Flowers blue.

Distribution: Pakistan, Kashmir.

Grows in temperate parts of Himalaya from 2300-2800 m. July-August.

8. **D. pyramidale** Royle Pl. 2

Stout plants from 55-100 cm or more tall, glabrous to hairy. Petiole of lower leaves 25-40 cm long. Leaves round, 20-25 mm broad, 3-5-angled, into broad cuneate segments; segments incised-serrate. Pedicels up to 5 cm long. Spur of upper sepal slender, up to 4 mm broad. Sepals purple-blue. Upper petals dark, almost black.

Distribution: Pakistan, Kashmir, Tibet and Nepal.

Common in alpine and sub-alpine areas of Himalaya specially in Swat, Hazara and Kashmir from 2100-3600 m. July-August.

9. **D. nordhagenii** Wendelbo Pl. 2

Plants from 15-30 cm tall, pubescent-glandular. Petioles of lower leaves 4-10 cm long; leaves 30-45 mm wide, 3-5-lobed, lobes crenate. Flowers in racemes. Pedicels 20-60 mm long. Sepals purple-blue. Spur of upper sepal 10 mm, curved. Follicles 4-5 in number, hairy.

Distribution: Endemic to Chitral (Pakistan).

Not common, confined to alpine areas of Chitral from 3600-4500 m. July-August.

4. Aquilegia 4 spp. *Columbine*

Perennial herbs with compound leaves and inner petals with 5 incurved spurs. Flowers solitary or several.

+	Spur of petals straight, 15-18 mm. Flowers white or cream, 30-50 mm long.	10. **A. fragrans**
–	Spur of petals curved, 5-12 mm long. Flowers purplish-blue or pale yellow, up to 30 mm long.	11. **A. pubiflora**

10. **A. fragrans** Benth. Pl. 2

Plant up to 70 (-75) cm tall. Basal leaves bi-triternate; ultimate segment lobes cuneate to obovate. Flowers purple-white, nodding, 30-50 mm long, scented. Follicles 5-9 in number, pubescent.

Distribution: Pakistan eastward to Uttar Pradesh (India).

A delicate plant with large attractive flowers. Found as an undergrowth and in open places from 3000-3700 m in Chitral, Hazara, Gilgit and Kashmir. July-August.

11. **A. pubiflora** Wall. ex Royle Pl. 2

An attractive plant up to 65 cm tall. Basal leaves biternate. Petals 10-15 cm long, ultimate segment lobes often green above, pale beneath. Flowers purplish-blue or lemon, 2.5-4 cm long, spurred. Follicles hairy.

Distribution: Pakistan, Kashmir eastward to west Nepal.

Common in mixed coniferous forest in temperate Himalaya from 2100-3500 m, extending west up to Chitral. Forms with lemon or whitish-yellow flowers are less common. July-August.

5. Paraquilegia 1 sp.

Perennial, caespitose herbs with regular flowers. Inflorescence few-flowered. Stamens included. Follicles oblong, glabrous.

12. **P. anemonoides** (Willd.) Ulbr. Pl. 2

Plant forming dense cushions. Leaves compound. Leaflets tri-fid to tripartite with lobed or entire segments. Scapes slender, up to 3 cm tall. Flower 2-3 cm broad, white.

Distribution: Altai, Pamir, Tien Shan, Afghanistan, Pakistan, Kashmir, Tibet and south west China.

Found in open cushion and rosette plant formations in the moist alpine belt from 3200-4300 m in Chitral, Kaghan, Baltistan and Kashmir. June-July.

6. Clematis 11 spp.

Shrubby climbers. Leaves opposite, ternate or pinnate. Flowers solitary or in axillary panicles. Sepals 4, petal-like. Petals absent. Fruit a collection of achenes, overtopped by the feathery persistent style.

1.	+	Leaves ternate. Large petal-like sepals white, tips not recurved.	13. **C. montana**
	–	Leaves pinnate or bipinnate. Petal-like sepals white to pale yellow, tips recurved.	2
2.	+	Leaflets lanceolate. Sepals yellow, sometimes purplish tinged to outside, spreading.	15. **C. orientalis**
	–	Leaflets broadly ovate-cordate. Sepals bell shaped, whitish-yellow.	14. **C. connata**

13. **C. montana** Buch.-Ham. ex DC. Pl. 2

An extensive climber, with large flowers up to 7 cm across. Leaflets 3, ovate, margin toothed. Achenes plumose.

Distribution: Afghanistan, Pakistan eastward to south west China, Taiwan.

Very common in temperate Himalaya from 2000-3000 m in the forest zone. Conspicuous in Murree and adjacent hills in Spring. April-June.

14. **C. connata** DC. Pl. 3

Leaflets 3-7 in number, 3-9 cm long, ovate-cordate, margin toothed. Flowers more or less nodding, pale yellowish-white, bell-shaped.

Distribution: Pakistan, Kashmir, Bhutan to south west China.

In forests from 1500-2700 m in Dir, Swat, Hazara, Murree hills and Kashmir. July-August.

15. **C. orientalis** L. Pl. 3

Leaves pinnate or bi-pinnate. Leaflets lanceolate, often basally with 1-2 shorter lobes. Flowers yellow, purplish tinged to the outside, in lax panicles, spreading, ovate-lanceolate, tip recurved. Achenes silky hairy.

Distribution: Widely distributed from east Mediterranean through west and south west Asia, south Siberia, Pakistan, Mongolia, Tibet, China.

Found in arid places in northern areas of Pakistan, from 2000-4100 m. Common in Hunza valley. August-September.

7. Anemone 12 spp.

Perennial herbs with long stalked radical lobed leaves. Flowers quite showy, in clusters. Sepals 5 or more, petal-like. Petals absent.

1.	+	Erect robust plants from 30-80 cm tall. Basal leaves 5-7 lobed; irregularly toothed.	16. **A. vitifolia**
	–	Caespitose or small plants. Leaves not as above.	2
2.	+	Slender erect plants up to 20 cm tall. Flowers 10-18 mm broad. Usually found in forests.	19. **A. falconeri**
	–	Caespitose plants. Flowers 20-50 mm broad. Usually found in alpine or sub-alpine places.	3
3.	+	Rhizome elongated, not tuberous. Petal-like sepals 25-30 mm long, white.	17. **A. rupicola**
	–	Rhizome short and tuberous. Petal-like sepals 20-25 mm long, white to purplish-blue.	18. **A. obtusiloba**

16. **A. vitifolia** Buch.-Ham. ex DC. Pl. 3

Leaves up to 20 cm broad, shallowly 5-lobed, like in the grape. Flowers 4.5 cm broad, woolly. Achenes in large heads, embedded in dense wool.

Distribution: Afghanistan, Pakistan east to north Burma, south west China.

A tall species commonly found in forests from 1700-3000 m in Hazara and Murree hills. July-September.

17. **A. rupicola** Camb. Pl. 3

A low growing rock loving plant with showy solitary or paired white flowers 40-50 mm broad. Fruit forming a round woolly head with embedded achenes.

Distribution: Afghanistan, Pakistan east to Tibet, south west China.

Fairly common in alpine belt on stony outcrops from 3000-4200 m in Chitral, Hazara, Gilgit, Baltistan. June-August.

18. **A. obtusiloba** D. Don Pl. 3

A low growing perennial of open slopes. Leaves deeply 3-lobed; lobes toothed. Flowers are white on inner surface of petals-bluish on outer surface. Achenes hairy.

Distribution: Pakistan to south Tibet, Burma.

Common in alpine pastures and meadows from 2100-4100 m in Chitral, Swat, particularly in Murree Hills, Kaghan and Kashmir. June-August.

19. **A. falconeri** Thoms. Pl. 3

A slender erect perennial up to 20 cm tall. Radical leaves many, long petioled. Flowers 10-18 mm broad, white. Fruit forming a more or less globose head, hairy.

Distribution: Pakistan, Kashmir.

In forests from 1800-2800 m in Swat, Hazara and Murree hills. April-May.

8. Ranunculus 26 spp. *Buttercup*

Annual or perennial herbs with long petioled and lobed leaves. Flowers in panicles. Petals 5, yellow. Stamens many. Fruit a collection of achenes.

1.	+	Perennial robust plants 30-60 cm tall. Leaves up to10 cm broad. Flowers 17-25 mm broad.	20. **R. laetus**
	–	Annual plants generally much smaller in size. Leaves up to 4 cm broad. Flowers 5-12 mm broad.	2
2.	+	Leaves 3-lobed; lobes crenate. Petioles up to 15 cm long. Achenes muricate.	21. **R. muricatus**
	–	Leaves sub-entire or 3-parted into oblanceolate to linear lobes. Achenes spiny or tuberculate.	22. **R. arvensis**

20. **R. laetus** Wall. ex Hook. f. & Thoms. Pl. 4

Basal leaves hairy, suborbicular-reniform, deeply 2-3-lobed; lobes coarsely toothed. Flowers bright yellow, 17-25 mm broad. Achenes 2-2.5 mm broad, suborbicular.

Distribution: Afghanistan, Pakistan, east to Tibet, south west China.

The commonest and largest of our buttercups from 1200-2700 m in temperate Himalaya in moist places. April-June.

21. **R. muricatus** L. Pl. 4

A glabrous annual with solitary hollow stems. Flowers yellow, 8-12 mm broad. Achenes 5-6 mm long, ovate, compressed, sides muricate, rarely smooth.

Distribution: Atlantic and south Europe west and south west Asia, Crimea, Caucasus, south Siberia, Pakistan, India, Kashmir.

Widely distributed throughout the country. Common as an annual spring weed near damp places from 500-2000 m. March-May.

22. **R. arvensis** L. *Corn Buttercup* Pl. 4

Similar to *R. muricatus,* but narrower leaflets and sepals which are not reflexed.

Distribution: Temperate Europe and Asia.

Common spring weed of fields, widely distributed in the country from 200-2200 m. April-May.

9. Batrachium 2-3 spp.

Sometimes included in the genus *Ranunculus,* but differing in the aquatic habit and finely dissected leaves.

23. **B. trichophyllum** (Chaix.) van der Bosche Pl. 4
Syn.: *Ranunculus trichophyllus* Chaix.

Aquatic plants. Leaves divided into filiform segments. Fruiting stalks 1-5 cm long. Flowers white, c. 10 mm broad. Fruitlets c. 1.5 mm long, clustered on a subglobose receptacle.

Distribution: Temperate Europe and Asia, Africa, north America.

Found in streams of northern Punjab, N.W.F.P. and northern Balochistan. March-April.

10. Caltha 1 sp. *Marsh Marigold*

Herbaceous clump-forming plants with long petioled reniform-cordate leaves; upper leaves sessile, amplexicaule. Flowers white (in our plant) rarely yellow, solitary. Sepals 5-6, petaloid. Petals absent.

24. **C. alba** Jacq. ex Camb. *White Marsh Marigold* Pl. 4

A stout, glabrous 20-60 cm tall branched plant growing in clumps. Leaves showy, 4-18 cm broad, more or less fleshy, margin crenate. Flowers in few flowered corymbs, white. Mature follicles 8-10 mm long.

Distribution: Pakistan, Kashmir to Bhutan.

Common and gregarious in open places along streams from 2700-4300 m in alpine Himalaya. May-August.

11. Trollius 1 sp. *Globe Flower*

Herbs with deeply divided 5-veined leaves. Flowers yellow (in our species), solitary. Sepals many, caducous.

25. **T. acaulis** Lindley Pl. 4

Radical leaves suborbicular, long petioled, 5-lobed; lobes oblong-obovate, toothed. The glossy flowers are solitary, bright yellow, 5-6 cm broad.

Distribution: Pakistan, Kashmir to west Nepal.

Open screes in alpine belt from 3000-4100 m in Himalaya and Karakoram. May-June.

PAEONIACEAE *Peony Family*

Herbaceous plants, sometimes with a woody base. Leaves large, divided. The large flowers white or pale pink. Sepals 5. Petals large. Stamens many.

Paeonia 1 sp.

Characters as that of the family.

26. **P. emodi** Wall. ex Royle *Mamekh, Himalayan Peony* Pl. 5

Glabrous perennial 50 cm or more tall. Leaves ternate or biternate, glabrous. Flowers solitary, white or pale pink, 8-9 cm broad. Follicles ovoid-globose, glabrous or with yellowish-white tomentum, c. 2.5 cm long, woody, splitting at maturity. Seeds 7-8 mm broad, arillate; aril scarlet-red.

Distribution: Afghanistan, Pakistan, Kashmir eastward to west Nepal.

Gregarious in moist shady slopes in forests in Chitral, Swat, Kaghan, Hazara and Kashmir from 2000-3200 m. Toxic to grazing animals. May-June.

BERBERIDACEAE

Barberry Family

Branched shrubs, often spinose. Leaves in tufts, simple. Flowers bisexual, 6-merous, regular, in condensed racemes or clusters. Stamens opposite the petals.

Berberis 20 spp.

Barberry

Leaves with 3-5-branched spines at the base, margin spiny-toothed. Flowers bright yellow or yellow-orange. Perianth segments in three whorls, subequal to unequal. Sepals petal-like. Fruit a berry.

+	Leaf spines on long shoots 3-7-fid. Stems pale yellowish. Leaves spinosely toothed. Flowers yellow-orange.	27. **B. baluchistanica**
–	Leaf spines on long shoots 1-3-fid. Stems whitish-grey. Leaves subentire, with few spines only. Flowers pale yellow.	28. **B. lycium**

27. **B. baluchistanica** Ahrendt *Karoskai* Pl. 5

Shrub up to 2.5 m tall. Stems reddish-brown. Long shoots with spines 3-7-fid. Leaves thickish, margin subentire to subserrate. Flowers up to 10 mm broad, yellow-orange. Berries black, 6 x 3 mm.

Distribution: Endemic to Balochistan (Pakistan).

The roots are said to be medicinal and used for tanning. Dry slopes between 2000-2700 m in Quetta valley and Ziarat area. March-May.

28. **B. lycium** Royle *Kashmal* Pl. 5

Shrub with whitish-grey stems. Spines 1-3-fid, straw coloured. Leaves similar to the above species but with fewer marginal spines. Flowers 6-8 mm broad, pale yellow. Berries 7-8 x 5 mm.

Distribution: Pakistan, Kashmir to east Nepal.

Forests from 1000-2900 m in Himalaya. The blue-black berries are edible. A brown extract 'Rasaunt' from root and stem is used as a cooling agent and as an eye lotion. April-June.

PODOPHYLLACEAE *May Apple Family*

Sometimes included in the *Ranunculaceae* or *Berberidaceae*. Holds an intermediate position between the two.

Podophyllum 1 sp.

Perennial herbs with large palmately lobed, petiolate leaves and showy large white flowers. Berry pulpy, many-seeded.

29. **P. hexandrum** Royle *Ban Kakri, May Apple* Pl. 5

Perennial herb up to 40 cm tall. Leaves long petioled, palmate, deeply 3 (-5) lobed. Lamina 12-22 cm broad. Flowers white, 50 mm across. Berry oblong-elliptic, 3-4 x 2-2.5 cm, red. Seeds 2-3 mm broad.

Distribution: Afghanistan, Pakistan, Kashmir to south west China.

In forests from 2200-4200 m throughout the Himalaya. Shade loving. Usually much exploited for commercial use throughout its distribution range. It is included in the CITES list of endangered plants. The trade is restricted. The rhizome contains the alkaloid podophyllin which is a palliative in the treatment of cancer. April-May.

NYMPHAEACEAE *Water Lily Family*

An aquatic family with a world-wide distribution. Leaves large, handsome, usually floating, long petioled. Flowers solitary, very showy.

+	Leaves exserted above water, on long petioles. Fruit on maturity floating.	**1. Nelumbo**
–	Leaves floating. Fruit on maturity under water.	**2. Nymphaea**

1. Nelumbo 2 spp.

30. **N. nucifera** Gaertn. *Kanwal, Water Lily, Lotus* Pl. 5

An aquatic plant with large orbicular leaves present above water, 70-80 cm broad, attached to a submerged stalk up to 1.5 m. Flowers 18-22 cm broad, white-pink. Fruit flat-topped, up to 10 cm broad, with many embedded seeds.

Distribution: Pakistan, India, north China.

Widely naturalized up to 700m in Punjab and Sindh. The seeds and young roots are edible. July-August.

2. **Nymphaea** 3 spp. *Water Lily*

Like *Nelumbo*, but leaves floating.

31. **N. mexicana** Zucc. Pl. 5

Leaves floating, suborbicular, up to 15 cm long. Flowers yellow, 9-11 cm broad.

Distribution: Naturalized in many places.

A hybrid. Cultivated and naturalized in several places in Sindh and Punjab. February-March.

PAPAVERACEAE *Poppy Family*

Annual or perennial herbs, often with milky parts. Leaves alternate, usually divided. Flowers often large and showy, 4-merous. Fruit a capsule dehiscing by pores or valves.

1.	+	Stamens 4 in number. Inner petals 3-partite.	**1. Hypecoum**
	–	Stamens many. Inner petals not 3-partite.	2
2.	+	Fruit linear-cylindrical, many times longer than broad.	**2. Roemeria**
	–	Fruit globose to subglobose.	3
3.	+	Style absent. Stigma disc-like. Capsule dehiscence by pores.	**3. Papaver**
	–	Style present, Stigma not disc-like. Capsule dehiscence by valves.	4
4.	+	Style conspicuous. Stigma globose. Seeds rugose.	**4. Meconopsis**
	–	Style not conspicuous. Stigma globular not forming a mass. Seeds pitted.	**5. Argemone**

1. **Hypecoum** 2 spp.

Low growing annual herbs with 2-4-pinnatisect leaves. Flowers 4-merous. Inner petals 3-partite. Stamens 4, opposite petals. Fruit linear.

32. **H. pendulum** L. Pl. 6

Small annual herbs with 2-3-pinnatisect leaves; ultimate segments narrow. Inflorescence a 2-4-flowered axillary or terminal cyme. Flowers 7-10 mm broad, yellow. Sepals 2. Petals 4, the inner ones 2-3-lobed. Fruit linear, deflexed, many-seeded.

Distribution: Europe, north Africa, Iran, Afghanistan, Pakistan.

Weed of cultivated fields from 600-2000 m in Balochistan, N.W.F.P. and Punjab. Two varieties occur; the var. *parviflorum* differs in the narrower elliptic inner petals and glands which are absent from base of filament of the stamen. March-May.

2. Roemeria 2 spp.

Evil smelling annual herbs with a yellowish juice. Leaves bi-or tripinnate. Flowers red or violet. Stigmas 3-4, capitate. Fruit not horned.

+	Flowers red. Capsule glabrous.	33. **R. refracta**
–	Flowers violet. Capsule not glabrous.	34. **R. hybrida**

33. **R. refracta** (Stev.) DC. Pl. 6

Annual hispid herb up to 30 (-35) cm tall. Leaves bi-pinnatisect into narrow segments. Petals 20-22 mm long, suborbicular to obovate, with a black spot at the base. Capsule 20-35 cm long, linear, glabrous.

Distribution: Turkey, Iran, Afghanistan, Pakistan.

Dry areas from 1500-1800 m in south Waziristan and north Balochistan. Also as a weed. April May.

34. **R. hybrida** (L.) DC. Pl. 6
subsp. **dodecandra** (Forssk.) Maire

Plant parts not as hispid as in the previous species. Leaves bi-or tripinnatisect. Capsule 20-75 mm long, setosely hairy.

Distribution: Egypt, Jordan, Iran, Russia, Afghanistan, Pakistan.

Distinguished from the subsp. *hybrida* by the violet flowers, broader leaf segments and capsules which are densely setose; peduncles generally erect. Sandy places or low hills. Found from 1400-1800 m. February-April.

3. Papaver 10 spp.

Annual or perennial herbs with a white or yellow sap. Leaves usually pinnately lobed. Flowers solitary, axillary or terminal; long peduncled, brightly coloured. Buds apically horned or not. Fruit a capsule, dehiscing by pores, smooth or not.

1. + Flowers yellow. Perennials. 35. **P. nudicaule**

 – Flowers red to scarlet or orange. Annuals. 2

2. + Capsule setosely hairy, ovoid or broadly so. 3

 – Capsule oblong, glabrous. 36. **P. dubium**

3. + Flower buds 2-horned. Capsule more or less ovoid. 38. **P. pavoninum**

 – Flower buds not horned. Capsule subglobose. 37. **P. hybridum**

35. **P. nudicaule** L. *Yellow Poppy* Pl. 6

Perennial tufted setosely hairy herb. Leaves radical, pinnately 3-lobed; lobes further subdivided. Flowers solitary, arising from a leafless scape, yellow to yellow-orange, 30-50 (-60) cm broad. Capsule oblong, 15-18 x 7-8 mm, hairy.

Distribution: North and central Asia, Afghanistan, northern Pakistan, Kashmir.

A variable species in size and flower colour. Usually found in alpine rocky places from 3300-4800 m. June-August.

36. **P. dubium** L. *Post, Long headed Poppy* Pl. 6

Annual up to 20 cm tall with deeply divided leaves. Flowers red, 3-4 (-7) cm broad. Each petal with a dark dot in the center. Capsule oblong, 15-20 x 4-8 mm, smooth.

Distribution: Temperate Europe and Asia, Afghanistan, Pakistan, Kashmir to central Nepal.

A conspicuous weed of fields from the plains to 1800 m. March-June.

37. **P. hybridum** L. *Wild Poppy* Pl. 6

Similar to the above species, but flowers generally darker and with a black spot at the petal base and bristly capsules.

Distribution: Europe, north Africa, south west and central Asia, Afghanistan, Pakistan.

Common as a weed in fields up to 900 m. February-April.

38. **P. pavoninum** Schrenk Pl. 6

Similar to *P. hybridum,* but flowers generally orange-red and larger, and capsules more slender. The flower buds are generally 2-horned, which is not the case in *P. hybridum.*

Distribution: Turcomania, Iran, Afghanistan, Altai, Pakistan.

A variable species in leaf and flower size. A more common species than *P. hybridum,* often in wheat fields from plains to 1900 m. February-April.

4. Meconopsis 2 spp.

Resembling *Papaver*, but capsule valvular and flowers with a distinct style.

39. **M. aculeata** Royle *Blue Poppy* Pl. 7

A perennial prickly herb 40-65 cm tall and with bristly, pinnately divided radical leaves. Scapes many-flowered, stout. Flowers blue or pale blue (with age), 5-7 cm broad. Capsule prickly, 15-18 x 8-10 mm.

Distribution: Pakistan, Kashmir.

A rare plant in Pakistan with large attractive blue flowers. Alpine Himalaya. Found in Hazara district on rocks and steep places from 3000-3800 m. July-August.

5. Argemone 2 spp.

Prickly annual herbs exuding a yellow sap. Leaves divided. Flowers terminal, solitary. Capsules prickly, dehiscing by valves.

40. **A. mexicana** L. *Pila Dhatura, Prickly Poppy* Pl. 7

Perennial prickly herb. Flowers large, attractive, yellow. Leaves variegated green-white, prickly.

Distribution: Native of Mexico. Naturalized in most warm countries.

Found in the plains in waste places and as a noxious weed up to 900 m. The plant exudes a yellow juice that is used in jaundice and dropsy. The seeds are reputed to be medicinal as well. February-May.

FUMARIACEAE *Fumitory Family*

Closely related to *Papaveraceae*, differing in the plants without latex (sometimes watery juice), spurred corolla, dimorphous anthers.

+	Annuals with 2-4 -pinnatisect leaves. Fruit a 1-2-seeded nutlet.	**1. Fumaria**
–	Perennials with less cut leaves. Fruit a many-seeded capsule.	**2. Corydalis**

1. Fumaria 3 spp.

41. **F. indica** (Hausskn.) Pugsley *Pit Papra, Fumitory* Pl. 7

A diffuse, prostrate much branched herb. Leaves much dissected, greyish-green. Flowers pale-pink, diffused with crimson, borne on short racemes up to 2 cm long. Nutlets globose, 1.5-2 mm broad.

Distribution: West Asia. Introduced elsewhere.

Common as a weed in fields from the plains to 1800 m. Plant parts used in local medicine and the herbage of some value as fodder. March-May.

2. Corydalis c. 22 spp.

Characterized by zygomorphic spurred flowers which are 2-lipped.

1.	+	Radical leaves absent. Flowers pink or pink-purple.	42. **C. diphylla**
	–	Radical leaves present. Flowers yellow.	2
2.	+	Stem leaves many. More or less a bushy perennial. Leaves fleshy; pinnae fan-shaped. Sepals 3-5 mm long.	43. **C. flabellata**
	–	Plants tufted or prostrate. Leaves not fleshy. Sepals c. 1 mm long or obsolete.	3
3.	+	Plants prostrate. Racemes 3-5 cm long, 5-10 - flowered. Flowers 6-10 mm.	45. **C. stewartii**
	–	Plants erect, tufted. Racemes 5-10 cm long, 10-30 - flowered. Flowers 20-25 mm long.	44. **C. govaniana**

42. **C. diphylla** Wall. Pl. 7
Syn.: *C. rutifolia* auct. non (Sibth & Sm.) DC.

A slender herb with pretty pink or pink-purple flowers. Leaves few, paired, pinnately dissected. Flowers pink-purple, 3-10 per raceme. Capsule 6-8 mm long.

Distribution: Pakistan, Kashmir eastward to north Nepal.

Found on moist shady slopes of Himalaya from 1800-3100 m in mixed coniferous forest zone. April-June.

43. **C. flabellata** Edgew. Pl. 7

A tufted bushy perennial up to 80 cm tall, with thickish fan-shaped leaves. Flowers yellow, 12-18 mm long. Sepals 3-5 mm long. Capsule 12-22 x 2-5 mm.

Distribution: Pakistan, east to Nepal, Tibet.

Conspicuous plant on open screes in alpine belt of Himalaya from 2000-3000 m. August-September.

44. **C. govaniana** Wall. Pl. 7

An erect tufted perennial up to 60 cm tall. Racemes 5-10 cm long, 10-30-flowered. Flowers including spur 20-25 mm long, yellow. Capsule 10-12 x 3-4 mm.

Distribution: Pakistan, India to east Nepal.

In relatively moist alpine habitat in Swat, Hazara and Kashmir area from 2100-4200 m. Its roots are said to be medicinal. June-August.

45. **C. stewartii** Fedde *Mamiri* Pl. 7

A slender, prostrate herb with ternate leaves. Flowers pale yellow, 6-10 mm long, in racemes 3-5 cm long. Capsule linear oblong, 12-15 x 1.5-2 mm.

Distribution: Central Asia, Pakistan, Kashmir.

Found in the moist sub-alpine and lower alpine habitat in Himalaya from 1800-3500 m. July-August.

BRASSICACEAE *Mustard Family*

A large family represented by 92 genera and 250 species in Pakistan. Including economically important crop species cultivated for vegetables and oil seeds throughout the country.

Annual or perennial herbs with regular flowers. Sepals and petals 4, free. Stamens 6. Fruit a short silicula or long siliqua, often opening from below by 2-valves.

1. + Fruit with a beak-like projection, biarticulate, with the upper and lower parts dissimilar. 2
   – Fruit without a beak-like projection, not biarticulated. 3

2. + Fruit many seeded. **1. Diplotaxis**
   – Fruit one seeded. **2. Crambe**

3.	+	Flowers white or pink, or pink-purple, (sometimes orange-yellow to yellowish in **Farsetia**).	4
	–	Flowers yellow or yellowish-orange (also white in **Draba**).	9
4.	+	Leaves pinnate or deeply lobed.	5
	–	Leaves entire or shallowly lobed (sometimes subentire in **Malcomia**).	8
5.	+	Sepals not spreading. Fruit usually hairy.	6
	–	Sepals spreading. Fruit glabrous.	7
6.	+	Plants of wet habitats. Racemes lax in fruit. Fruit more or less cylindrical.	**3. Nasturtium**
	–	Plants mesophytic, but never standing in water. Racemes many-flowered. Fruit compressed.	**4. Cardamine**
7.	+	Fruit not constricted. Dehiscent.	**6. Malcomia**
	–	Fruit constricted breaking into one seeded parts.	**5. Chorispora**
8	+	Fruit suborbicular to obovate-oblong, compressed, margin winged, apex notched.	**9. Thlaspi**
	–	Fruit linear to oblong, compressed or subcylindrical, not winged, apex not notched.	**10. Farsetia**
9.	+	Leaves divided or sometimes entire. Flowers of medium size, yellow or orange.	**7. Erysimum**
	–	Leaves entire to toothed. Flowers small, white to yellow.	9
10.	+	Plants glabrous. Basal leaves elliptic-oblong, upper cordate and sessile.	**8. Isatis**
	–	Plants hairy. Leaves not as above.	**11. Draba**

1. Diplotaxis 1 sp.

A European or Mediterranean genus with large cream-violet to pinkish flowers and long slender fruit.

46. **D. griffithii** (Hook. f. & Thoms.) Boiss. *Barani Muli* Pl. 8

A branched hispid annual up to 60 cm tall. Basal leaves lyrate pinnatisect. Flowers racemose, 12-15 mm broad, pale pink. Fruit 30-50 x 1.5-2.5 mm, compressed.

Distribution: Afghanistan, Pakistan.

A desert annual, locally abundant especially Punjab Salt Range, Kalla Chitta hills, Nushki, Panjgur, Qila Saifullah from 500-1700 m. February-March.

2. Crambe 1 sp.

Stout perennial herbs with very large leaves. Fruit biarticulate, one seeded indehiscent.

47. **C. cordifolia** Steven *Filgosh, Elephant Ear* Pl. 8
subsp. **kotschyana** (Boiss.) Jafri

Plants up to 180 cm tall with large tap root. Basal leaves large, rounded, thick textured up to 40 cm in diameter. Flowers small, white.

Distribution: Iran, Afghanistan, Turkmenistan, Pakistan.

Conspicuous plant in dry hilly areas of Waziristan and Balochistan up to 2400 m. Leaves and roots edible. April-June.

3. Nasturtium 2 spp.

Plants of aquatic habitats with pinnate leaves. Flowers small, white. Fruit subcylindrical, glabrous.

48. **N. officinale** R. Br. *Water Cress* Pl. 8

Branched floating perennial up to 100 cm long, rooting at the nodes. Leaves lyrate-pinnate with 5-9-leaflets. Flowers 15-30 per raceme, 3.5-4 mm broad. Fruit oblong, 10-20 x 2-2.5 mm.

Distribution: Temperate Europe and Asia. Introduced elsewhere.

Very common in shallow streams and water courses from the plains to 2100 m. March-June.

4. Cardamine 5 spp.

Glabrous to hairy annual or perennial herbs with pinnate leaves and linear compressed fruit, the valves of which coil and open upwards to eject the minute seeds.

49. **C. loxostemonoides** O. E. Schulz Pl. 8

A procumbent perennial up to 25 cm long. Flowers 12-15 mm broad, lilac or pink, 5-15 per raceme. Fruit 2.5-3 x 1.5 mm, glabrous.

Distribution: Endemic to Himalaya.

Usually found in rock crevices or screes between 3000-3500 m in the alpine areas. Common in Hazara area. May-June.

5. Chorispora 5 spp.

Annual or perennial herbs with divided or entire leaves. Flowers medium to large, in various shades from pink to white. Fruit linear, beaded.

50. **C. sabulosa** Camb. Pl. 8

A suberect branched perennial. Leaves pinnatifid or toothed. Flowers 10-12 mm broad, pink or violet. Fruit 15 x 2 mm, constricted, giving a beaded appearance, glandular hairy.

Distribution: Afghanistan, Kashmir to Uttar Pradesh.

Usually in alpine places from 3200-4600 m. Pods are used as a pot herb by local people. June-August.

6. Malcolmia 7 spp.

Annual hairy herbs. Leaves deeply divided to subentire. Flowers in lax racemes, small to large, in shades of pink. Fruit linear, subcylindrical, glabrous to hairy.

51. **M. cabulica** (Boiss.) Hook. f. & Thoms. Pl. 8

Annual or biennial branched herb, 12-28 cm tall. Basal leaves in rosettes, pinnately lobed to toothed. The crinkly flowers showy, deep pink to reddish, 12-15 mm broad, racemes 8-15 cm long. Fruit 20-25 x 2 mm, linear, strigosely hairy.

Distribution: Afghanistan, Pakistan.

Found in dry open places up to 600 m in Sindh, Balochistan, N.W.F.P. and Punjab. Often carpeting gravelly areas with red-pink flowers. March.

7. Erysimum 14 spp.

Annual, biennial or perennial herbs with hairy parts. Leaves pinnatifid to entire and toothed. Flowers medium, yellow or orange.

52. **E. melicentae** Dunn. Pl. 8

Biennial or perennial herbs up to 90 cm tall. Basal leaves in rosettes 40-90 x 6-12 mm, toothed. Flowers showy, 12-15 mm broad, yellow to orange. Fruit 30-65 x 1.5 mm, linear, cylindric, hairy.

Distribution: Pakistan, Kashmir.

In open sunny situations. Common in Swat, Kaghan valley from 2000-3000 m. May-July.

8. Isatis 7 spp.

Branched annual, biennial or perennial herbs, with glabrous parts. Stem leaves cordate to ovate-oblong, entire. Flowers small, yellow. Fruit compressed, winged, 1-seeded.

53. **I. costata** C.A. Meyer Pl. 9

Much branched annual up to 100 cm or more tall with clasping leaves. Flowers about 3.5 mm broad. Fruit oblong-elliptic, 12-15 x 3.5-6 mm.

Distribution: Afghanistan, Pakistan, India.

In open places, from 1300-2000 m in Balochistan, Kurram valley, Swat, Gilgit and Kashmir. April-June.

9. Thlaspi 8 spp. *Pennycress*

Leaves entire. Flowers white to pink. Fruit suborbicular or ovate-oblong, compressed, winged.

54. **T. andersonii** (Hook. f. & Thoms.) O. E. Schulz Pl. 9

A subfleshy perennial 12-15 cm tall with stem leaves cordate, amplexicaul. Flowers white or pale pink, 6-7 mm broad. Fruit 6-8 x 2 mm, elliptic-oblong.

Distribution: Pakistan, Kashmir.

In forests from 2100-3200 m in the alpine belt of Himalaya. May-July

10. Farsetia 4 spp.

Undershrubs or perennial herbs with narrow leaves and oblong compressed fruit. Flowers short pedicelled. Petals longer than sepals.

55. **F. jacquemontii** Hook. f. & Thoms. Pl. 9

Branched plants up to 40 cm tall. Flowers pink or orange-yellow. Fruit 20-48 x 3-4 mm. Seeds winged, c. 2.5 mm broad.

Distribution: Afghanistan, Pakistan, India.

Desert areas and dry hills up to 1300 m throughout the country. March-May.

11. Draba 23 spp.

Perennial often tufted plants with yellow or white flowers. Fruit ovate to suborbicular to elliptic, compressed.

56. **D. oreades** Schrenk Pl. 9

Plant cushion-like, mat forming. Leaf rosettes 5-6 mm broad. Stems short, leafless. Flowers yellow, c. 4 mm broad. Leaves linear-lanceolate, up to 1 cm long, margin hairy.

Distribution: Central Asia, Pakistan, India to south west China.

Open screes from 3200-4300 m in alpine zone of Himalaya and Karakoram. June-July.

CAPPARIDACEAE

Caper Family

Closely related to the *Brassicaceae*, but differs in the glandular hairy parts and stamens which are never in groups.

\+ Sepals united below to form a distinct tube. Flowers without a cupular disc around the floral axis. **1. Capparis**

– Sepals not united or only slightly so at the base. Flowers with a cupular disc around the floral axis. **2. Maerua**

1. Capparis 6 spp.

Generally shrubs with simple leaves, showy flowers and baccate fruit.

\+ Fruit ellipsoid. Flowers solitary, white. 57. **C. spinosa**

– Fruit more or less subglobose. Flowers one to many, brick-red or pink-red. 58. **C. decidua**

57. **C. spinosa** L. *Caper* Pl. 9

A low growing shrub. Leaves ovate to elliptic or orbicular, 15-40 x 10-30 mm. Flowers 25-50 mm, white. Fruit 20-40 x 10-20 mm, red when ripe.

Distribution: Tropical and sub-tropical countries of both hemispheres.

A plant of dry areas, browsed by goats etc. conspicuous in the dryer far northern regions of Pakistan, but also found in Balochistan. Fruit used for pickling. April-May.

58. **C. decidua** (Forssk.) Edgew. *Karildelha* Pl. 9

Low, much branched shrub or small trees with crooked, spiny and leafless branches. Flowers 10-20 mm broad, brick-red or pink-red. Fruit subglobose, 12-18 mm broad, smooth, red when ripe.

Distribution: Tropical Africa, Arabia to India.

A common shrub of arid areas up to 700 m. Widespread in Las Bela, Thal, Tharparkar and Cholistan. The nectariferous flowers attract many bird species. The young fruit and flower buds are pickled. May-June.

2. Maerua 2 spp.

Shrubs or small trees with usually simple leaves. Flowers 1-5 in the axil of upper leaves. Stamens many. Fruit cylindrical to round.

+ A scrambling shrub. Flowers many, in corymbose racemes. 59. **M. arenaria**

– Flowers solitary or 2-4. A small tree. 60. **M. crassifolia**

59. **M. arenaria** (DC.) Hook. f. & Thoms. Pl. 10

Flowers greenish-white. Fruit cylindrical, 20-75 x 10-15 mm, beaded, somewhat twisted.

Distribution: Pakistan, India, Sri Lanka.

Arid, sandy or stony places up to 1300 m. March-April.

60. **M. crassifolia** Forssk. Pl. 10

Tree up to 5 m tall. Leaves somewhat fleshy, ovate-oblong, 5-15 x 2-8 mm. Fruit irregularly cylindrical, 25-50 x 5-8 mm, constricted.

Distribution: North Africa, Egypt eastward to south Balochistan (Pakistan).

In dry arid sandy places up to 300 m. February.

RESEDACEAE

Mignonette Family

Herbs or shrubs with usually simple, stipulate leaves. Flowers irregular. Sepals and petals 4-7. Fruit a capsule or berry.

+ Straggling shrubs with petals reduced or absent. Fruit a berry or capsule. **1. Ochradenus**
– Herbs with petals well developed. Fruit capsular. **2. Reseda**

1. Ochradenus 2 spp.

Dioecious glabrous spinescent shrubs with flowers in terminal spikes or racemes. Leaves entire, often whorled. Sepals 5-6 in number.

61. **O. baccatus** Delile Pl. 10

Straggling shrubs up to 2.5 m tall with greenish-yellow branches. Leaves linear. Flowers small, greenish-yellow, in terminal stiff racemes up to 15 cm. Berries 4-5 mm broad, white.

Distribution: Libya, Egypt, Ethiopia, Somaliland, Socotra, Middle East to south Iran and Pakistan.

Found in hot dry places up to 1000 m especially Balochistan and lower Sindh. The plants are browsed by animals and fruit is edible. January-April.

2. Reseda 4 spp.

Annual or perennial herbs. Leaves sometimes divided. Petals entire or divided, lower part appendaged. Capsule indehiscent, open at the top, tip 3-4-lobed.

62. **R. pruinosa** Delile Pl. 10

Perennial herb. Stems pruinose to papillose. Leaves linear-lanceolate, uppermost ternate. Capsule 3-4 mm broad, subglobose, 3-lobed at the top.

Distribution: Egypt, Arabia, Middle East, Iran, Afghanistan, Pakistan, India.

In dry rocky places, stream beds etc., up to 700 m in Salt Range, Sindh, Kohistan, Balochistan and N.W.F.P. hills. October-June.

VIOLACEAE *Violet Family*

Annual or perennial herbs with zygomorphic spurred flowers. Leaves simple, alternate. Flowers capsular, dehiscing by 3 valves.

Viola 16 spp. *Violet*

Plants with underground rhizomes or stolons. Leaves ovate-cordate to reniform, stipulate. Petals 5, the lateral pair larger than other and forming a spur.

1.	+	Flowers yellow.	63. **V. biflora**
	–	Flowers lavender to pink or shade of blue.	2
2.	+	Plant parts glabrous. Spur 1-2.5 mm long. Leaves ovate to spathulate.	64. **V. stocksii**
	–	Spur 4-8.5 mm long. Leaves ovate-cordate, reniform or more or less triangular in outline.	3
3.	+	Plants stoloniferous; leaves ovate-cordate. Ovary hairy, stigma club-like.	65. **V. canescens**
	–	Plants not stoloniferous. Leaves more or less triangular. Ovary glabrous, stigma 3-lobed.	66. **V. betonicifolia**

63. **V. biflora** L. *Yellow Violet* Pl. 10

A perennial or annual decumbent plant with a slender rootstock. Leaves reniform, margin crenate. Stipules 2, ovate or oblong, entire. Flowers 8-12 mm long; spur 2.5-4 mm, straight.

Distribution: Europe, Siberia, central Asia, Pakistan, Kashmir to Nepal, south west China, Japan, north west America.

Uncommon, found in humid sub-alpine open places and forest zone, from 3700-4000 m. June-August.

64. **V. stocksii** Boiss. Pl. 11

Slender annual 3-15 cm. Leaves spathulate to ovate. Stipules linear-lanceolate, 3-9 x 1-2 mm. Flowers small, 2-5 mm long, bluish-white; spur 1-2.5 mm long, straight. Capsule 5-6 mm long, glabrous.

Distribution: Iran, Afghanistan, Pakistan, India.

A species of the plains and drier areas up to 1200 m. March-April.

65. **V. canescens** Wall. ex Roxb. Pl. 11
Syn.:*V. serpens* var. *canescens* (wall. ex Roxb.) Hook .f. & Thoms.

Prostrate stoloniferous plant with pubescent leaves. Flowers 12-18 mm long, pale violet or violet. Spur 3-4 mm long, straight to slightly curved. Capsule globose, hairy.

Distribution: Pakistan, Kashmir to Bhutan.

A forest species on shady banks etc. from 950-2600 m in Himalaya. March-June.

66. **V. betonicifolia** Smith Pl. 11
Syn.: *V. caespitosa* D. Don

A glabrous perennial with slender unbranched roots. Leaves triangular, base truncate. Stipules free, dentate. Flowers 5-10 (-12) mm long, purplish to lilac. Capsule 8-10 mm, oblong, glabrous.

Distribution: Pakistan, India, Kashmir to south west China, Australia.

Generally found along water courses or shady banks from 900-2800 m. Common in the Murree hills. March-May.

POLYGALACEAE

Milkwort Family

Herbs, shrubs or trees with simple, alternate leaves. Flowers irregular, racemose. Fruit a capsule.

Polygala 8-10 spp.

Milkwort

Annual or perennial herbs with flowers in terminal or lateral racemes. Petals 3, somewhat united below. Capsule compressed, 2-seeded.

67. **P. abyssinica** R. Br. ex Fresen. Pl. 11

Slender perennial herbs up to 45 cm tall. Leaves linear-lanceolate. Petals obovate, 3-veined, pink. Capsule obovate, 3-4 x 2.5 mm, winged.

Distribution: Africa, Afghanistan, Pakistan, Kashmir to Uttar Pradesh. (India)

Common in dry rocky places up to 2400 m in north Balochistan, north Punjab and N.W.F.P., March-September.

CARYOPHYLLACEAE

Carnation Family

A large cosmopolitan family of mostly herbs with simple and opposite leaves (sometimes alternate or whorled) and swollen stem nodes. Flowers regular. Sepals and petals generally 5. Fruit capsular or berry-like.

1.	+	Sepals free.	2
	–	Sepals united.	5
2.	+	Leaves stipulate.	**1. Spergularia**
	–	Leaves exstipulate.	3
3.	+	Capsule cylindric, dehiscing by teeth that are reflexed.	**2. Cerastium**
	–	Capsule globose to ovoid, dehiscing by erect teeth.	4
4.	+	Capsule teeth 6 in number, less than half the body length.	**3. Arenaria**
	–	Capsule teeth 4-6 in number, about half the body length.	**4. Stellaria**
5.	+	Bracteoles present.	**5. Dianthus**
	–	Bracteoles absent.	6
6.	+	Calyx broadly 5 winged.	**6. Vaccaria**
	–	Calyx not winged.	7
7.	+	Leaves and bracts subulate and spiny. Calyx tip spinose.	**7. Acanthophyllum**
	–	Leaves, bracts and calyx not as above.	8
8.	+	Calyx nerves with hyaline intervals between. Capsule dehiscing by 4 valves.	**8. Gypsophila**
	–	Calyx ribs without hyaline intervals. Capsule dehiscence by 6-10 teeth.	**9. Silene**

1. Spergularia 3 spp.

Closely related to *Spergula* but differing in the united stipules.

Annual or perennial herbs with whorled leaves. Stipules scarious, united. Sepals and petals 5. Sepals free. Capsule 3-valved.

68. **S. marina** (L.) Griseb. Pl. 11

Stem branched. Leaves linear, more or less fleshy. Stipules triangular. Petals pink, shorter than sepals, ovate. Capsule 4 mm long ovoid.

Distribution: The Mediterranean, south west and east Asia, Iran, Turkmenistan.

Common in neglected fields etc. February-March.

2. Cerastium 8 spp. *Chickweed*

Annual or perennial herbs, often glandular. Petals 5, tip bilobed. Styles 3-5. Capsule cylindrical, membranous, dehiscing by 6 or 10 teeth.

69. **C. pusillum** Ser. Pl. 11

Perennial up to 10 cm long with glandular parts. Leaves lanceolate, up to 16 x 5 mm. Pedicels much exceeding the flower length. Sepals 5-6 mm, elliptic. Petals white, bilobed, exceeding the sepals. Styles 5. Capsule teeth 10.

Distribution: East and west Siberia, central Asia, Pakistan, Kashmir.

Found in the subalpine and alpine zones from 3100-4200 m. July-August.

3. Arenaria 7 spp. *Sandwort*

Annual or perennial herbs, often caespitose with narrow opposite leaves. Sepals and petals 5. Petals white, usually entire. Styles 3. Capsule teeth twice as many as the style.

70. **A. griffithii** Boiss. Pl. 12

A low growing perennial with erect spreading flowering stems. Leaves subulate, 8-12 mm long. Petals obovate. Capsule exceeding calyx length.

Distribution: Afghanistan, Pakistan, Kashmir to Uttar Pradesh (India).

Rocky places from 2000-3300 m. June-July.

4. Stellaria 11 spp. *Stitchwort*

Annual or perennial usually few flowered herbs. Petals 2-lobed. Styles 2 or 3. Capsule. ovoid, splitting by 4-6 valves.

71. **S. media** (L.) Vill. Pl. 12

Annual with prostrate or decumbent stems. Leaves 7-20 x 3-10 mm, elliptic to ovate-elliptic. Pedicels slender. Petals white. Styles 3.

Distribution: Cosmopolitan.

Common weed of cultivated areas and waste places. Plains to 3900 m. April-August.

5. Dianthus 10 spp. *Pink*

Annual or perennial herbs with linear leaves. Flowers bracteolate, white to shades of pink. Calyx 5-toothed. Petal limb incised. Styles 2. Capsule dehiscing by 4 teeth.

72. **D. anatolicus** Boiss. Pl. 12

Tufted perennial with linear leaves. Bracteoles 4-6. Flowers solitary, white to pink. Petal about 15 mm, toothed at the apex.

Distribution: West Asia, Pakistan to Kashmir.

In drier regions of Chitral, Gilgit and Kaghan beyond monsoon influence. Common in open sunny places from 2400-3400 m. June-August.

6. Vaccaria 1 sp.

Annual glabrous herbs. Petals 5. Stamens 10. Styles 2. Capsule dehiscing by 4 valves.

73. **V. hispanica** (Miller) Rausch. *Cow Basil* Pl. 12

Plants erect, up to 50 cm tall. Leaves ovate-lanceolate to lanceolate. Flowers 3-7 per inflorescence, pink. Pedicels slender. Petals 15-20 mm, limb entire or with a toothed apex. Capsule subglobose, 8-10 mm.

Distribution: West Europe to west Asia.

Commonly a weed of cultivated fields from the plains to 2700 m. February.

7. Acanthophyllum 5 spp.

Tufted perennials often cushion-like with narrow spiny leaves and bracts. Flowers white or pink, solitary or in compact heads. Petals 5. Stamens 20. Capsule ovoid, 1-seeded.

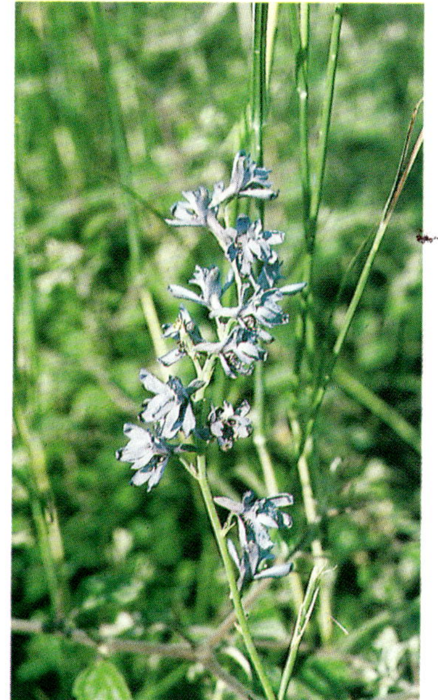

6. *Delphinium kohatense*, Attock hills, S. A. Sultan

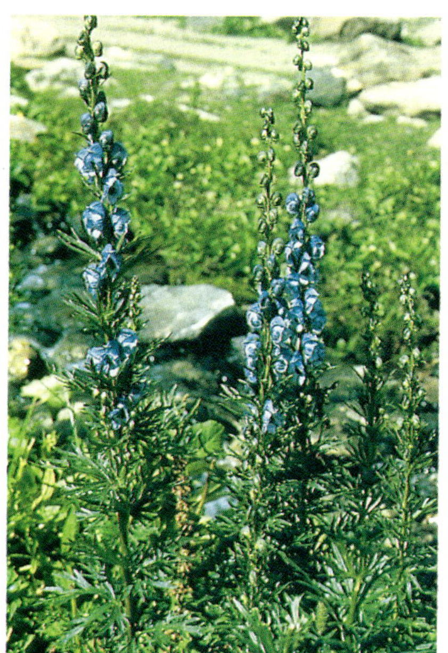

3. *Aconitum violaceum*, Besal-Kaghan valley, T. J. Roberts

1. *Aconitum laeve*, Lowari pass, T. J. Roberts

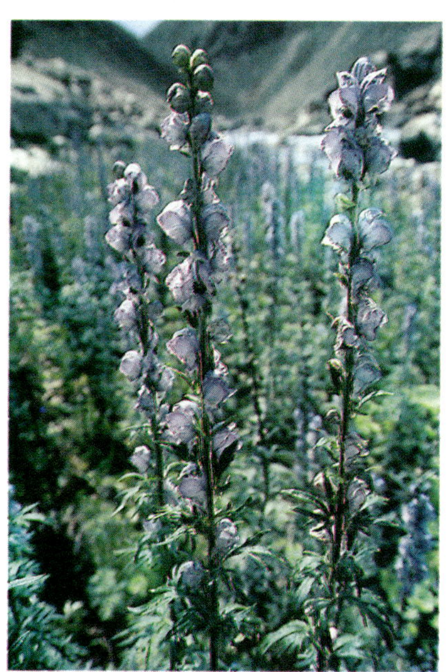

2. *Aconitum chasmanthum*, Kaghan, Fritz Berger

5. *Delphinium uncinatum*, Shangla pass, T. J. Roberts

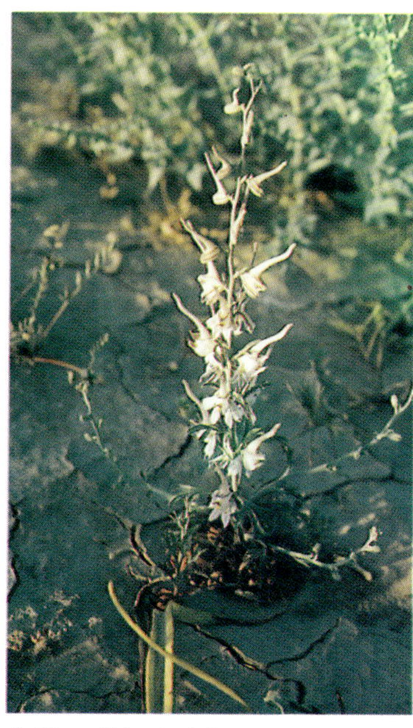

4. *Consolida stocksiana*, Quetta, Rubina A. Rafiq

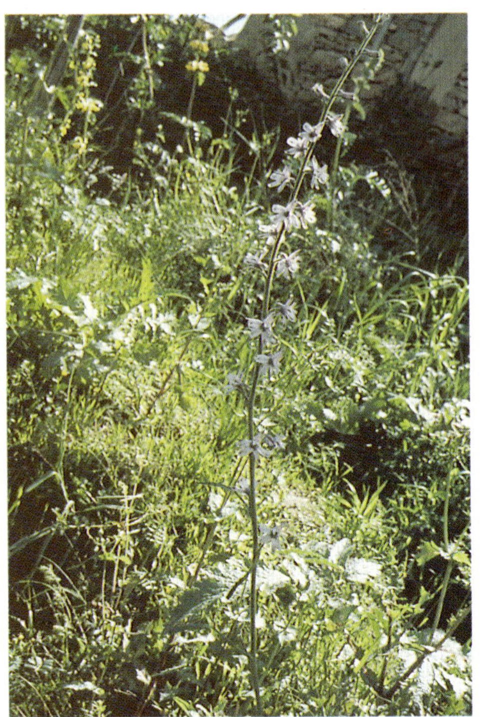

7. *Delphinium roylei,*
Lower Swat, T.J. Roberts

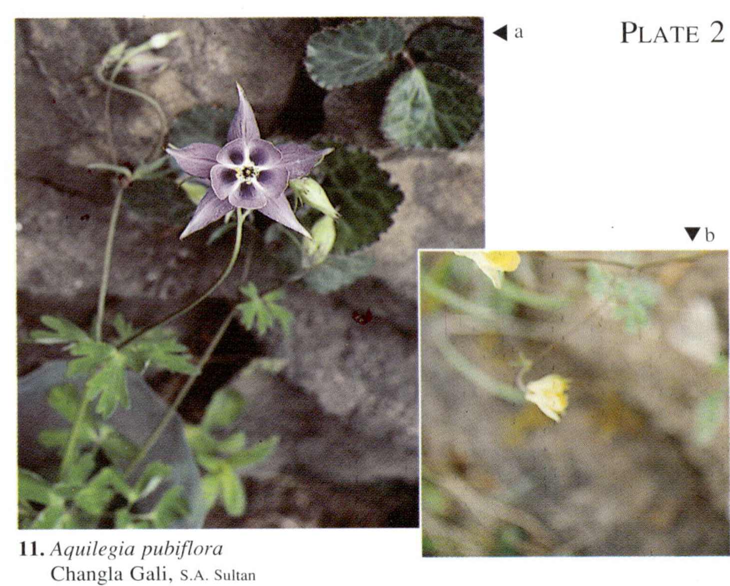

◄a ▼b

11. *Aquilegia pubiflora*
Changla Gali, S.A. Sultan

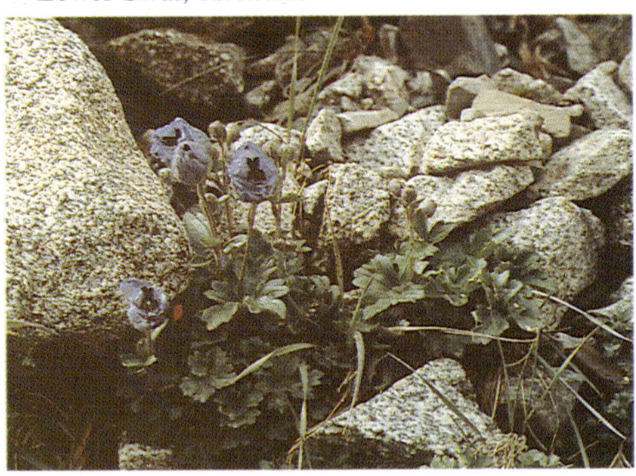

9. *Delphinium nordhagenii,*
Khunjerab pass, Mark Mallalieu

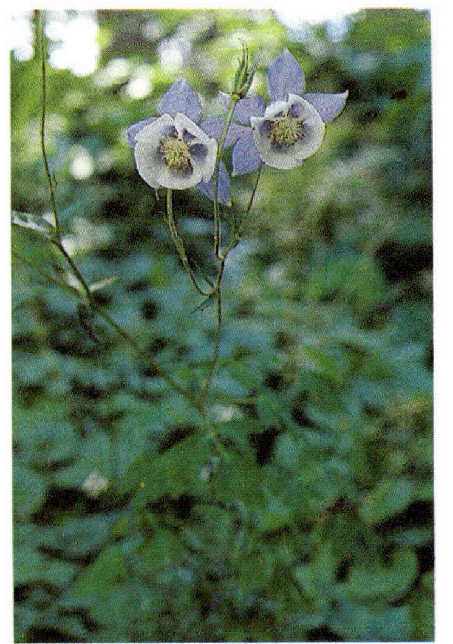

10. *Aquilegia fragrans,*
Kaghan, Rubina A. Rafiq

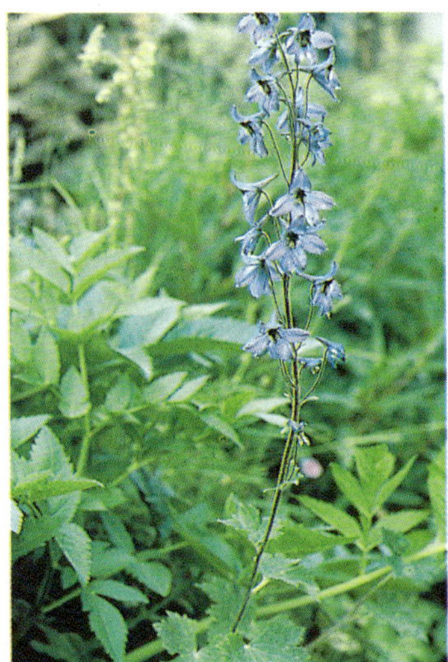

8. *Delphinium pyramidale,*
Babusar valley, Chilas, T.J. Roberts

13. *Clematis montana* Murree hills, S.A. Sultan

12. *Paraquilegia anemonoides*
N.W. Himalaya, Chris Chadwell

15. Clematis orientalis Hunza, Y. J. Nasir

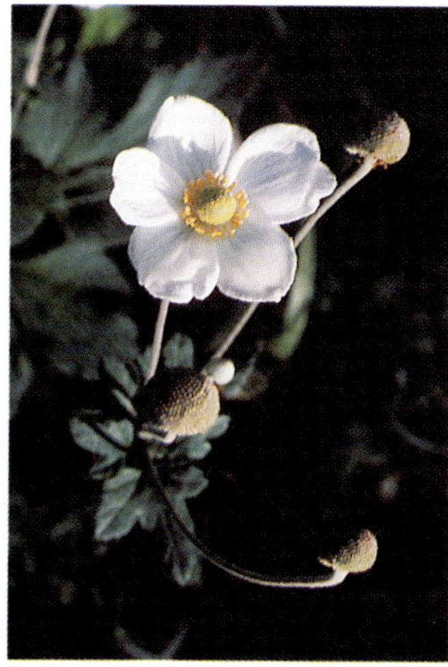

16. *Anemone vitifolia* Murree hills, Y. J. Nasir

17. *Anemone rupicola* Musa ka Musala, Kaghan, T. J. Roberts

18. *Anemone obtusiloba* Upper Kaghan, S. A. Sultan

14. *Clematis connata* Thandiani, Hazara, Y. J. Nasir

19. *Anemone falconeri* Dunga Gali, Y. J. Nasir

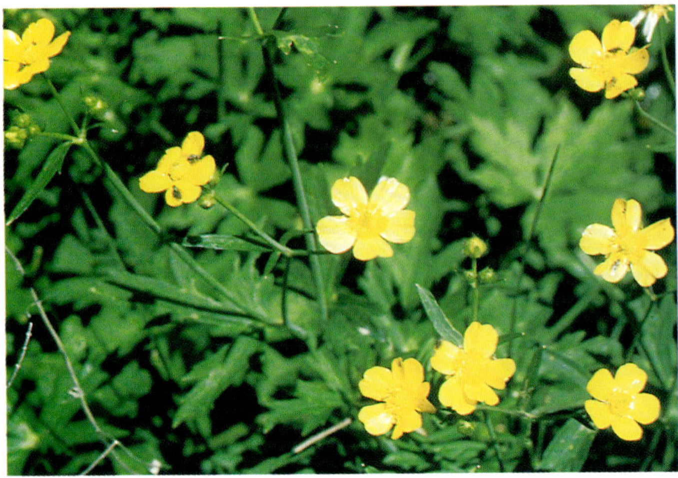

20. *Ranunculus laetus* Haripur, Hazara, S. A. Sultan

21. *Ranunculus muricatus* Margalla hills, S. A. Sultan

22. *Ranunculus arvensis* Margalla hills, S. A. Sultan

23. *Batrachium trichophyllum* Quetta, Rubina A. Rafiq

24. *Caltha alba* Naran, T. J. Roberts

25. *Trollius acaule* Musa ka Musalla, T. J. Roberts

29. *Podophyllum hexandrum*
a: Gabral, Kalam, Fritz Berger
b: Mahodand, Swat, Fritz Berger

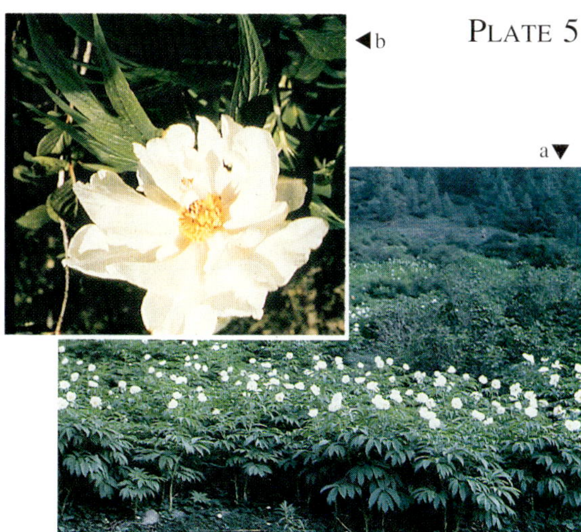

26. *Paeonia emodi* a: Sharan, Kaghan, T. J. Roberts
b: Manshi Forest, T. J. Roberts

30. *Nelumbo nucifera* Gujar Khan, S. A. Sultan

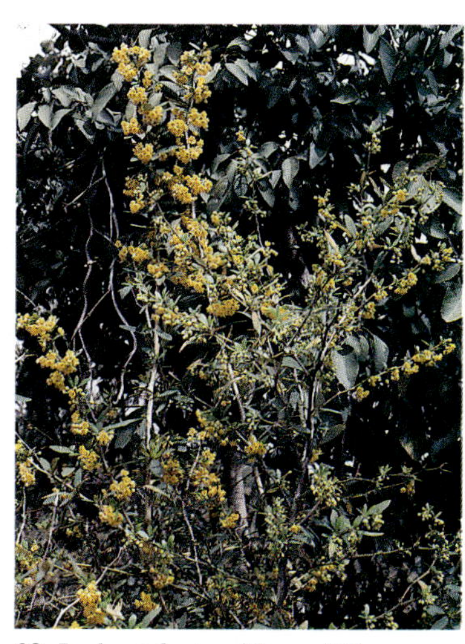

28. *Berberis lycium* Murree hills, Y. J. Nasir

27. *Berberis baluchistanica* Ziarat, Balochistan, T. J. Roberts

31. *Nymphaea mexicana* Karachi, Y. J. Nasir

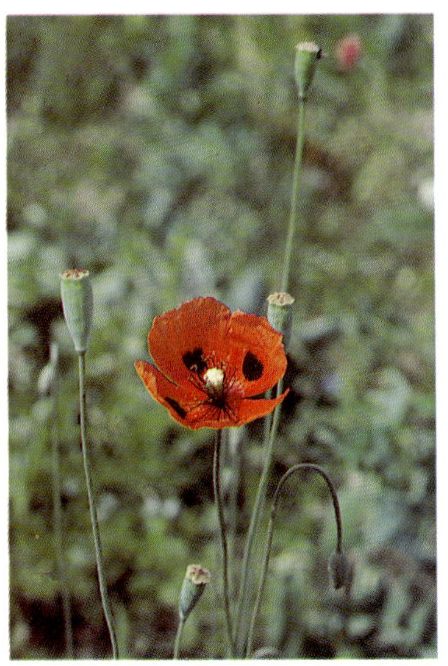

36. *Papaver dubium* Taxila, S. A. Sultan

32. *Hypecoum pendulum* Margalla hills, S. A. Sultan

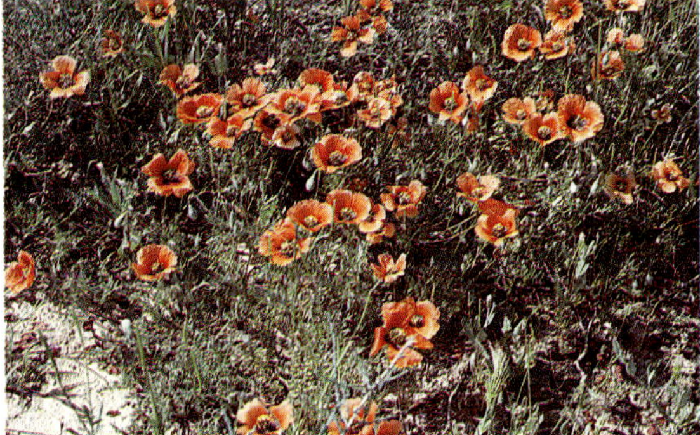

33. *Papaver nudicaule* Gilgit, Mark Mallalieu

38. *Papaver pavoninum* Swat, Fritz Berger

34. *Roemeria refracta* Quetta, Rubina A. Rafiq

37. *Papaver hybridum* Margalla hills, S. A. Sultan

35. *Roemeria hybrida subsp. dodecandra* Quetta, Rubina A. Rafiq

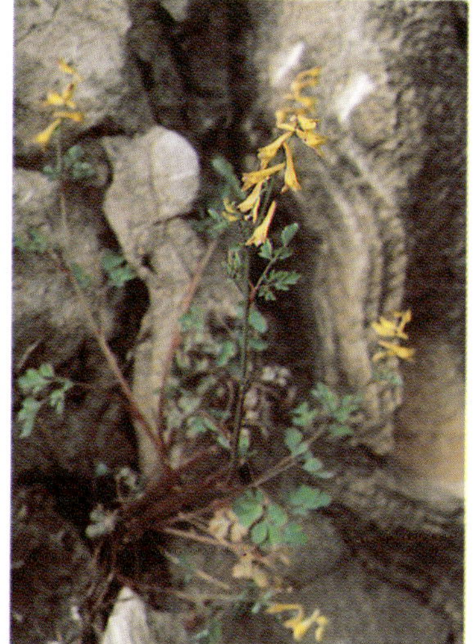

45. *Corydalis stewartii* Thandiani, Y. J. Nasir

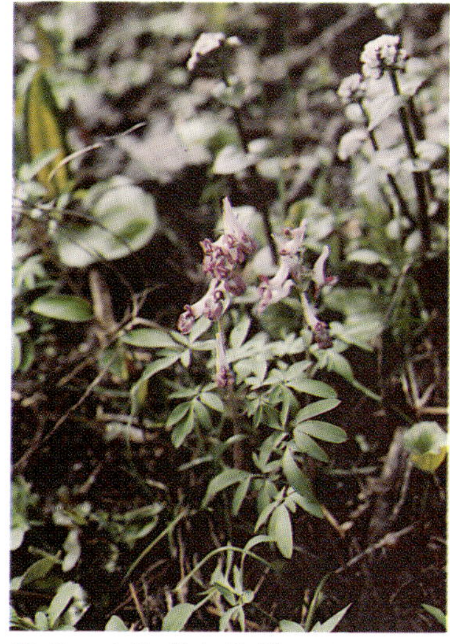

42. *Corydalis diphylla* Shogran, Rubina A. Rafiq

41. *Fumaria indica* Taxila, Rubina A. Rafiq

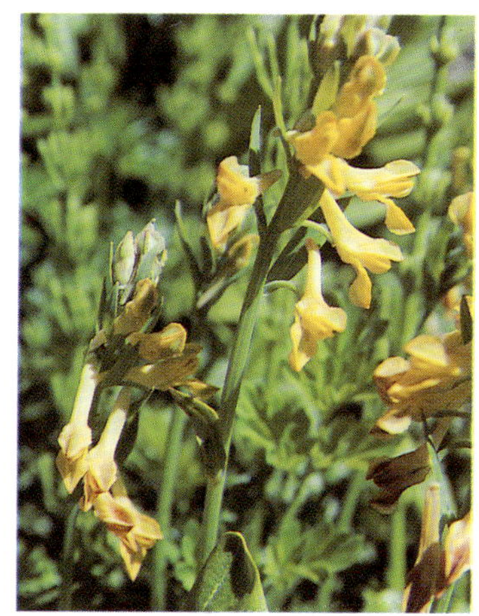

44. *Cordyalis govaniana* Mahodand, Swat, Fritz Berger

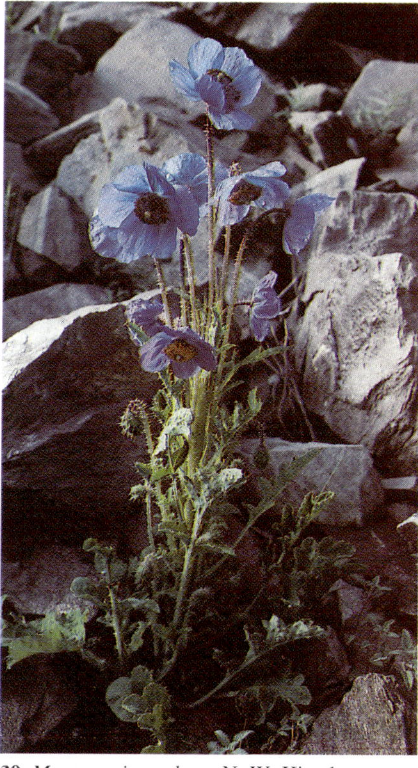

39. *Meconopsis aculeata* N. W. Himalaya, Chris Chadwell

43. *Corydalis flabellata* Hunza, Y. J. Nasir

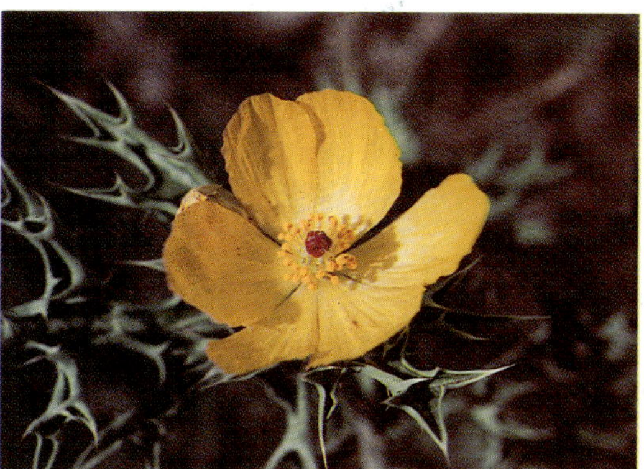

40. *Argemone mexicana* Lahore, Rubina A. Rafiq

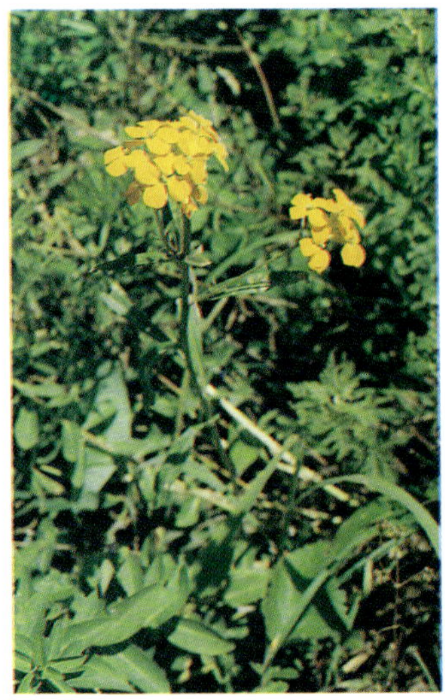

52. *Erysimum melicentae* Kalam, Y. J. Nasir

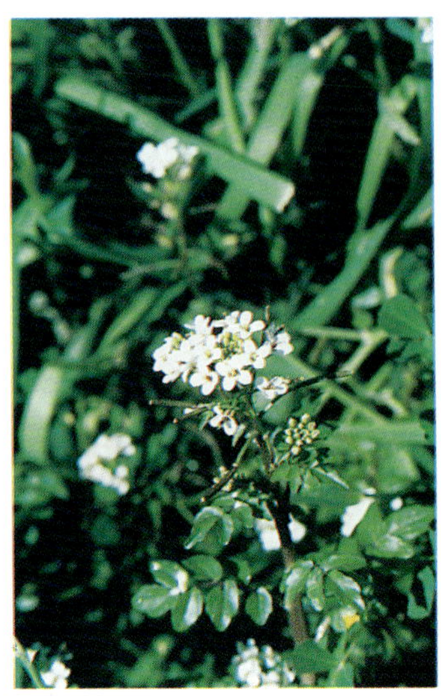

48. *Nasturtium officinale* Haripur, S. A. Sultan

46. *Diplotaxis griffithii* Kala Chita hills, S. A. Sultan

49. *Cardamine loxostemonoides* Kaghan, T. J. Roberts

47. *Crambe cordifolia subsp. kotschyana* Quetta, Rubina A. Rafiq

50. *Chorispora sabulosa* Hunza, Y. J. Nasir

51. *Malcomia cabulica* Taxila, Rubina A. Rafiq

55. *Farsetia jacquemontii* Karachi, Y. J. Nasir

58. *Capparis decidua* Salt Range, Rubina A. Rafiq

54. *Thlaspi andersonii* Swat, Fritz Berger

56. *Draba oreades* Deosai, Fritz Berger

53. *Isatis costata* Lower Swat, S. A. Sultan

57. *Capparis spinosa* Phandar, Gilgit, T. J. Roberts

62. *Reseda pruinosa* Makran, Y. J. Nasir

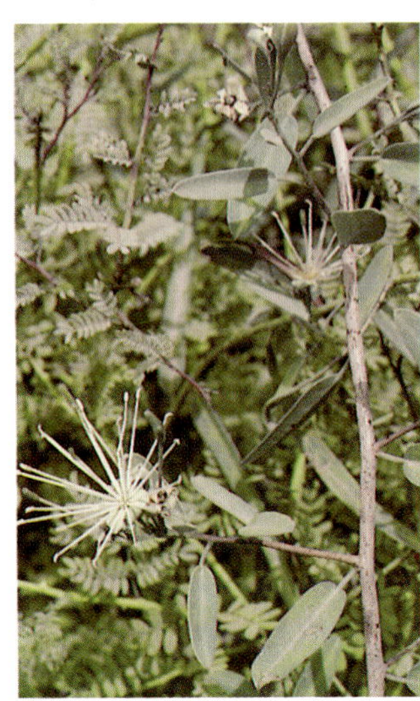

59. *Maerua arenaria* Karachi, Y. J. Nasir

63. *Viola biflora* Kaghan, Y. J. Nasir

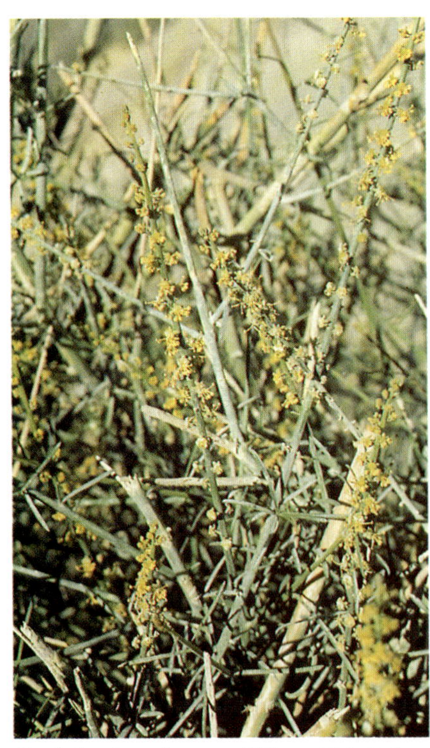

61. *Ochradenus baccatus* Makran, Y. J. Nasir

60. *Maerua crassifolia* Makran, Y. J. Nasir

68. *Spergularia marina* Lahore, Y. J. Nasir

67. *Polygala abyssinica* Margalla hills, S. A. Sultan

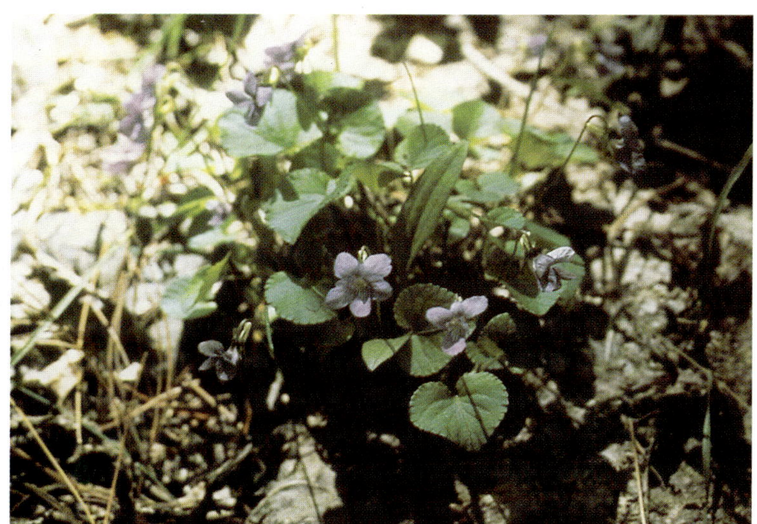

65. *Viola canescens* Hazara, Rubina A. Rafiq

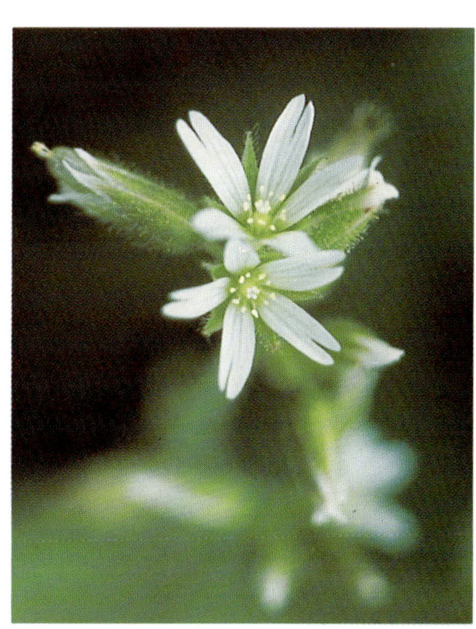

69. *Cerastium pusillum* Haripur, S. A. Sultan

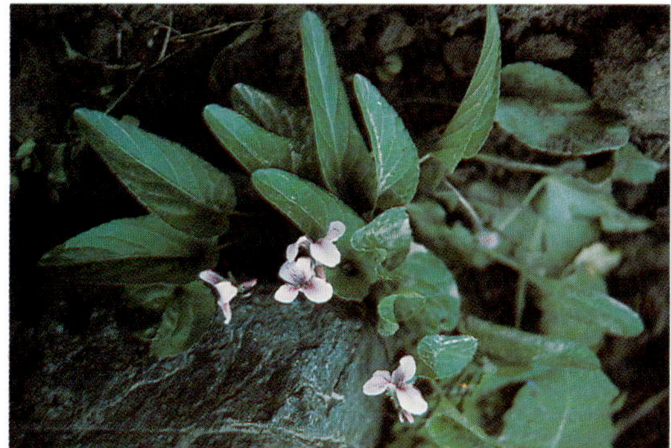

66. *Viola betonicifolia* Swat, Fritz Berger

64. *Viola stocksii* Kala Chitta hills, S. A. Sultan

70. *Arenaria griffithii* Kalam, Swat, Y. J. Nasir

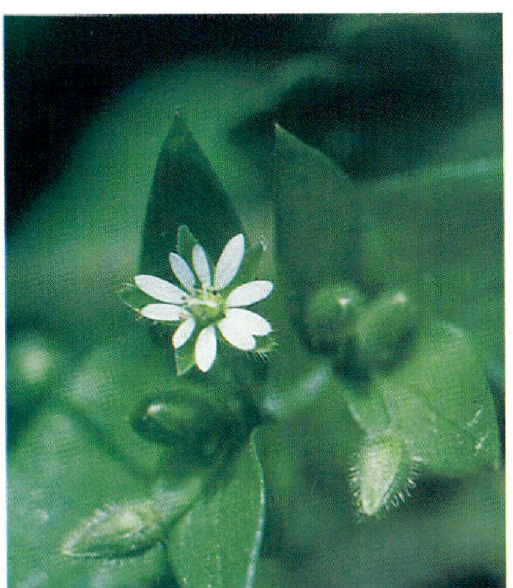

71. *Stellaria media* Islamabad, S. A. Sultan

74. *Acanthophyllum squarrosum* Ziarat, Rubina A. Rafiq

75. *Gypsophila cerastioides* Thandiani, Y. J. Nasir

73. *Vaccaria hispanica* Attock, S. A. Sultan

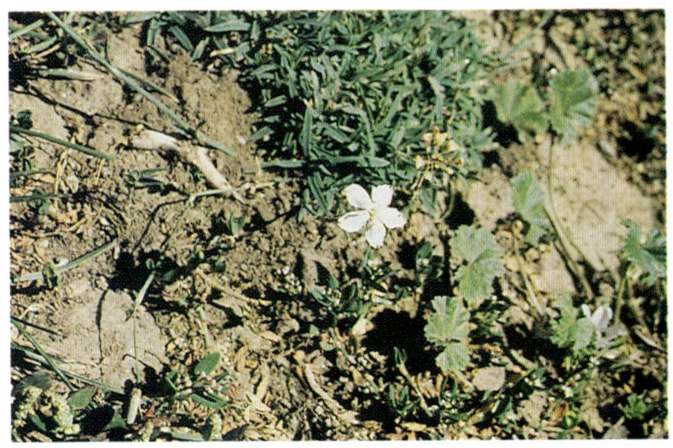

72. *Dianthus anatolicus* Kaghan, Rubina A. Rafiq

77. *Silene gonosperma* Kaghan, Y. J. Nasir

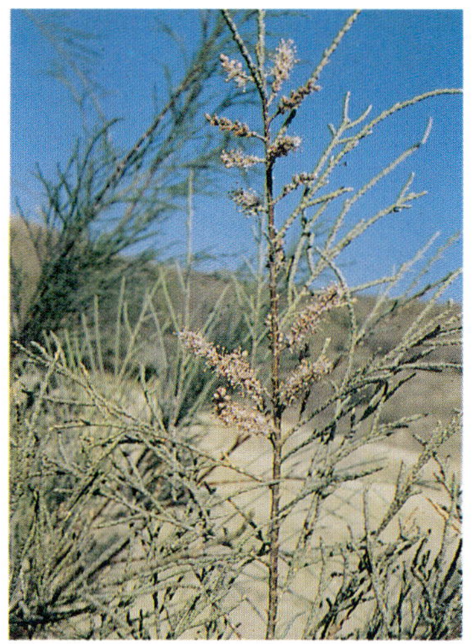

82. *Tamarix aphylla* Mekran, Y. J. Nasir

79. *Silene vulgaris* Gabral, Swat, Y. J. Nasir

80. *Reaumuria stocksii* Mekran, Y. J. Nasir

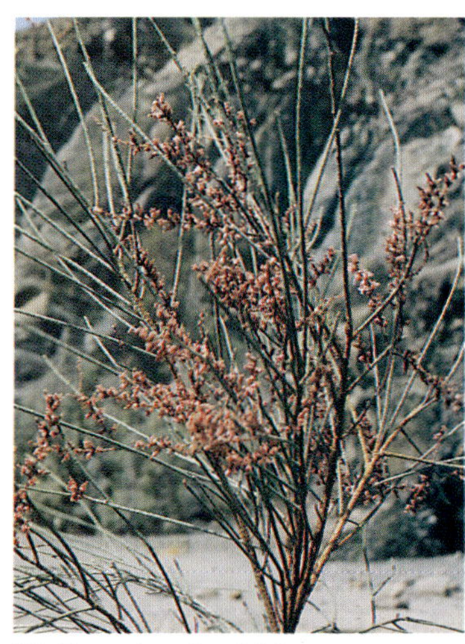

81. *Tamarix stricta* Mekran, Y. J. Nasir

78. *Silene viscosa* Shandur, Chitral, T. J. Roberts

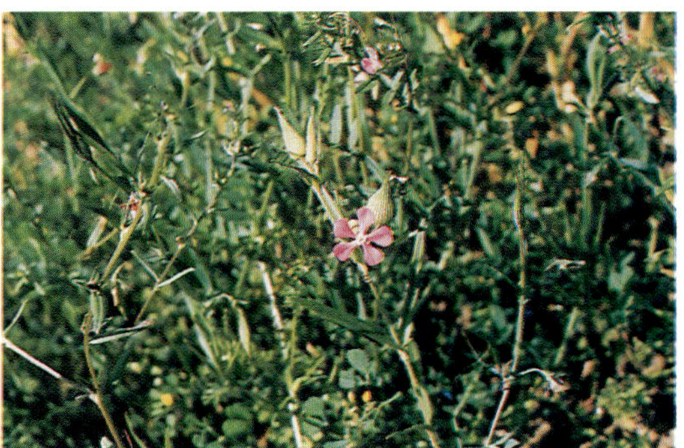

76. *Silene conoidea* Margalla hills, Rubina A. Rafiq

87. *Hypericum dyeri* Changla Gali, S. A. Sultan

85. *Hypericum perforatum* Kaghan, Y. J. Nasir

83. *Tamarix indica* Islamabad, Y. J. Nasir

86. *Hypericum oblongifolium* Hazara, Rubina A. Rafiq

84. *Tamaricaria elegans* Gabral, Swat, Y. J. Nasir

89. *Senra incana* Karachi, Y. J. Nasir

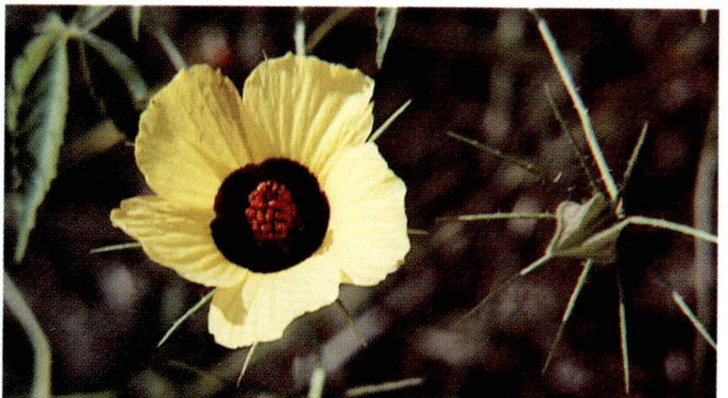

92. *Hibiscus caesius* Margalla hills, S. A. Sultan

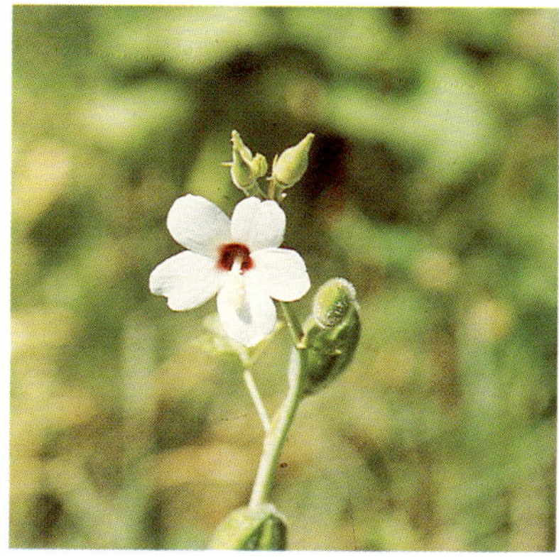

90. *Hibiscus lobatus* Margalla hills, S. A. Sultan

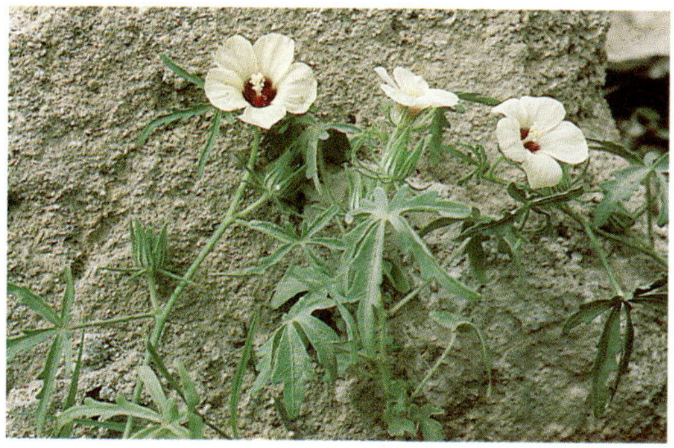

93. *Hibiscus aristivalvis* Karachi, Y. J. Nasir

91. *Hibiscus trionum* Shogran, Rubina A. Rafiq

94. *Malva neglecta* Mokshpuri, Y. J. Nasir

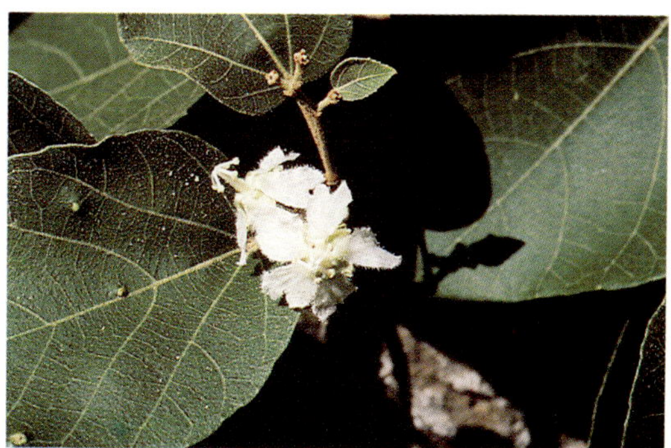

88. *Kydia calycina* Margalla hills, Y. J. Nasir

b ▲

◀ a

98. *Abutilon bidentatum* a, b: Changa Manga forest, Rubina A. Rafiq

100. *Sida pakistanica* Karachi, Y. J. Nasir

99. *Sida cordata* Rawalpindi, Y. J. Nasir

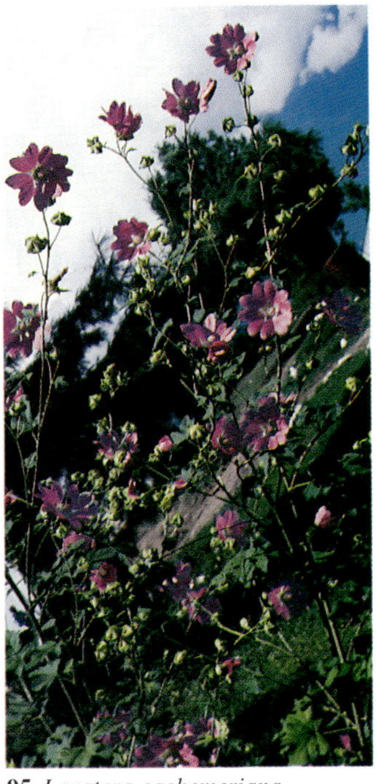

95. *Lavatera cachemeriana* Shogran, Rubina A. Rafiq

96. *Malvastrum coromandelianum* Islamabad, Rubina A. Rafiq

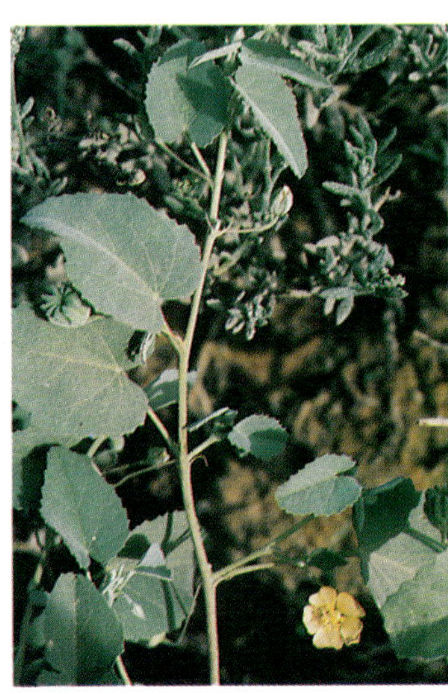

97. *Abutilon fruticosum* Karachi, Y. J. Nasir

74. **A. squarrosum** Boiss. Pl. 12

Caespitose tufted plant up to 20 cm tall. Flowers 1-3, in terminal corymbs, light pink. Petals narrow.

Distribution: Caucasus, Iran, Afghanistan, Pakistan.

Stony places from 1500-2500 m especially Balochistan. April-May.

8. Gypsophila 10 spp.

Annual or perennial glandular herbs with narrow leaves. Flowers many, ebracteate. Calyx tubular. Petals 5. Stamens 10. Styles 2. Capsule globose to oblong, 2-valved.

75. **G. cerastioides** D. Don Pl. 12

Perennial branched herbs with decumbent to ascending stems. Leaves obovate to subspathulate. Petals white, with purplish streaks, 7-9 mm long, notched.

Distribution: Pakistan, Kashmir to Bhutan, Assam.

Common in open places, slopes or rocks from 1900-3900 m. April-July.

9. Silene 28 spp. *Campion*

Annual or perennial herbs. Flowers solitary or not. Calyx tubular, 5-lobed. Petals with a distinct limb and claw. Stamens 10, in two whorls. Styles 3 or 5. Capsule dehiscing by 6 or 10 valves.

1.	+	Annuals. Flowers pink.	76. **S. conoidea**
	–	Perennials. Flowers white, greenish-white or dull purple.	2
2.	+	Flowers 1-2. Calyx with prominent purple or purplish-green veins.	77. **S. gonosperma**
	–	Flowers more than 5. Calyx nerves not purplish streaked.	3
3.	+	Plants sticky. Petal limb c. 8-9 mm long; claw 20 mm or more long.	78. **S. viscosa**
	–	Plants glabrous. Petal limb 5-6 mm long; claw 12-13 mm.	79. **S. vulgaris**

76. **S. conoidea** L. *Takla* Pl. 13

Plants up to 35 cm tall. Leaves linear-lanceolate. Petal tips entire or bilobed, pink. Capsule conical, 12-18 mm long.

Distribution: Cosmopolitan.

Common in cultivated fields from 1500-2200 m. March-April.

77. **S. gonosperma** Pl. 13
subsp. **Himalayensis** (Rohrb.) Bocquet

Slender plants up to 30 cm tall. Stem usually solitary. Calyx inflated with prominent purplish-tinged veins. Petals slightly exceeding calyx.

Distribution: Central Asia, Afghanistan, Pakistan, Kashmir to south west China.

Found in open rocky or stony places from 3300-5000 m. July.

78. **S. viscosa** (L.) Pers. Pl. 13

A robust plant up to 50 cm tall, with glandular sticky parts. Leaves 10-100 mm long, ovate-lanceolate to ovate. Calyx cylindrical, nerves greenish. Petals white, much incised. Capsule 12-14 mm long, ovoid, included in the calyx.

Distribution: Europe, Caucasus, Iran, Pakistan, north India.

Common in rocks from 2100-4200 m in Balochistan, Chitral and Gilgit. April-May.

79. **S. vulgaris** (Moench) Garcke *Bladder Campion* Pl. 13

Stems weak, erect to ascending, glabrous. Calyx inflated, 14-20 mm, greenish-yellow, many nerved. Flowers white or greenish-white. Petals bifid. Capsule 10-11 mm long, mouth constricted.

Distribution: North west Africa, Europe, Asia, Arabia, Pakistan, Kashmir to central Nepal.

Found from 1500-3100 m in Chitral, Swat, Gilgit, Hazara and Murree hills. June-July.

TAMARICACEAE *Tamarisk Family*

Shrubs or small trees with scale-like leaves. Flowers small, sometimes unisexual, pink to violet or white. Petals 4 or 5. Stigmas 3 or 4. Capsule conical or pyramidal, angled, dehiscing by 3-valves.

A widely represented family in Pakistan characteristic of saline soils.

1.	+	Flowers solitary axillary or terminal. Stamens many. Seeds hairy.	**1. Reaumuria**
	–	Flowers in racemes or spikes. Stamens up to 10 in number. Seeds glabrous.	2
2.	+	Leaves scale-like, styles distinct, seeds not beaked.	**2. Tamarix**
	–	Leaves not scale-like. Style absent. Seeds beaked.	**3. Tamaricaria**

1. Reaumuria 3-4 spp.

Prostrate herbs or undershrubs. Leaves somewhat fleshy. Flowers showy, bisexual. Sepals and petals 5. Stamens up to 70. Capsule globose to ellipsoid, dehiscing by 3-5 valves.

80. **R. stocksii** Boiss. Pl. 13

A profusely branched annual prostrate herb. Leaves linear 8-20 mm long. Flowers pink or pink-purple. Petals 5, free, 8-11 mm long. Capsule brown.

Distribution: South east Iran, Makran (Pakistan).

An attractive herb when in flower. Sandy places and low hills up to 200 m in Makran area. February-March.

2. Tamarix c. 24-26 spp. *Tamarisk*

A difficult genus taxonomically, characterized by woody habit, scale-like leaves and small flowers dispersed in simple or branched racemes. Stamens 4-10. Stigmas 3 or 4, rarely more or less. Capsule 3-5 - angled, conical or pyramidal, dehiscing by 3-valves.

1.	+	Stamens 7-10, in 2 series.	81. **T. stricta**
	–	Stamens 4-5, in one series.	2
2.	+	Trees or shrubs to 13 m. Leaves and bracts boat like. Capsule pyramidal, 2.5-3.5 x 1-5 mm.	82. **T. aphylla**
	–	Small shrubs rarely tree-like, 2-6 m tall. Leaves and bracts not or slightly boat-like. Capsule conical, c. 6 mm long.	83. **T. indica**

81. **T. stricta** Boiss. Pl. 13

Shrub, rarely a tree, 3-5 m tall. Leaves closely sheathing the stem, 3-4 mm long. Petals 5, pink to purplish-pink. Stamens 7-10; 5 short alternating with 5 long. Capsule conical, 5 x 1.5 mm.

Distribution: South Iran, Balochistan.

Usually dry river beds up to 900 m. In Balochistan, this species in autumn tends to turn golden-orange and sheds its leaves—a character rather unusual for the family. February-April.

82. **T. aphylla** (L.) Karst. *Farwan, Frash* Pl. 13

Flowers pinkish-white, bisexual. Usually a tree with reddish-brown or grey glabrous bark.

Distribution: Africa, Middle East, Afghanistan, Pakistan, India.

The largest of our tamarisks, which is also commonly planted. Growth is fast but tree is short-lived. Widely distributed throughout the country. The tamarisk forests are important source of wood and fuel in arid, sandy and saline areas where few other trees can grow.

83. **T. indica** Willd. *Pilchi, Gaz* Pl. 14

Shrubby plant with reddish-brown glabrous bark. Leaves ovate, 1-2 mm long, broader at the base. Bracts triangular, acuminate. Petals 5, 1-2 x 1 mm, obovate narrowed towards base.

Distribution: Afghanistan, Pakistan, India, Sri Lanka, Bangladesh. January-October.

One of the commonest shrubs of dry sandy and saline areas of Sindh, Punjab and Balochistan, also found in desert areas of N.W.F.P. Ascends to 3600 m in Tibet. Abundant on river banks, usually growing in association with *Salvadora oleoides* (wan).

3. Tamaricaria 1 sp.

A monotypic genus mainly central Asiatic, related to *Myricaria*, but with stamens that are free and pollen grains with reticulate sculpturing.

84. **T. elegans** (Royle) Qaiser & Ali Pl. 14
Syn.: *Myricaria elegans* Royle

A branched shrub up to 3 m with dark reddish-brown spreading branches. Leaves 5-16 x 4-5 mm. Flowers pinkish-white, in racemes up to 20 cm long. Capsule about 10 x 3 mm.

Distribution: Central Asia, northern Pakistan, Kashmir, Tibet.

Usually along streams from 2700-3700 m in Chitral, Gilgit and Kaghan. July-August.

GUTTIFERAE

St. John's Wort Family

A family of tropical trees, shrubs or herbs. Leaves simple, opposite. Flowers bisexual, regular. Sepals and petals 4-5. Stamens many, free or united in groups. Fruit a capsule, berry or drupe.

Hypericum 9 spp.

St. John's Wort

Glabrous perennial or annual herbs (in our species). Usually with gland dotted herbage. Leaves opposite, entire. Flowers showy, yellow. Sepals and petals 5. Fruit capsular.

1.	+	Flowers 5-7 cm broad. Leaves sessile, oblong to elliptic.	86. **H. oblongifolium**
	–	Flowers up to 3.5 cm broad.Leaves shortly petiolate, narrower.	2
2.	+	Shrubs. Leaves not gland-dotted. Styles 5.	87. **H. dyeri**
	–	Perennial herbs. Leaves gland-dotted. Styles 3.	85. **H. perforatum**

85. **H. perforatum** L. Pl. 14

Herbs up to 100 cm tall. Leaves varying from ovate to elliptic-oblong or linear. Inflorescence usually in many flowered subcorymbs. Petals oblanceolate. Stamens in groups of 3. Capsule 4-8 mm long.

Distribution: Europe, west Asia, Pakistan, Kashmir to Uttar Pradesh (India), Australia, south Africa.

A common slender herb in open places from 1800-3000 m. Said to be poisonous to livestock. July-September.

86. **H. oblongifolium** Choisy *Pinli* Pl. 14
Syn.: *H. cernuum* Roxb. ex D. Don

A branched shrub up to 1.5 m tall, with pendant branches. Leaves 25-80 x 8-30 mm. Inflorescence few-flowered. Petals obovate. Stamens in groups of 5. Capsule 9-17 mm long, ovoid.

Distribution: Pakistan to central Nepal.

Usually on steep rocky banks from 900-2200 m, fairly common in foothill Himalaya and Salt Range. March-April.

87. **H. dyeri** Rehder Pl. 14

Similar to the above species but with smaller flowers and narrower sepals and leaves.

Distribution: Pakistan, Kashmir eastward to Nepal.

On rocky cliffs from 1500-2300 m in Hazara, Swat and Dir especially Murree hills and Galis.

MALVACEAE

Cotton or Hibiscus Family

A mainly tropical and sub-tropical family of herbs, shrubs or trees characterized by stipulate, and usually alternate leaves. Stamens many, united to form a staminal tube. Fruit a capsule or schizocarpic, rarely a berry.

1.	+	Trees. Epicalyx segments 3, cordate.	**1. Kydia**
	–	Shrubs or perennial herbs. Epicalyx segment if present linear or ovate.	2
2.	+	Flowers violet or reddish-violet.	**2. Senra**
	–	Flowers yellow, white, pink or shades of it.	3
3.	+	Styles 5. Seeds usually kidney shaped.	**3. Hibiscus**
	–	Style 1, rarely divided. Seeds not kidney shapped.	4
4.	+	Epicalyx segments free.	**4. Malva**
	–	Epicalyx segments basally connate.	5
5.	+	Epicalyx segments 3. Seeds smooth or transversely ribbed.	**5. Lavatera**
	–	Epicalyx segments if present 6-9. Seeds radially ribbed.	6
6.	+	Epicalyx present.	**6. Malvastrum**
	–	Epicalyx absent.	7
7.	+	Capsule follicular, mericarps usually 2 or more seeded.	**7. Abutilon**
	–	Capsule not follicular, mericarps 1-seeded.	**8. Sida**

1. Kydia 1 sp.

Trees with unisexual or unisexual and bisexual flowers on the same plant. Flowers solitary or in panicles. Epicalyx segments 3. Capsule woody, subglobose.

88. **K. calycina** Roxb. *Pulian* Pl. 15

Monoecious trees up to 15 m tall. Leaves up to 10 x 15 cm, entire or 3-5-angled, base truncate or subcordate. Flowers white, in panicles. Epicalyx segments 3. Petals 8-10 x 4-5 mm, obovate, clawed, margin and base densely hairy. Capsule about 5 mm broad.

Distribution: Pakistan, Kashmir to Bhutan, India, Burma, Thailand and south west China.

Found in the sub-Himalayan tract east of the Indus from 900-1200 m. March-April.

2. Senra 1 sp.

Undershrubs with generally cordate to 3-lobed leaves and solitary axillary flowers. Sepals and petals 5. Capsule dehiscing by 5 valves.

89. **S. incana** Cav. Pl. 15

A sub-prostrate, spreading, branched perennial herb with softly hairy parts. Leaves cordate-orbicular, 3-lobed. Flowers axillary, solitary, violet with dark purple centre. Capsule ovoid or subglobose, 6 x 5 mm.

Distribution: Arabia, Ethiopia, Egypt, Pakistan.

Common in Sindh and lower Balochistan in sandy waste places or along field edges. October.

3. Hibiscus 9 spp.

Perennial or annual herbs or undershrubs, usually stellately hairy. Flowers axillary, usually of various colours, solitary. Epicalyx segments 3 or more. Styles 5. Fruit capsular. Seeds usually kidney shaped.

1.	+	Flowers whitish or creamish, 1.5-2 cm broad. Epicalyx segments absent or obsolete.	90. **H. lobatus**
	–	Flowers yellowish-white to creamish, with a dark centre, 3-5 cm broad. Epicalyx segments present.	2
2.	+	Calyx inflated in fruit.	91. **H. trionum**
	–	Calyx not inflated in fruit.	3
3.	+	Flowers yellow with a dark brown centre. Epicalyx segments 15-35 mm long. Seeds glabrous.	92. **H. caesius**

– Flowers pale yellow to creamish with a dark brown centre. Epicalyx segments 7-10 mm long. Seeds tomentose. 93. **H. aristivalvis**

90. **H. lobatus** (J.A. Murray) O. Kuntze Pl. 15

Annual herb up to 1m tall. Leaves 3-5–fid or partite, but not deeply divided. Flowers axillary solitary, 15-20 mm broad. Capsule oblong or ovoid, 10 x 5-7 mm.

Distribution: Tropical Africa, Asia.

Plains and foothill areas up to 900 m. Near water courses, open places. August-September.

91. **H. trionum** L. Pl. 15

A straggling annual with hispid parts. Leaves 3-5-partite, each segment further toothed or lobed. Epicalyx segments 7-12, with stiff hairs. Corolla white to pale yellow with a dark centre, 15-30 mm broad. Anthers yellow. Capsule 10-15 x 6-12 mm, oblong, hairy.

Distribution: South Europe, south Africa, Asia, Australia.

Found from plains to 2100 m. Flowering mostly throughout the year.

92. **H. caesius** Garcke Pl. 15

A suberect branched herb up to 1 m tall. Leaves palmately 3-5-partite, usually stellately hairy. The pretty flowers comprise of yellow corolla with dark centre and about 50 mm broad. Capsule similar to above species.

Distribution: Tropical and subtropical regions of Old World.

Found up to 900 m in low hilly areas of Punjab and N.W.F.P. August-September.

93. **H. aristivalvis** Garcke Pl. 15

A prostrate or subprostrate herb with hairy parts. Leaves 3-7-partite, lower surface hairy. Conspicuous flowers pale yellow to creamish with dark centre, about 30 mm broad. Capsule subglobose, 6-7 mm broad.

Distribution: Tropical and subtropical Africa, Asia.

Common in Sindh. Found from plains to 700 m. February.

4. **Malva** c. 8 spp. *Mallows*

Annual or perennial herbs or undershrubs. Leaves stipulate. Flowers axillary. Epicalyx of

3 free narrow segments. Calyx 5-lobed. Petals 5, notched at the apex. Fruit small, of 1-seeded mericarps.

94. **M. neglecta** Wallr. *Saunchal* Pl. 15

Prostrate perennial herb, with stellate pubescent parts. Leaves kidney shaped, 10-25 (-30) x 10-50 mm, long petioled. Flowers usually in groups of 3 or 4. Epicalyx segments narrow. Petals pale pink, apex notched. Capsule 5-6 mm broad, depressed.

Distribution: Cosmopolitan.

Found from plains to 4200 m. April-May, earlier in warmer places.

5. Lavatera 1 sp.

Annual or perennial herbs with hairy parts. Epicalyx segments connate at the base. Flowers axillary solitary or in fascicles or racemes. Seeds solitary.

95. **L. cachemiriana** Camb. *Wild Hollyhock* Pl. 16

Perennial branched herbs up to 2.5 m tall. Leaves suborbicular in outline, 5-angled, 25-100 x 30-60 mm. Epicalyx segments 3, fused at base, enlarging in fruit. Corolla 4-7 cm broad, lilac pink; petals notched at apex. Capsule 10-12 mm broad. Mericarps 1-seeded.

Distribution: Pakistan, Kashmir to Nepal.

Found from 1600-3200 m. July-September.

6. Malvastrum 1 sp.

Herbs or undershrubs with elliptic-ovate lobed or entire leaves. Bracts 3, linear. Flowers yellow. Mericarps indehiscent, strigose. Seeds glabrous.

96. **M. coromandelianum** (L.) Garcke Pl. 16

A branched undershrub up to 1 m tall. Leaves ovate, 10-50 mm long, coarsely serrate. Flowers usually in fascicles, yellow. Petals 5-8 x 4-5 mm. Fruit globose, mericarps 8-14 in number.

Distribution: Tropical and subtropical regions of both new and old worlds.

Widely naturalized in waste places, roadsides etc. from plains to 2400 m. Flowering mostly throughout the year.

7. Abutilon c. 12-14 spp.

Perennial herbs or undershrubs with stellately hairy parts. Leaves toothed, lobed. Epicalyx absent. Flowers yellow to orange-yellow. Fruit subglobose, splitting into 1-seeded mericarps.

+ Pedicels very slender. Mericarps 8-10 in number.
Corolla 8-12 mm broad. 97. **A. fruticosum**

– Pedicels not slender. Mericarps 13-16 in number.
Corolla c. 15 mm broad 98. **A. bidentatum**

97. **A. fruticosum** Guill. & Pers. Pl. 16

Perennial herb up to 1 m tall or undershrub. Leaves hairy, cordate at base. Flowers pale yellow. Fruit subcylindric, 7-9 x 5-6 mm; mericarps pubescent, 8-10, 3-seeded.

Distribution: Tropical Africa, Arabia, Pakistan, India.

Fairly common in lower Sindh and parts of Punjab up to 500 m. February-March.

98. **A. bidentatum** A. Rich Pl. 16

A spreading shrub with greenish stems and broadly ovate leaves. Corolla pale yellow, about 15 mm broad. Mericarps 13-16.

Distribution: Tropical and subtropical Africa, Arabia, Pakistan, India, China.

A very variable species in leaf and fruit size. Usually in arid dry places in Punjab and Sindh plains. Flowering mostly throughout the year.

8. Sida 9 spp.

Perennial herbs or undershrubs with pubescent parts. Leaves entire or not. Epicalyx usually absent. Flowers yellow, orange, white or pinkish. Carpels up to 10 in number.

+ Mericarps 5 in number. Pedicels 15-25 cm long.
Corolla 7-8 mm broad. 99. **S. cordata**

– Mericarps 7-8 in number. Pedicels up to 10 mm
long. Corolla 15-20 mm broad. 100. **S. pakistanica**

99. **S. cordata** (Burm.f.) Boiss. Pl. 16

Semiprostrate branched herb with stellate pubescence. Stems green or sometimes dark brown-purplish. Leaves 10-50 x 10-40 mm, lanceolate to broadly ovate, crenate to serrate. Corolla pale yellow. Fruit depressed globose, 3-4 mm broad.

Distribution: Tropical and subtropical parts of the old World.

A widespread subtropical weed found in the plains to 2200 m. Flowering throughout the year.

100. **S. pakistanica** Sultanul Abedin Pl. 16

Leaves 10-35 x 5-30 mm, ovate to elliptic. Flowers yellow. Petals 8-10 x 5-6 mm. Fruit depressed globose, enclosed in calyx. Mericarps 3-4 mm long.

Distribution: Socotra, Pakistan, India.

Fairly common in parts of lower Sindh, Balochistan and Punjab. Flowering mostly throughout the year.

BOMBACACEAE *Bombax Family*

A tropical family of trees with alternate, simple or compound deciduous leaves. Flowers large and showy, often appearing before the leaves. Sepals and petal 5. Fruit capsular.

Bombax 1 sp.

Generally large trees with trunk armed with prickles. Leaves digitately compound, long stalked; leaflets 5-9 in number. Inflorescence usually terminal and solitary. Capsule subwoody, dehiscing by 5-valves, woolly within.

101. **B. ceiba** L. *Simbal, Silk Cotton Tree* Pl. 17

A large to medium sized tree. Bark of trunk often with conical excrescences. Leaves compound; leaflets 5-7 in number, elliptic to elliptic-lanceolate. Flowers large, red, more or less fleshy, appearing on the branches before leaves appear and the nectariferous flowers are particularly attractive to birds. Capsule 10-13 cm long, woody, dehiscing by 5 valves.

Distribution: Pakistan, Kashmir to Bhutan, India, south China.

Widely naturalized. Plant parts are medicinal. The wood is used in match and box industry. The 'fluff' from the capsule is used in stuffing pillows and cushions. Found from the plains to 1200 m. December-March.

STERCULIACEAE *Sterculia or Cocoa Family*

A family of trees or shrubs, with alternate, simple or compound leaves. Flowers regular. Fruit a dry dehiscent or indehiscent capsule. Mostly ornamentals.

Melhania 3 spp.

Undershrubs or perennial herbs. Flowers bisexual, yellow, Petals present. Carpels generally united.

102. **M. futteyporensis** Munro ex Masters Pl. 17

Shrubby, up to 60 cm tall, with hoary-pubescent parts. Leaves elliptic-lanceolate or ovate-lanceolate, 2-8 x 1-5 cm, serrate-crenate. Inflorescence 2-4 flowered. Flowers 15-20 mm broad.

Distribution: Pakistan, India.

Usually found in dry rocky places up to 1400 m. March-December.

TILIACEAE *Linden Family*

A family of trees and shrubs with alternate simple leaves. Flowers in cymes or panicles. Sepals and petals 4-5. Carpels 2-5, usually united. Fruit capsular, dehiscent or a drupe or berry.

Grewia c. 9 spp.

Flowers bisexual, yellow or white, 5-merous. Sepals free, thickish, mostly coloured. Petals usually shorter than sepals, clawed. Drupe yellow or red, sometimes 4-lobed.

1.	+	Petals 10-15 mm long. Flowers solitary or in leaf opposed cymes.	103. **G. tenax**
	–	Petals 2.5-7 mm. Flowers in axillary cymes.	2
2.	+	Shrub or small tree with large ovate-lanceolate leaves, 4-12 cm long, pale green on under surface. Flowers creamish-white.	104. **G. optiva**
	–	Shrub with smaller ovate-oblong to narrow oblong leaves, 10-24 mm long ,whitish-grey on under surface. Flowers yellow.	105. **G. damine**

103. **G. tenax** (Forssk.) Fiori *Kango, Gwangi* Pl. 17

Shrub up to 2.5 m tall. Leaves ovate to elliptic-ovate to suborbicular, 5-40 x 4-35 mm, serrate. Flowers white, 20-25 mm broad. Drupe 2-4 - lobed, orange-yellow to pale reddish.

Distribution: Tropical and subtropical north Africa eastward to south Iran, Pakistan, India, Sri Lanka.

A xerophytic shrub widespread throughout drier and warmer parts of the country from Makran to Azad Kashmir to 600 m. February-August.

104. **G. optiva** Drum. ex Burret *Dhaman* Pl. 17

A small tree or large shrub 3-4 m tall. Stem and branches minutely hairy. Leaves 4-12 x 3-7 cm, ovate or broadly so. Flowers in extra-axillary cymes of up to 6, creamish-white to the inside, 3-3.5 cm broad. Drupe 2-4-lobed.

Distribution: Pakistan, eastward to Nepal.

Generally found in the sub-Himalayan tracts. A useful tree. The wood is used for furniture etc. The green bark is used by women to wash hair. Leaves and shoot are used as fodder and the fruit is eaten. April-September.

105. **G. damine** Gaertn. *Bihul* Pl. 17

Shrub with oblong to narrow oblong leaves up to 25 mm long; undersurface of leaves greyish-white. Flowers yellow, up to 16 mm broad, in axillary clusters of 4-5. Drupe 6-8 mm broad, more or less glabrous.

Distribution: Tropical Africa, Pakistan, west and south India, Nepal.

From Sindh up to N.W.F.P. and Salt Range on dry arid plains and hills up to 1400 m. July-September.

LINACEAE *Flax Family*

Herbs or shrubs with simple, entire leaves. Flowers regular, in cymose or racemose clusters. Fruit capsular or a drupe.

+ Herbs with narrow sessile leaves. Capsule 10-locular. **1. Linum**

– Shrubs with broad elliptic and petiolate leaves. Capsule 6-8-locular. **2. Reinwardtia**

1. Linum 2 spp.

Flowers ebracteate, 5-merous, yellow, red or blue. Stamens and carpels 5.

106. **L. corymbulosum** Reichenb. Pl. 17

An annual herb up to 30 cm tall. Stem solitary or few branched, glabrous. Leaves 8-12 x 4 mm. Flowers yellow, cymose. Capsule subglobose, c. 4 mm long.

Distribution: South Europe, Canary Islands, east Africa, Afghanistan, Pakistan, India.

Common in the plains and hills as far north as Hunza up to 2200 m. March-May.

2. Reinwardtia 1 sp.

Shrubs or undershrubs with elliptic to oblanceolate, petiolate leaves. Flowers 5-merous. Capsule 6-8-locular

107. **R. indica** Dumort. *Basant* Pl. 18

Shrub 1-1.5 m tall, often with dangling branches. Leaves 25-90 x 8-35 mm. Petiole up to 12 mm long. Flowers 30-40 mm broad, bright yellow. Capsule 10 mm broad, globose.

Distribution: Pakistan, Kashmir to Bhutan, India, Burma, Thailand, Indo-China, China.

Found in the sub-Himalayan tracts from 500-1900 m. In spring a showy shrub covered with bright yellow flowers. February-April.

MALPIGHIACEAE *Hiptage Family*

A family of trees or climbers with entire, opposite, exstipulate leaves. Flowers in axillary or terminal racemes or panicles. Sepals and petals 5. Stamens 10. Carpels 3, united. Fruit schizocarpic or a drupe.

Hiptage 1 sp.

Climbers with generally handsome, irregular and bracteate flowers. Petals 5. Ovary 3-lobed. Style filiform.

108. **H. benghalensis** (L.) Kurz *Endra* Pl. 18

A vigorous woody liana. Leaves elliptic, 90-200 x 35-85 mm, glabrous. Flowers creamish-white, yellow in the centre, in axillary and terminal racemes up to 24 cm long. Ovary pubescent. Style 13-16 mm long.

Distribution: Pakistan (sub tropical Himalaya), Assam, Sri Lanka, Burma, Indo-China, south west China, Andaman Isls.

An attractive climber with fragrant flowers often planted in gardens. The leaves are used in rheumatism and skin diseases. Forest shrubberies up to 800 m especially Murree foot hills, Kahuta, Margalla hills. March-April.

ZYGOPHYLLACEAE

Caltrop Family

Annual or perennial herbs with opposite and generally fleshy leaves. Flowers regular or irregular, solitary or in axillary cymes. Sepals and petals 5. Carpels 4-5, syncarpous. Fruit capsular, rarely a drupe or berry.

1.	+	Stipules spiny.	**1. Fagonia**
	–	Stipules not spiny.	2
2.	+	Leaves alternate, simple, the lamina multifid. Stamens 12 or more.	**2. Peganum**
	–	Leaves opposite, compound. Stamens 8-10.	3
3.	+	Prostrate herb. Leaves pinnate. Stamens of unequal length. Fruit hairy.	**3. Tribulus**
	–	Shrub or suffruticose herbs. Leaves 1-3–foliolate. Stamens of one size. Fruit subglabrous to glabrous.	**4. Zygophyllum**

1. Fagonia 6 spp.

Annual or perennial much branched herbs. Flowers regular, 5-merous, pink or purplish-pink. Petals 5, free, clawed, caducous. Stamens 10, free. Capsule 5-angled, more or less subglobose in outline, dehiscing into 5 one-seeded cocci.

109. **F. indica** Burm. f. Pl. 18

Perennial or annual glabrous or subglabrous low growing herb. Stipules spiny. Leaves unifoliolate, basal one trifoliate. The pretty flowers are pink to pink-purple, about 10 mm broad. Petals 5 x 3 mm. Capsule subglobose, 3-4 mm broad.

Distribution: North east tropical Africa eastward to Pakistan, subtropical India.

A much branched spiny desert plant. Widespread in dry places in the plains and hilly tracts up to 500 m. Flowering most of the year.

2. Peganum 1 sp.

Perennial herbs. Flowers solitary, regular. Sepals and petals 5. Stamens 12 or more. Capsule 3-4-locular.

110. **P. harmala** L. *Harmal* Pl. 18

Glabrous branched herb up to 50 cm tall. Leaves pinnately dissected into linear segments.

Flowers white or creamish, about 30 mm broad. Stamens 15. Capsule 6-8 mm broad, trigonal.

Distribution: North Africa, Europe, Russia, Afghanistan, Pakistan, Kashmir.

A common plant of waste places and roadsides from 200-2500 m. Widespread from southern Sindh, Balochistan up to Hunza and Baltistan. The seeds and leaves are medicinal. April-October.

3. Tribulus 4 spp.

Annual or perennial herbs. Leaflets 3-10 pairs. Flowers solitary axillary, yellow. Sepals and petals 5. Stamens 10, in 2 whorls. Fruit splitting, winged or spinose, hairy.

111. **T. terrestris** L. *Gokhru, Tirkundi* Pl. 18

Fruit spinose, 8-10 x 4-8 mm, pubescent. Flowers 10-15 mm broad. Leaves 25-50 cm long.

Distribution: Cosmopolitan.

Found in sandy barren places or as a weed widespread from Balochistan, N.W.F.P. to Gilgit and Baltistan up to 2800 m. Plant parts are medicinal. Flowering mostly throughout the year.

4. Zygophyllum 6 spp.

\+ Perennial shrubs 20-50 cm tall. Leaves compound. Flowers yellowish-white, 6-8 mm broad. 112. **Z. propinquum**

– Annual up to 20 cm tall. Leaves simple. Flowers yellow, up to 5 mm broad. 113. **Z. simplex**

112. **Z. propinquum** Decne. Pl. 18

Stem much branched. Leaves pinnately bifoliolate; leaflets 5-10 mm long, cylindrical. Capsules 10-13 x 4 mm, soon falling off.

Distribution: North Africa, Saudi Arabia, Jordan, Kuwait, south Iran, Afghanistan, Pakistan.

Fairly common in Sindh and lower Balochistan in sandy soils, up to 700 m. Seeds said to be anthelmintic. September-June.

113. **Z. simplex** L. *Alethi, Pulani* Pl. 18

A delicate much branched glabrous herb with succulent bright green leaves. Flowers yellow. Capsule deflexed, top shaped.

Distribution: Cape Verde Isls., north Africa, Iran, Pakistan, India.

Common in sandy saline plains in Sindh and lower Balochistan. The plant is browsed by camels. The leaves and seeds are medicinal. August-May.

GERANIACEAE

Geranium Family

Annual or perennial herbs with leaves variously divided. Flowers regular, usually in cymes. Sepals and petals 5, free. Stamens twice as many as the petals. Capsule splitting into 1-seeded mericarps.

+	Leaves not longer than broad. Fertile stamens 10. Stylar axis in fruit elastically coiling upwards.	**1. Geranium**
–	Leaves longer than broad. Fertile stames 5. Stylar axis not so.	**2. Erodium**

1. Geranium 22 spp.

Cranes Bill

Leaves stipulate. The brightly coloured flowers are 5-merous. Fruit of 5 mericarps which remain attached to the elastically coiling stylar axis.

1.	+	Annual or biennials of the plains and hills up to 2300 m.	2
	–	Perennial of mountainous areas from 2100-4800 m.	3
2.	+	Flowers 12-15 mm broad, pink with a dark centre.	114. **G. ocellatum**
	–	Flowers 6-7 mm broad, lilac.	115. **G. rotundifolium**
3.	+	Flowers mauve to pinkish-blue. Leaves 5-angled. Stipules egg-shaped, 8-20 x 6-12 mm.	116. **G. wallichianum**
	–	Flowers white or pale rose-lilac. Leaves suborbicular in outline; stipules linear, 5-7 mm long.	117. **G. collinum**

114. **G. ocellatum** Camb. *Bhand* Pl. 19

A diffuse slender annual or biennial with pretty pink flowers with black centres. Leaves suborbicular, 8-50 mm broad.

Distribution: East Africa, Afghanistan, Pakistan, Kashmir to Nepal, north Assam, west China.

Found in open places, shady banks etc. from 800-1800 m in *Pinus roxburghii* zone. March-April.

115. **G. rotundifolium** L. Pl. 19

Low growing annual with glandular hairy, often reddish stem. Leaves reniform. Flowers small, 6-7 mm broad, pretty pink.

Distribution: West and central Europe, Siberia, Turkey, Iran, west and east Mediterranean, Africa, Afghanistan, Pakistan, temperate and tropical Himalaya.

Fairly common in the plains in spring. Found up to 3000 m. Plant parts medicinal. March-April.

116. **G. wallichianum** D. Don ex Sweet *Rattanjot* Pl. 19

Prostrate to decumbent perennial, with diffuse spreading branches. Leaves 5-palmatisect into ovate-rhombic segments. Stipules conspicuous. Peduncles up to 13 cm long, usually 2-flowered. Flowers 32-40 mm broad, purplish-blue to mauve.

Distribution: Afghanistan, Pakistan, Kashmir to Bhutan.

Common and gregarious in forest undergrowth from 1600-3300 m. The large flowers make this one of the more showy geraniums in the Himalaya forests. Plant parts are used in tooth aches. July-September.

117. **G. collinum** Steph. ex Willd. Pl. 19

A prostrate branched perennial with deeply divided suborbicular or orbicular leaves, 15-80 mm broad. Peduncles up to 15 cm long, ascending recurved. Flowers c. 3-4 cm broad. Sepals with a 1-2 mm awn. Petals with a ciliate claw.

Distribution: Central and south Russia, west Siberia, Romania, Turkey, central Asia, Caucasus, Turkey, Iran, Afghanistan, Pakistan, Kashmir.

The species is variable in leaf size, shape, incision, degree of hairiness and flower colour. White colour forms are occasional. Usually found along damp courses or ditches from 2800-4800 m in dryer north western regions such as Swat, Chitral and Gilgit. July-August.

2. Erodium 7 spp.

Annual or perennial herbs with lobed or divided leaves. Sepals and petals 5. Stamens 5, alternating with 5 staminodes. Beak of fruit (capsule) plumose within on dehiscence. Mericarps 2.

+	Leaves crenate to pinnatifid, not deeply lobed.	118. **E. malacoides**
–	Leaves 1-2 - pinnatisect.	119. **E. cicutarium**

118. **E. malacoides** (L.) L'Herit ex Aiton Pl. 19

Prostrate branched annual. Leaves oblong or oblong-ovate, 15-37 x 10-30 mm, crenate or more or less serrate. Flowers 3-6, on peduncles up to 3 cm long. Beak of fruit 2.5-3.5 cm long. Mericarps 4.5 mm long.

Distribution: North and south America, north Africa, Mediterranean, Turkey, Syria, Arabia, Iraq, Iran, Caucasus, central Asia, Pakistan.

A spring annual with small pretty pink flowers. Found as an undergrowth in shady places from 700-1000 m. March-April.

119. **E. cicutarium** (L.) L'Herit ex Aiton Pl. 19

Similar, but with larger and deeply divided leaves. Flowers 2-8, reddish-purple.

Distribution: Europe, north Africa, south west and central Asia, Arabia, Caucasus, Siberia.

Common in hills from 700-2400 m. Replaces *E. malacoides* at higher elevations. March-April.

OXALIDACEAE *Wood Sorrel Family*

Shade loving annual or perennial herbs with 3-foliolate leaves. Flowers regular, 5-merous, quite showy. Stamens 10, basally connate. Fruit capsular.

Oxalis 5 spp.

Bulbous or rhizomatous creeping herbs. Filaments 5 long, alternating with 5 short ones.

1.	+	Flowers yellow.	2
	–	Flowers pink or mauve.	3
2.	+	Stem creeping. Bulbils absent. Scape barely raised above foliage. Flowers 6-7 mm long.	120. **O. corniculata**

– Stem reduced. Bulbils present. Scapes well exserted above the foliage. Flowers 20 mm long. 121. **O. pes-caprae**

3. + Leaflets triangular in outline. Pedicels glabrous. 122. **O. latifolia**

– Leaflets obcordate. Pedicels hairy. 123. **O. corymbosa**

120. **O. corniculata** L. *Khatti Buti* Pl. 20

Leaflets 4-12 x 8-30 mm,, obcordate. Flowers solitary or in axillary 2-5-flowered umbels. Capsule up to 2.5 cm long, subcylindric, pubescent.

Distribution: Cosmopolitan.

Found in waste places, cultivated areas from plains to 2700 m. The leaves can be used as a vegetable. March-December.

121. **O. pes-caprae** L. *Yellow Sorrel* Pl. 20

A more robust plant than the preceding with attractive lemon yellow flowers on well developed scapes.

Distribution: Native to south Africa. Widely naturalized elsewhere.

Found as a weed or in shady places in the plains up to 900 m. January-April.

122. **O. latifolia** Kunth Pl. 20

A stoloniferous herb with underground bulbils. Leaflets 15-35 (-40) x 30-70 mm, triangular in outline. Flowers subumbellate, pale to deep pink. Petals 8-10 mm long.

Distribution: Native to tropical south America.

Occurs as a weed of cultivated areas and near water courses from 1200-2100 m. July-September.

123. **O. corymbosa** DC. Pl. 20

A plant similar to *O. latifolia* but with obcordate leaves. Flowers mauve.

Distribution: Native to south America. Widely naturalized elsewhere.

Found in the plains to 800 m. March-April.

BALSAMINACEAE

Balsam Family

A handsome family related to *Geraniaceae*, but differing in the irregular spurred flowers and the anthers that form a hood or ring around the ovary.

Impatiens 12 spp. *Balsam*

Perennial or annual herbs. The pretty shaped flowers are solitary axillary or disposed in subcorymbs, subumbels or racemes. Sepals 3, the lowermost petaloid and spurred. Petals 3-5, the anterior (lower) the largest, lateral petals often lobed.

1. + Flowers white, 4-5 (-6) mm long, almost spurless. 124. **I. brachycentra**

– Flowers pink-white, yellow or a mixture of both, more than 12 mm long. Spur well developed. 2

2. + Flowers yellow. 125. **I. edgeworthii**

– Flowers pink, lilac or pink-yellow. 3

3. + Flowers pink and yellow. 126. **I. bicolor**

– Flowers pink to lilac or shade of it, sometimes suffused with yellow. 4

4. + Robust plants 100 cm or more tall. Flowers 25-30 mm long, generally darker coloured. 127. **I. glandulifera**

– Slender plants up to 50 cm tall. Flowers 8-20 mm long, pink or pink-white. 5

5. + Capsule 14-19 mm long, somewhat clavate, nodding. Throat of flower spotted brown. 128. **I. thomsonii**

– Capsule 10-12 mm long, broadly linear. Throat of flower some-times suffused with yellow or white. 129. **I. flemingii**

124. **I. brachycentra** Kar. & Kir. *Spurless Balsam* Pl. 20

The smallest flowered balsam. Spur obsolete, about 1 mm long, sometimes tinged yellow. Lower petal 3-4 mm long. Capsules erect, 12-22 mm long.

Distribution: Central Asia, Sinkiang, Pakistan, Kashmir to Uttar Pradesh (India).

Common and gregarious during monsoons from 1800-3000 m in moist temperate Himalaya. July-August.

125. **I. edgeworthii** Hook. f. Pl. 20

A much branched leafy annual up to 60 cm tall, with yellow flowers. Lower sepal with spur 20-30 mm long. Lower petal 7-8 x 10-14 mm, apex 2-lobed. Capsules 20-30 mm long, broadly linear, erect.

Distribution: Pakistan, Kashmir to Nepal.

Gregarious and very common during monsoons preferring shady banks, waste places from 1800-3000 m especially Murree and Hazara hills, Swat and Kaghan. July-September.

126. **I. bicolor** Royle Pl. 20

Flowers pink-yellow. Lower sepal broadly conical with a short spur. Lower petal broad, orbicular, 6-8 x 12-14 mm, pink. Capsule cylindric, 20-30 mm long, erect.

Distribution: Pakistan, Kashmir to west Nepal.

One of the more spectacular flowered balsams in our area. Fairly common in open places from 1700-3000 m in Swat, Kaghan and Kashmir. More adapted to lower elevations and drier locations. July-August.

127. **I. glandulifera** Royle *Himalayan Balsam* Pl. 21

Robust plants up to 2 m tall with succulent stems. Leaves 50-150 x 15-60 (-65) mm, elliptic-ovate to lanceolate, serrate. Lower sepal saccate, abruptly ending in an incurved spur 5-6 mm long. Lower petal more or less orbicular, 6-8 x 11-12 mm. Capsule broadly clavate, 14-20 mm long, nodding.

Distribution: Pakistan, Kashmir.

In damp places, forest clearings, field borders, roadside ditches from 1600-4300 m in temperate Himalaya from Hazara to Kashmir. Common in Kaghan valley; often growing gregariously . Flowers showy pink-purplish. July-August.

128. **I. thomsonii** Hook. f. Pl. 21

A slender annual up to 50 cm tall, with serrated elliptic-lanceolate leaves and lilac flowers. Lower sepal conical, with spur 8-11 mm long. Capsule 14-19 mm long, nodding.

Distribution: Pakistan, Kashmir to Uttar Pradesh.

Found in the dry inner Himalaya from 1700-4000 m. Usually as an undergrowth. July-August.

129. **I. flemingii** Hook. f. Pl. 21

Flowers lilac, suffused with some yellow. Lower sepal with spur 20-25 mm long, conical.

Spur slender, straight. Lower petal 5-6 x 10-12 mm. Capsule broadly linear, 10-12 mm long, erect.

Distribution: Pakistan, Kashmir.

The long spurs of the flowers add to the beauty of this plant. Usually found along pathways, embankments from 1500-3200 m. July-August.

RUTACEAE

Citrus Family

Trees or shrubs with glandular parts. Leaves simple or compound. Flowers solitary axillary or in terminal cymes or panicles. Sepals and petals 4-5. Fruit a capsule, berry or drupe.

1.	+	Shrubs or small trees, often prickly.	**1. Zanthoxylum**
	–	Perennial herbs or semishrubs.	2
2.	+	Leaves simple. Flowers small, in corymbose cymes.	**2. Haplophyllum**
	–	Leaves compound. Flowers large and showy, in racemes.	**3. Dictamnus**

1. Zanthoxylum 1 sp.

Shrub with stem and branches prickly. Leaves compound. Fruit subglobose, splitting.

130. **Z. armatum** DC. *Timbar* Pl. 21

Leaves imparipinnate, aromatic, gland dotted. Petiole and midrib winged. The small yellow flowers are unisexual. Male: stamens 6-8 in number. Female: ovary 3-lobed. Capsule red and splitting in two when ripe.

Distribution: Pakistan, Kashmir to south west China, Taiwan, Philippines.

Found in the sub-Himalayan tracts from 600-1500 m. The twigs are used as a substitute for tooth brushes and stems in making walking sticks. March-April.

2. Haplophyllum 7 spp.

Stems branched, often woody below. Sepals and petals 5. Stamens 10. Capsule 3-5-lobed, dehiscent.

+	Flowers reddish. Petals 7-9 mm long.	131. **H. erythraeum**

– Flowers yellow. Petals 4-5 mm long. 132. **H. tuberculatum**

131. **H. erythraeum** Boiss. Pl. 21

Glabrous herbaceous perennial, up to 25 cm tall, branching from the base. Flowers rusty red, in lax corymbs. Petals 7-9 mm long. Capsule 4.5-5 mm broad, surface beset with minute appendages.

Distribution: Afghanistan, north Balochistan (Pakistan).

Found in sandy and stony places up to 1900 m in hilly areas from Balochistan to Chitral. April-May.

132. **H. tuberculatum** (Forssk.) Juss. Pl. 21

Perennial herb up to 30 cm tall. Leaves 7-40 x 2-15 mm, lanceolate to oblong. Flowers yellow, corymbose. Capsule 3-4 mm broad, glabrous or not.

Distribution: North Africa, south west Asia.

Arid sandy places or low hills up to 1400 m, particularly in Balochistan and Sindh Kohistan. March-September.

3. Dictamnus 1 sp.

Perennial herbs with imparipinnate, gland dotted leaves. Capsule woody, 5-lobed.

133. **D. albus** L. *Burning Bush, Gas Plant* Pl. 22

Erect herbaceous perennial up to 90 cm tall. Leaflets 35-100 x 12-35 mm long, elliptic to ovate-lanceolate. The tall spikes of pale pink flowers are striped darker. Petals 25-30 mm long, oblong, clawed. Capsule 18-20 mm broad.

Distribution: Temperate Eurasia, Pakistan, Kashmir to Uttar Pradesh, (India).

Plant parts are fragrant; also secrete a volatile oil which can become inflammable in hot weather. Not common. In forest shrubberies and damp meadows from 2200-3300 m in temperate Himalaya in Swat Kohistan, Kaghan and Kashmir. June-July.

CELASTRACEAE

Spindle-tree Family

Trees or shrubs with simple leaves. Flowers small, sometimes unisexual, solitary or dispersed in cymes. Sepals and petals 4-5. Fruit capsular or not.

Maytenus 3 spp.

Small trees or shrubs, usually spiny. Leaves alternate, exstipulate. Stamens 4-5. Capsule 3-angled, 1-4-seeded.

134. **M. royleanus** (Wall. ex Lawson) Cufodont. *Pataki* Pl. 22

Spines 10-25 mm long, axillary, not bearing leaves or flowers. Leaves 10-50 x 7-25 mm, ovate to obovate. Flowers in axillary cymes which are shorter than the leaves. Petals 2-2.5 mm long, whitish. Capsule 5-7 mm broad.

Distribution: Afghanistan, Pakistan, India.

Common on hot dry slopes in Salt Range and sub-Himalayan tracts up to 1500 m. Flowering mostly throughout the year.

RHAMNACEAE

Buckthorn Family

Trees, shrubs or climbers with simple, coriaceous leaves. Flowers regular, in axillary cymes, corymbs or panicles, sometimes unisexual. Petals 5. Stamens 10. Fruit usually a drupe.

1.	+	Weak plants climbing by means of tendrils.	**1. Helinus**
	–	Shrubs or trees.	2
2.	+	Fruit with 1 pyrene.	**2. Zizyphus**
	–	Fruit with 3 pyrenes.	**3. Rhamnus**

Helinus 1 sp.

Unarmed climbers with greenish-yellow flowers. Ovary 3-locular, style 3-fid. Fruit subglobose with 1-seeded, three cocci.

135. **H. lanceolatus** Wall. ex Brandis Pl. 22

Glabrous plant climbing by means of tendrils. Leaves 20-60 x 8-20 mm, 3-nerved. Flowers 3-9 in number. Capsule 5-7 mm broad.

Distribution: Pakistan, Kashmir to Nepal.

The sub-Himalayan tracts and adjacent plains to 900 m. In shrubberies. April.

Zizyphus c. 5 spp. *Ber*

Thorny shrubs or small trees with stipules modified into spines. Leaves 3-nerved, coriaceous. Flowers 5-merous, greenish-yellow. Drupe fleshy, oblong to subglobose; pyrenes 1-seeded.

136. **Z. oxyphylla** Edgew. *Phitni* Pl. 22

A shrub or small tree with slender reddish-brown branches and glabrous, ovate-lanceolate and acuminate leaves. The edible drupes 8-10 mm broad, subglobose, bright red, tuning darker with age.

Distribution: Pakistan, Kashmir, India.

Widespread in the drier foothill zone. Forest shrubberies. Found from 600-1800 m in Himalaya. June-September.

Rhamnus 6 spp. *Buckthorn*

Evergreen or deciduous trees or small shrubs. Flowers small, greenish or greenish-yellow, sometimes unisexual. Stamens 4-5. Drupe fleshy or dry and 1-seeded.

137. **R. triquetra** (Wall.) Brandis *Girgithan* Pl. 22

Medium sized tree with tomentose shoots. Leaves subcoriaceous, 6-13 x 2-7 cm, ovate to elliptic, under surface velvety. Lateral pairs of nerves 7-8. Flowers 2-3 mm broad. Drupes 4-5 mm long, 3-lobed.

Distribution: Pakistan, Kashmir to Nepal.

Typical of dry hilly tracts up to 1800 m in N.W.F.P. and Margalla Hills. July-August.

VITACEAE *Grape Family*

Mostly lianas or woody creepers with simple or palmately lobed leaves. Tendrils often present. Flowers small, sometimes unisexual. Fruit a 1-4-seeded berry.

Tetrastigma 1 sp.

138. **T. serrulata** (Roxb.) Planch. Pl. 22

A scrambler with digitately 5-palmate leaves, climbing by means of tendrils. Leaflets elliptic-serrate. Flowers small, greenish, in umbel-like clusters.

Distribution: Pakistan eastward to Nepal, west China.

A rare plant found only in the Margalla hills above Islamabad from 800-1100 m. August-September.

SAPINDACEAE *Soap Nut Family*

Trees, shrubs or climbers with simple or compound, alternate leaves. Flowers small, unisexual, regular or irregular, disposed in panicles or cymes. Sepals and petals 3-5, sometime absent. Stamens 5-10. Fruit a nut, berry or capsule, sometimes winged.

+	Young shoots sticky. Stems unarmed. Petals absent. Capsule winged.	**1. Dodonaea**
–	Young shoots not sticky. Stems armed. Petals present. Capsule not winged.	**2. Stocksia**

1. Dodonaea 1 sp.

Dioecious shrubs. Leaves simple. Fruit capsular, 2-6-valved, winged.

139. **D. viscosa** (L.) Jacq. *Sanatha* Pl. 22

Evergreen shrub to 4 m tall, with resinous parts. The bright green shining leaves oblanceolate, 20-70 x 7-22 mm. Male flowers: Stamens 6-8, anthers red. Female flowers: Ovary 3-angled. Capsule 12 x 15 mm, winged.

Distribution: Pantropical.

A component of the dry scrub vegetation of the sub-Himalayan tracts and Salt Range. Often dominant and gregarious on exposed dry slopes up to 900 m. Excellent as a hedge plant and valued for fuel wood. February-March.

2. Stocksia 1 sp.

Spinose shrubs with simple entire leaves. Petals 4, free, without scales. Stamens 8. Fruit capsular.

140. **S. brahuica** Benth. *Koh Tor, Mountain Peach* Pl. 22

Shrub 1.5-3 m tall, branched. Branches modified into spines up to 6 cm long. Leaves narrow, 12-26 mm long. Flowers small, greenish-yellow. Capsule 25-32 x 30 mm, inflated and bladder-like, trigonal, reddish-pink.

Distribution: Iran, Afghanistan, Pakistan.

In dry open places in Balochistan and southern N.W.F.P from 1300-1900 m. The silvery bark and conspicuous pinkish-red fruit make this an attractive bush. April-May.

HIPPOCASTANACEAE *Horse Chestnut Family*

Trees or shrubs with opposite and palmate compound leaves. Inflorescence with upper flowers male. Sepals 5, united. Petals 4-5. Fruit capsular, dehiscing by 3-valves. Seeds few, large.

Aesculus 1 sp. *Horse Chestnut*

Deciduous trees with pyramidal clusters of yellowish white flowers. Petals 4-5, clawed. Capsule leathery, with 1-3 seeds.

141. **A. indica** (Wall. ex Camb.) Hook. f. *Bankhor* Pl. 23

Deciduous trees 15-20 m or more tall, with longitudinally flaking bark. Leaflets 5-7, elliptic to oblanceolate, 9-18 x 2-7 cm, acuminate. Panicles erect, 15-30 cm long. Flowers white, tinged yellow. Capsule somewhat ovoid, 4-4.5 cm long, leathery. Seed 25-35 mm broad, dark brown, glossy.

Distribution: Afghanistan, Pakistan, Kashmir to central Nepal.

Both cultivated and wild from 1200-3300 m. In parts of the Neelam valley (Azad Kashmir) it is the dominant forest tree up to 2200 m. The leaves are used as fodder and bark is medicinal. April-May.

ACERACEAE *Maple Family*

Deciduous trees or shrubs with palmately lobed or entire leaves. Flowers regular, sometimes unisexual, dispersed in fascicles or corymbs. Sepals and petals 4 or 5. Fruit a samara.

Acer 6 spp. *Maple*

Deciduous trees with simple or palmately compound leaves. Flowers bisexual or unisexual. Sepals and petals 4-5. Fruit a samara.

1.	+	Leaves simple, entire, persistent.	143. **A. oblongum**
	–	Leaves palmately lobed, deciduous.	2
2.	+	Leaf margin entire; lobes 5-7.	142. **A. cappadocicum**
	–	Leaf margin serrate; lobes 3-5.	3
3.	+	Inflorescence terminal. Tall trees with generally 5-lobed leaves.	144. **A. caesium**
	–	Inflorescence lateral. Smaller trees with generally 3-lobed leaves.	145. **A. pentapomicum**

142. **A. cappadocicum** Gleditsch *Kilpattar* Pl. 23

Trees up to 15 m tall with brown smooth bark. Leaves 5-7 palmately lobed; lobes acuminate, margin entire. Flowers yellowish-green, 7-8 mm broad. Samaras 8-3.5 cm long, wings divergent. Nutlets compressed.

Distribution: North east Turkey, Caucasus, north Iran, Afghanistan, Pakistan, Kashmir to Bhutan, Assam.

Found in forests or along streams from 1800-2500 m especially in lower Kaghan valley but not common in Murree hills. Farming tools are made from its wood. April.

143. **A. oblongum** Wall. ex DC. *Mark* Pl. 23

Medium sized evergreen trees up to 12 m tall. Leaves ovate-lanceolate, 5-15 x 2-6 cm. Flowers greenish-white, 8-9 mm broad. Samaras 20-30 mm long, wings divergent.

Distribution: Pakistan, Kashmir to south west China, Burma, south east Asia.

A tree of moist ravines in the sub-Himalayan tracts from 600-2000 m. Occasionally cultivated. February-March.

144. **A. caesium** Wall. ex Brandis *Tarkanna* Pl. 23

Trees up to 20 m or more tall. Leaves palmately (3-) 5 lobed, 7-18 cm broad. Petiole up to 15 cm long. Flowers yellowish-green, 4-5 mm broad. Samara 3-5 cm across, glabrous, wings more or less erect. Nutlets dark brown.

Distribution: The Himalaya

The commonest of the maples. Found from 2000-3500 m in Himalayan moist temperate forest especially in the Murree hills and Swat. March-May.

145. **A. pentapomicum** J. L. Stewart ex Brandis Pl. 23

A smaller tree up to 5 m tall. Leaves 3-lobed; lobes triangular, pale green beneath. Flowers greenish, c.6 mm broad. Samaras 2-3 cm across; wings divergent. Nutlets gibbous.

Distribution: Turkestan, Afghanisan, Himalaya.

Found from 1300-2400 m. Wood used for making tools. March-April.

ANACARDIACEAE *Mango Family*

Trees or shrubs with resiniferous parts often including turpenes. Leaves alternate, simple or compound. Flowers regular, unisexual or bisexual, disposed in racemes. Petals 3-5, sometimes absent. Fruit a drupe, fleshy or dry.

1.	+	Petals absent.	**1. Pistacia**
	–	Petals present.	2
2.	+	Shrubs. Leaves simple.	**2. Cotinus**
	–	Trees. Leaves compound.	**3. Lannea**

1. Pistacia 3 spp.

Leaves imparipinnate compound. Flowers unisexual, small. Male: Stamens 3-6, Female: Bracteate, styles 3. Fruit a 1-seeded drupe.

146. **P. chinensis** Bunge *Kakar Singhi* Pl. 23
subsp. **integerrima** (J. L. Stewart) Rech. f.

Dioecious trees up to 20 m tall. Leaflets 7-9 in number, lanceolate, serrate, nerves prominent. Male panicles drooping, the female erect. Drupe 5-6 mm broad, rugose, greyish-brown.

Distribution: Afghanistan, Pakistan, Kashmir to west Nepal.

Found both cultivated in plains and wild in sub-Himalayan tracts from 400-1900 m in the spring. The newly emerging leaves are bright bronze red, making an attractive show in the spring. The wood is used for furniture. March-April.

2. Cotinus 1 sp.

Dioecious shrubs with male and female flowers on separate plants. Leaves simple. Flowers in axillary or terminal panicles.

147. **C. coggyria** Scop. *Bhan, Wig or Smoke Tree* Pl. 24

A small bushy shrub to 2 m tall. The pale green leaves 25-70 x 20-50 mm, elliptic-oblong or suborbicular. Female flowers slightly smaller than male. Sterile flowers many, on long hairy reddish peduncles. Drupe laterally compressed, 4-5 mm long, minutely hairy.

Distribution: South Europe, south west Asia, Pakistan, Kashmir to central Nepal, north China.

Fairly common and gregarious from 760-1550 m in the foothill zone. Much affected by browsing domestic stock.

3. **Lannea** 1 sp.

Monoecious or dioecious trees with imparipinnate alternate leaves. Flowers small, paniculate, precocious. Petals 4. Stamens 8. Drupe small, compressed.

148. **L. coromandelica** (Houtt.) Merrill *Kamlai* Pl. 24

A deciduous medium sized tree up to 10 m tall. Leaflets 5-9 in number, ovate to ovate-oblong, hairy on the undersurface. Flowers greenish-yellow, the male in compound racemes. Drupe 10-12 mm long, oblong, compressed, red.

Distribution: Pakistan, Kashmir to Bhutan, Assam, Burma, China, Indo-China, Sri Lanka, Andaman Isls., Malaysia.

Found in the sub-Himalayan tracts up to 1300 m. March.

MORINGACEAE *Horseradish Family*

Deciduous trees with compound leaves. Flowers irregular, bisexual, borne in axillary panicles. Sepals 5, united. Petals 5, free. Carpels 3, united. Fruit a cylindrical beaded pod dehiscing by 3 valves.

Moringa 1 sp.

149. **M. oleifera** Lamk. *Sohanjna* Pl. 24

Trees with 3 pinnate leaves. Leaflets 12-20 mm long, obtuse. Flowers 20-25 mm broad, scented. Pod up to 40 cm long, beaded, pendulous.

Distribution: Pakistan, India, Sri Lanka, Malabar Isls.

Wild in the sub-Himalayan tracts. A quick growing tree cultivated in the plains. The root is a substitute for horseradish. The flowers and young fruit are used as a pot-herb; an oil (ben oil) from the seeds is used as a lubricant in fine-machinery. January-April.

PAPILIONACEAE *Pea Family*

A large family of trees, shrubs, climbers and herbs with usually compound leaves and irregular flowers. Calyx of 5 united sepals. Petals 5, free or united. The uppermost (standard) largest, two laterals, and two innermost (keel) united along margin. Stamens 10, free or united. Fruit dehiscent or indehiscent or jointed and breaking up into 1-seeded parts.

1. + Trees. 2
   – Shrubs or herbs. 3

2.	+	Flowers red. Keeled petal much shorter than the standard petal.	**1. Erythrina**
	–	Flowers pinkish-orange to orange, deep brown at the base. Keeled petal equalling or longer than standard petal.	**2. Butea**
3.	+	Vigorous woody climbers or stragglers.	**3. Pueraria**
	–	Herbs or shrubs, if climbers than herbaceous and weak.	4
4.	+	Fruit inflated on each face.	**4. Astragalus**
	–	Fruit not inflated on each face.	5
5.	+	Procumbent or decumbent unarmed herbs, trailers or weak climbers.	6
	–	Erect herbs or shrubs (sometimes prostrate in **Indigofera** and **Onobrychis**).	11
6.	+	Leaves paripinnate, the rachis ending in a tendril or bristle.	**5. Lathyrus** **6. Vicia**
	–	Leaves imparipinnate, the rachis ending in a terminal leaflet.	7
7.	+	Leaflets 3.	9
	–	Leaflets 5 or more.	8
8.	+	Leaflets 5.	**15. Lotus**
	–	Leaflets more than 5.	**16. Chesneya**
9.	+	Stamens with both long and short anthers.	**9. Argyrolobium**
	–	Stamens all alike.	10
10.	+	Leaflets toothed or serrate.	**12. Trifolium** **13. Trigonella**
	–	Leaflets entire or lobed.	**10. Vigna** **11. Shuteria**

11.	+	Leaves simple or 3-foliolate.	**14. Crotalaria**
	–	Leaflets 5 or more.	12
12.	+	Fruit inflated or sac-like.	**17. Oxytropis** **18. Colutea** **19. Caragana**
	–	Fruit not inflated, mostly compressed.	13
13.	+	Fruit indehiscent, suborbicular, prickly or crested.	**7. Onobrychis**
	–	Fruit not as above.	14
14.	+	Hairs 2-branched, stamens apiculate.	**8. Indigofera**
	–	Hairs if present, not as above. Stamens not apiculate.	15
15.	+	Plants spinose.	**20. Ebenus**
	–	Plants not spinose.	**21. Sophora** **22. Taverniera** **23. Tephrosia** **24. Desmodium**

1. Erythrina 1 sp.

Leaves 3-foliolate. Flowers showy, in axillary or terminal clusters. Calyx 2-lipped. Pod beaded.

150. **E. suberosa** Roxb. *Dhauldak, Coral tree* Pl. 24

Trees to 15 m tall. Branches sometimes prickly. Leaves long petioled. Flowers 1-4 in number, bright scarlet red. Standard 4-5 cm long, larger than the keel or laterals.

Distribution: Pakistan, India, Burma, Indo-China.

Uncommon in the sub-Himalayan tracts. Occasionally cultivated. Found up to 900 m. April.

2. Butea 1 sp.

Flowers orange or flame coloured, in short racemes or panicles . Standard ovate, recurved. Stamens diadelphous. Pod 2-valved.

151. **B. monosperma** (Lam.) O. Kuntz. *Dhak, Flame of the Forest* Pl. 24

Deciduous tree up to 15 m tall. Flowers appearing before the leaves, silky pubescent to the touch. Standard 4-5 cm long. Pods up to 20 cm long, silky.

Distribution: Pakistan, Kashmir to Bhutan, Sri Lanka, Malaysia.

Wild in the foothills of Himalaya and Salt Range from 600-900 m. A spectacular tree when in flower. Occasionally cultivated. Gum extracted is medicinal. March-April.

3. Pueraria 1 sp.

Roots often tuberous. Leaf 3-foliolate. Flowers in axillary racemes or panicles. Standard with 2 inflexed appendages. Pod linear, many seeded.

152. **P. tuberosa** (Roxb. ex Willd.) DC. Pl. 25

An extensive climber/straggler with large tuberous roots. Young branches and leaves densely pubescent. Petiole up to 22 cm long. Pod 5-8 cm long, constricted, silky to the touch.

Distribution: Pakistan, Kashmir to central Nepal from 400-1100 m.

Foothill shrubbery from 400-1100 m, especially conspicuous in Margalla hills. The hanging racemes of bluish mauve flowers are reminiscent of *Wisteria*. March-April.

4. Astragalus c. 134 spp.

A large and taxonomically difficult genus in Pakistan. Herbs, shrubs or undershrubs with compound stipulate leaves. Flowers bracteate; bracteoles present or not. Calyx 5-toothed. Wing and keel clawed. Stamens 10, diadelphous; 9 united, 1 free stamen opposite standard. Pod 2-valved.

1.	+	Stipules leafy and prominent.	154. **A. graveolens**
	–	Stipules not leafy and prominent.	2
2.	+	Flowes yellow.	3
	–	Flowers pale cream or pink to lilac.	4
3.	+	Plants straggling, not spiny.	153. **A. pyrrhotrichus**
	–	Plants erect, spiny.	155. **A. psilocentros** 156. **A. stocksii**

4. + Herbaceous unarmed plants. 5

– Spinose shrubs. 159. **A. anisacanthus**

5. + Flowers pink. Stragglers. 158. **A. scorpiurus**

– Flowers pale yellow. Erect plants. 157. **A. retamocarpos**

153. **A. pyrrhotrichus** Boiss. Pl. 25

Perennial with yellow flowers borne on an axillary peduncled raceme. Leaf imparipinnate; leaflets 39-53, somewhat orbicular in outline. Standard 14-18 mm long. Fruit densely silky.

Distribution: Afghanistan, Pakistan, India.

From 700-1800 m, mainly N.W.F.P., Chitral. March-April.

154. **A. graveolens** Buch.-Ham. ex Benth. Pl. 25

A tall glabrous undershrub with imparipinnate leaves. Stipules 2-4 cm long, leafy. Flowers yellow or pale yellow, in axillary peduncled racemes 20-30 cm long. Pod elongated, 30-40 mm long.

Distribution: Afghanistan, Pakistan, Kashmir to west Nepal, Tibet, south west China.

An attractive plant with prominent stipules. In mountainous areas by roadsides, open places from 1000-2600 m in Punjab, N.W.F.P. and Balochistan. April-May.

155. **A. psilocentros** Fisch. Pl. 25

Erect spiny shrub up to 60 cm tall. Leaves paripinnate; leaflets 10-12 in number, 4-10 mm long, obovate. Stipules 2, connate. Flowers yellow, 1-2 in number. Fruit 14-17 mm long, pubescent.

Distribution: Pakistan, Kashmir to Himachal Pradesh.

Generally in drier hilly areas of Punjab, N.W.F.P. and Balochistan, ascending to 2700 m. March-May.

156. **A. stocksii** Bunge Pl. 25

Like *A. psilocentros* but a larger plant with fewer leaflets (6-8), spiny stipules and flowers in axillary racemes.

Distribution: Afghanistan, Pakistan.

Dry arid stony and sandy places from 1100-2000 m especially Sindh Kohistan and Balochistan. April-May.

157. **A. retamocarpus** Boiss. & Hohen. Pl. 25

An erect bushy perennial up to 100 cm tall. Leaves imparipinnate; leaflets 31-61 in number. Stipules 10-12 mm long. Flowers in short axillary racemes, greenish. Pod 8-11 mm long, glabrous.

Distribution: Iran, Pakistan, Russia, Mt. Turkmenia, Pamir Alai.

Common in field borders in Swat area. 600-1000 m. February-March.

158. **A. scorpiurus** Bunge Pl. 26

A prostrate herb with spreading branches. Leaves imparipinnate; leaflets 11-19 in number. Flowers red or reddish pink, in axillary pedunculate racemes. Pod 12-20 mm long, pubescent.

Distribution: Afghanistan, Pakistan.

A species of the plains and foothill areas, usually in sandy places, forming carpets of red flowers. A weedy plant in Kala Chita hills. March-April.

159. **A. anisacanthus** Boiss. Pl. 26

Small spiny shrub up to 1.3 m tall with pink flowers and inflated fruiting calyx. Fruit 5-6 mm long, 2-seeded, pubescent.

Distribution: Iran, Afghanistan, Pakistan.

In dry stony areas and low hills up to 1800 m in Balochistan and N.W.F.P. Fairly common. February-March.

5. Lathyrus 8 spp. *Matar, Pea*

Annual or perennial herbs with stems weak and usually winged. Leaves paripinnate, the rachis usually ending in a tendril. Stipules large. Flowers in axillary or solitary racemes. Pod compressed.

1.	+	Whole leaf modified into a tendril.	160. **L. aphaca**
	–	Rachis of leaf modified into tendril of bristle. Leaflets present, 5-15 in number.	2
2.	+	Flowers in terminal racemes, yellow.	161. **L. pratensis**
	–	Flowers solitary, red, orange or deep blue.	3

3. + Stem winged. 162. **L. sativus**

– Stem not winged. 163. **L. sphaericus**

160. **L. aphaca** L. Pl. 26

A weak trailer, with leaf modified into tendril and foliaceous paired stipules. Flowers lemon yellow.

Distribution: North Africa, Europe, west and central Asia, Pakistan, Kashmir, India, Nepal.

A common spring weed. Plains to 1300 m. February-April.

161. **L. pratensis** L. Pl. 26

A scrambling perennial with angled stems and paripinnate leaves; terminal leaflets modified into tendril. Peduncle exserted above the foliage. Pod 18-28 mm long, glabrous.

Distribution: Temperate Europe, Asia, Pakistan, Kashmir to central Nepal.

Widespread in open places in the forest zone from 1800-3100 m from Chitral to Murree hills. July-August.

162. **L. sativus** L. Pl. 26

A subglabrous annual with winged stems. Leaf paripinnate. Leaflets 3, terminal modified into a tendril. Flowers deep blue to red or white. Pod winged on one side.

Distribution: Widely cultivated and naturalized as a weed.

Often cultivated for fodder especially in Sindh, also found as a weed in cultivated areas from the plains to 2400 m. Human consumption of seeds said to cause paralysis. March-June.

163. **L. sphaericus** Retz. Pl. 26

Resembling the preceding species but stems and pods not winged. Flowers red or vermilion.

Distribution: Africa, Europe, west and central Asia, Pakistan, Kashmir to west Nepal.

Found in plains and hills, often as a field weed from 600-1800 m in Swat, Hazara and Murree hills. February-April.

6. Vicia 14 spp. *Vetch*

Resembling *Lathyrus*, but generally with larger number of leaflets and stems that are unwinged and hairless style.

+	Inflorescence many-flowered. Plants more or less bushy, suberect.	164. **V. bakeri**
–	Inflorscence 1-2 - flowered. Plants low growing, procumbent.	165. **V. sativa**

164. **V. bakeri** Ali Pl. 27

Annual with stalked axillary inflorescence of pink drooping flowers borne on long peduncles. Leaves pinnate, tendrillar.

Distribution: Pakistan, Kashmir to central Nepal.

In open places from 2100-3300 m particularly Kaghan valley and Murree hills. Flowers occasionally white. July-August.

165. **V. sativa** L. Pl. 27

A low growing annual decumbent herb, variable in its morphology. Flowers 1-2, pale pink to crimson or purple-violet. Leaflets 4-18, linear to lanceolate or broader. Pod 20-60 cm long, glabrescent with age.

Distribution: Cosmopolitan.

A common weed of cultivated fields. Also used as fodder. Plains to 2300 m. March-April.

7. Onobrychis 9 spp.

Undershrubs or herbs. Leaflets imparipinnate. Flowers bracteate, in axillary pedunculate racemes. Wings shorter than standard or keel. Stamens diadelphous. Fruit indehiscent, compressed, suborbicular in outline.

1.	+	Plants tufted, cushion-like, spinose. Flowers deep red.	166. **O. cornuta**
	–	Plants prostrate, not spinose. Flowers light pink or white with red streaks.	2
2.	+	Branches spreading. Leaflets 9-13, elliptic to ovate-oblong.	167. **O. stewartii**

– Branches short. Leaflets 3-5, suborbicular to obovate 168. **O. tavernierifolia**

166. **O. cornuta** (L.) Desv. Pl. 27

Perennial, cushion-like and very spiny plant with profuse pretty deep red flowers. Leaflets 5-11 in number, linear-lanceolate. Fruit somewhat round, compressed.

Distribution: Syria, Turkestan, Iran, Afghanistan, Pakistan.

A plant of dry stony places, slopes from 2500-3400 m especially in the juniper forest zone in Balochistan. May.

167. **O. stewartii** Baker Pl. 27

A prostrate straggler with branches up to 60 cm long. Flowers lilac, with darker streaks. Fruit c. 8 mm long, suborbicular, pubescent, back crested, the faces honey-combed.

Distribution: Pakistan, India.

Sandy places, rocky slopes up to 800 m in Balochistan, N.W.F.P. and Punjab. March-April.

168. **O. tavernierifolia** Stocks ex Boiss. Pl. 27

An annual. Stem reduced. Flowers on a peduncled raceme, white with red veins. Calyx much shorter than corolla. Fruit 10-12 mm long, orbicular, face deeply pitted.

Distribution: Iran, Afghanistan, Pakistan.

Stony or sandy places, low hills from 1300-1800 m. March-May.

8. Indigofera 24 spp.

Shrubs or herbs with pinnate compound stipulate leaves. Flowers in axillary racemes or spikes. Standard sessile. Stamens diadelphous. Fruit linear or globose, compressed.

1. + More or less erect shrubs. 2

   – Plants prostrate. 3

2. + Shrub up to 2 m tall. Fruit hairy. 169. **I. heterantha**

   – Shrub up to 90 cm tall. Fruit glabrous. 170. **I. atropurpurea**

3. + Leaves simple. Fruit 1-seeded. 171. **I. linifolia**

– Leaves imparipinnate. Fruit 4-6 seeded. 172. **I. argentea**

169. **I. heterantha** Wall. ex Brandis *Kainthi* Pl. 27

Shrub up to 2 m tall. Leaves imparipinnate. Leaflets 9-33 in number. Flowers red or reddish-purple, in axillary racemes. Pod 14-40 mm long, cylindric.

Distribution: Afghanistan, Pakistan, Kashmir to Bhutan, Sikkim and China.

Gregarious in hot dry open places from 1500-3000 m. May-June.

170. **I. atropurpurea** Buch.-Ham. ex Hornem. Pl. 28
Syn.: *I. cassioides* Rottl. ex DC.

A smaller shrub than the preceding one with pale pink or pink-red flowers. Flowers on a peduncled raceme up to 30 cm long. Pod up to 4 cm long, narrow, glabrous.

Distribution: Pakistan, Kashmir to Bhutan.

Found from 900-1500 m. Generally open slopes in the chir pine (*Pinus roxburghii*) zone. March-April.

171. **I. linifolia** (L. f.) Retz. Pl. 28

A prostrate tropical annual weed with simple linear leaves and red flowers. Pod 1-seeded, globose, pubescent.

Distribution: Widespread in tropical and subtropical regions of both hemispheres.

Throughout the plains and lower hills up to 1500 m. September-November.

172. **I. argentea** Burm. f. Pl. 28

An undershrub up to 70 cm tall. Leaves imparipinnate; leaflets 7-11 in number. Pods 10-13 mm. long, pubescent, 4-6-seeded.

Distribution: North Africa, Arabia, Sudan, south Iran, Pakistan.

A plant of dry sandy areas in the plains especially lower Sindh and Balochistan. Flowering mostly throughout the year.

9. Argyrolobium 3 spp.

Herbs with trifoliolate leaves. Flowers solitary or in short racemes or subumbels. Calyx 2-lipped. Standard suborbicular. Stamens monadelphous. Fruit narrow, compressed.

173. **A. roseum** (Camb.) Jaub. & Spach Pl. 28

Leaflets obovate. Flowers 5-6 mm long, red. Pod 20-30 cm long, pubescent.

Distribution: Iraq, Iran, Afghanistan, Pakistan, Kashmir to west Nepal.

Found in the plains and lower hills up to 1300 m. April-October.

10. Vigna c. 6 spp.

Climbing, prostrate herbs or semishrubs with 3-foliolate leaves. Flowers in axillary or terminal 1-many flowered subumbels or racemes. Calyx 5-lobed, 2-lipped. Standard with appendages. Stamens diadelphous. Pod linear or oblong.

174. **V. vexillata** (L.) A. Rich *Ud Salib* Pl. 28

A perennial vine. Leaflets 20-70 x 15-20 mm, lanceolate or ovate-lanceolate. Stipules 7-12 mm long. Flowers axillary, 2-6 in number, pink or purple-pink. Pod 6-12 x 3-4 mm, pubescent.

Distribution: Cosmopolitan.

Found in the lower hills in Punjab and N.W.F.P. from 700-2000 m especially in chir pine zone. August-September.

11. Shuteria 1 sp.

Climbing herbs with 3-foliolate leaves. Flowers in axillary racemes. Stamens 10 in two groups (9 + 1). Fruit linear, 2-valved.

175. **S. involucrata** (Wall.) Wight & Arnott Pl. 28

A weak climber with racemes of yellowish flowers. Leaflets obtuse, entire. Peduncle 4-7 cm long. Fruit 3-4 cm long, pubescent.

Distribution: Pakistan, India, Nepal.

Uncommon. Confined to sub-Himalayan tracts up to 1000 m. August-September.

12. Trifolium 8 spp. *Clovers & Trefoils*

Perennial or annual herbs with trifoliolate leaves. Flowers in pedunculate heads or racemes, sometimes solitary. Stamens diadelphous. Fruit included in calyx.

176. **T. repens** L. *Dutch Clover* Pl. 28

Flowers pink, in globose heads, about 25 mm broad. Peduncle exserted above the foliage. Pod linear.

Distribution: Temperate Europe and Asia, north Africa.

In open places, meadows from 1400-3400 m. April-August.

13. Trigonella 16 spp.

Annual with trifoliolate, stipulate leaves. Flowers solitary axillary, or in spikes or racemes. Stamens diadelphous. Fruit linear or oblong.

177. **T. monantha** C. A. Meyer Pl. 29
subsp. **incisa** (Benth.) Ali

Prostrate spreading annual. Leaflets 3-10 x 3-5 mm. Flowers yellow, in pedunculate clusters of 1-3.

Distribution: Afghanistan, Pakistan, India.

Found as a weed or in neglected fields from plains to 1600 m. March-April.

14. Crotalaria c. 9 spp.

Herbs or shrubs with simple or 3-foliolate leaves. Flowers in terminal or leaf opposed racemes. Stamens monadelphous. Anthers dimorphic. Pod turgid or inflated.

+ An undershrub. Branches densely pubescent. Leaves simple. Pod elongated. 178. **C. burhia**

– A herb. Branches sparsely pubescent. 179. **C. medicaginea**

178. **C. burhia** Buch. -Ham. ex Benth. Pl. 29

Low branched shrub up to 55 cm tall. Leaves 6-20 x 4-8 mm, oblong. Flowers 6-12 per raceme. Pod c. 8 x 3-4 mm, pubescent.

Distribution: Afghanistan, Pakistan, India.

Dry arid places especially N.W.F.P., Salt Range, Kala Chita hills and Balochistan up to 600 m. January-February.

179. **C. medicaginea** Lamk. Pl. 29

A perennial herb with subglabrous parts. Leaflets 3 in number, 5-8 x 3-5 mm, oblanceolate. Flowers 2-7 (-12) per raceme. Pod c. 4.5 mm broad.

Distribution: Afghanistan, Pakistan, Kashmir to west Nepal, India, Sri Lanka, Malaysia, Australia.

Common throughout plains to 1500 m. May-November.

15. Lotus 5 spp.

Annual or perennial herbs with 5 leaflets, the lower 2 taking the place of stipules. Flowers solitary or in axillary peduncles or umbellate. Stamens diadelphous. Pod dehiscent.

180. **L. corniculatus** L. *Bird's-foot Trefoil* Pl. 29

Prostrate to decumbent perennial herb. Leaflets obovate or narrowly so. Flowers yellow, 3-6 in number, borne on a peduncle up to 14 cm long. Standard 10-15 mm long. Pod 10-25 x 2.5 mm, cylindrical.

Distribution: Cosmopolitan, except New World.

Open places, meadows, from plains to 3300 m. April-August.

16. Chesneya 5 spp.

Prostrate herbs with imparipinnate stipulate leaves. Flowers 1-3, in axillary peduncles or umbels. Stamens diadelphous.

+ Flowers more than 15 mm long, creamish-white with keel reddish-pink. 181. **C. parviflora**

– Flowers 8-10 mm long, mauve or yellow. 182. **C. depressa**

181. **C. parviflora** Jaub. & Spach. Pl. 29

Pubescent prostrate herb. Leaflets 19-27 in number, opposite. Flowers creamish-white with keel tinged reddish-pink. Standard 15-18 mm long, pubescent.

Distribution: Iran, Pakistan.

In rocky or sandy plains up to 300 m, especially Balochistan. February-March.

182. **C. depressa** (Oliv.) Pop. Pl. 29

A prostrate plant with white-tomentose parts. Leaf imparipinnate; leaflets opposite to subopposite. Flowers solitary axillary, mauve or yellow. Standard c. 8 mm long.

Distribution: Pakistan, Kashmir.

Found from 1500-2400 m in the northern areas from Chitral to Gilgit. May-July.

17. Oxytropis c. 21 spp.

Herbs or undershrubs with imparipinnate leaves. Flowers in axillary racemes or spikes. Stamens diadelphous. Pod turgid.

183. **O. cachemiriana** Camb. Pl. 30

Flowers pink or pink-purple, in globular heads borne on peduncles 3.5-12 cm long. Leaves densely hairy. Pod 10-13 mm long, pubescent.

Distribution: Pakistan to Kashmir.

Dry places from 2300-4100 m in Chitral, Gilgit and Baltistan. July-August.

18. Colutea 3 spp.

Shrubs or small trees. Leaves imparipinnate. Flowers yellow, in racemes. Pod inflated with papery walls.

184. **C. paulsenii** Freyn Pl. 30

Shrubs 2-2.5 m tall, pubescent to subglabrous. Leaflets 5-11, elliptic. Flowers 3-5 in number, racemose. Standard 12-18 mm long. Pod 35-70 x 30-35 mm, tips acuminate.

Distribution: Russia, Pakistan, Kashmir.

In open places or along slopes etc. up to 2600 m in Gilgit. July-August.

19. Caragana 10 spp.

Generally shrubs. Leaves compound, stipulate, rachis ending in a spine. Flowers solitary or 2-3. Stamens diadelphous. Pod more or less cylindrical, many-seeded, glabrous or not.

185. **C. ambigua** Stocks Pl. 30

Low growing spinescent shrub with rachis modified into a spine. Leaflets 4-10 in number. Flowers yellow, up to 20 mm long. Pod curved sideways, 20 x 5 mm, minutely hairy.

Distribution: Pakistan.

Sandy and stony places from 1900-3000 m in Balochistan and south Waziristan. Quite floriferous. April-June.

20. Ebenus 1 sp.

Armed small shrubs with imparipinnate leaves. Flowers in pedunculate heads. Calyx 5-toothed. Standard slightly larger than the keel. Stamens 10, monadelphous.

186. **E. stellata** Boiss. Pl. 30

Perennial up to 50 cm tall. Leaflets linear-lanceolate, rachis spinose. Flowers yellowish-red. Peduncles 25-70 cm long. Pod compressed, oblong, 1-seeded.

Distribution: Arabia, Iran, Afghanistan, Pakistan.

Rocky places from 1500-2100 m; commonly in Balochistan and N.W.F.P. April-May.

21. Sophora 3 spp.

Perennial shrubs, rarely trees with compound leaves. Flowers in terminal racemes or spikes, creamish or yellow. Stamens free or united at the base. Fruit a lomentum.

+ Flowers yellow. Leaflets 19-37 in number, ovate or elliptic. 187. **S. mollis**

– Flowers cream. Leaflets 15-27 in number, oblong. 188. **S. alopecuroides**

187. **S. mollis** (Royle) Baker *Shamtastir* Pl. 30
subsp. **griffithii** (Stocks) Ali
Syn.: *S. griffithii* Stocks

A deciduous branched shrub up to 1.5 m tall. Young shoots silky white. Flowers yellow, in axillary racemes up to 12 cm long. Lomentum 6-12 cm long, beaded.

Distribution: Russia, Afghanistan, Pakistan, Kashmir to central Nepal.

A floriferous and gregarious bushy shrub which makes an especially brave show when it flowers during spring. Common throughout Balochistan in sandy or gravelly plains and hills from 1200-2100 m. Also found less commonly in Salt Range and Chitral. March-May.

188. **S. alopecuroides** L. Pl. 31

Herbaceous undershrub up to 1 m tall with pilose or densely pilose parts and creamish flowers on terminal spikes up to 22 cm long. Lomentum 3-14-seeded.

Distribution: Mediteranean, Soviet central Asia, Iran, Iraq, Turkey, Afghanistan, Pakistan (Balochistan, northern areas), China, Kansu.

Dry stony or sandy places, field edges from 400-2200 m in Balochistan, Gilgit, Baltistan and Kashmir. April-September.

22. Taverniera 4 spp.

Branched undershrubs with generally trifoliolate leaves. Flowers in axillary racemes. Stamens in one or two groups. Pod compressed, indehiscent, breaking up into 1-seeded parts.

+ Stems and shoots densely pubescent with white hairs. Leaves up to 8 mm long. 189. **T. spartea**

– Stems and shoots subglabrous, green. Leaves up to 26 mm long. 190. **T. cuneifolia**

189. **T. spartea** (Burm. f.) DC. Pl. 31

Perennial branched, densely pubescent shrubs up to 1.5 m tall, flowers pink with darker streaks. Leaves up to 8 mm long. Lomentum 1-2-seeded.

Distribution: Iran, Afghanistan, Pakistan.

Dry gravelly or sandy places up to 900 m in southern Balochistan. February-April.

190. **T. cuneifolia** (Roth) Arnott *Jethmad* Pl. 31

Similar to above species, but with less pubescence, greenish stems and larger leaves. Flowers pink with red streaks.

Distribution: Pakistan, India.

Dry stony and gravelly places up to 1600 m throughout foothill tracts. Flowers mostly all the year round.

23. Tephrosia c. 9 spp.

Undershrubs, shrubs or herbs with imparipinnate leaves. Flowers axillary or in short terminal racemes. Stamens 10, in one or two groups. Pod compressed, elongated.

191. **T. uniflora** Pers. Pl. 31

Low shrub with 3-7-oblanceolate leaflets. Flowers solitary, pink.

The plants is probably *T. uniflora* subsp. *petrosa* (Blatter & Hallberg) Gillett & Ali which it closely resembles.

Distribution: Pakistan, India, Arabia.

Dry places in lower Sindh, Balochistan and South Waziristan. February-March.

24. Desmodium 10 spp.

Shrubs, undershrubs or herbs with compound leaves. Flowers in terminal racemes or panicles. Stamens in one or two groups. Pod transversely jointed.

192. **D. elegans** DC. *Chamkat* Pl. 31

Shrub 2 m tall. Leaves trifoliolate. Flowers pale lilac, 10-12.5 mm long, in lax panicles up to 50 cm long. Pods 25-60 mm long.

Distribution: Afghanistan, Pakistan, Kashmir to Bhutan, China.

In pine forests from 1500-2500 m in Himalaya. July-September.

MIMOSACEAE *Mimosa Family*

A large family of tropical and subtropical trees or shrubs, represented by 4 genera and 8 species native to Pakistan. Many others are exotic and introduced.

1.	+	Stamens more than 10 in number.	2
	–	Stamens up to 10 in number.	3
2.	+	Stamens free.	**1. Acacia**
	–	Stamens united in one group.	**2. Albizzia**
3.	+	Flowers 3-6-merous. Anthers eglandular. Pods dehiscing by 2 valves.	**3. Mimosa**
	–	Flowers 5-merous. Anthers glandulose. Pods indehiscent, pulpy within.	**4. Prosopis**

1. Acacia c. 9 spp.

Prickly or spiny trees or shrubs with bipinnate leaves and stipules modified into spines. Flowers spicate or in globose heads.

Flowers small, 3-5-merous. Fruit linear to curved or twisted, sometimes woody.

1.	+	Flowers in spikes.	2
	–	Flowers in heads.	3
2.	+	Leaves with 10-30 pairs of pinnae.	193. **A. catechu**
	–	Leaves with 2-5 pairs of pinnae or fewer.	194. **A. modesta**
3.	+	Bark reddish-brown. Stem and branches densely spiny.	195. **A. jacquemontii**
	–	Bark grey to greyish-brown. Stem and branches not so.	4
4.	+	Generally large trees up to 20 m tall, with beaded pubescent pods. Ovary pilose.	196. **A. nilotica**
	–	Shrubs or small trees with smooth pods. Ovary glabrous.	5
5.	+	Pods flattened, dry.	197. **A. hydaspica**
	–	Pods turgid, pulpy within.	198. **A. farnesiana**

193. **A. catechu** (L. f.) Willd. *Khair, Katha* Pl. 31

A deciduous medium sized tree up to 10 m tall. Stipules hooked and pointed, 4-7 mm long; leaflets 14-56 pairs. Flowers in axillary spikes, 8-10 cm long, creamish. Pods flat, brown, dehiscing by 2-valves.

Distribution: Tropical and foothill Himalaya, Pakistan, India, Nepal, Sikkim, Assam, Burma.

A component of the sub-Himalayan tracts up to 1200 m. A useful plant which yields durable wood for construction purposes. A tannin 'Kath' is used in betel leaf for chewing. May-August.

194. **A. modesta** Wall. *Phulai* Pl. 32

A medium sized deciduous tree. Pinnae 2-5 in number; leaflets 3-5 pairs. Prickles stout,

105. *Grewia damine* Hab valley, Sindh, Khan Mohammad Khan

◀b

a▼

103. *Grewia tenax* a, b: Margalla hills, Rubina A. Rafiq

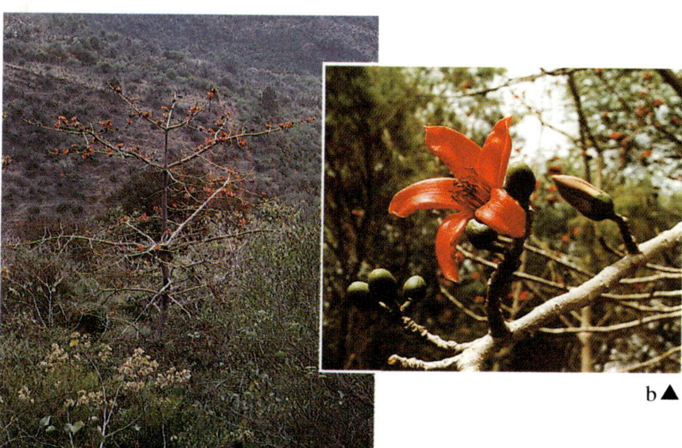

b▲

◀a

101. *Bombax ceiba*
a: Margalla hills, S. A. Sultan
b: Thatta distt, T. J. Roberts

102. *Melhania futteyporensis* Margalla hills, Rubina A. Rafiq

106. *Linum corymbulosum* Margalla hills, S. A. Sultan

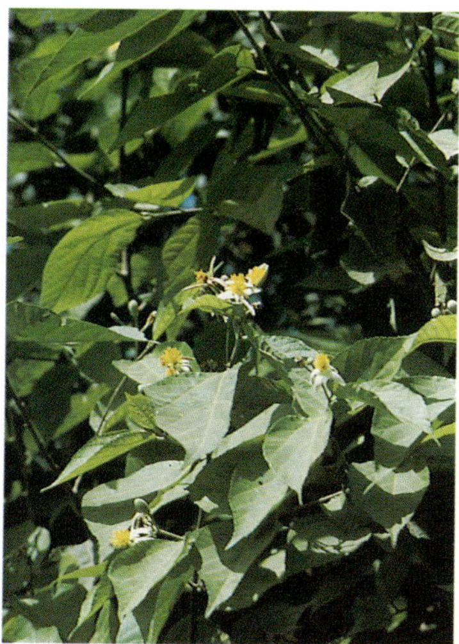

104. *Grewia optive* Margalla hills, S. A. Sultan

113. *Zygophyllum simplex* Karachi, Y. J. Nasir

107. *Reinwardtia indica* a: Balakot, T. J. Roberts b: Murree hills, Y. J. Nasir

108. *Hiptage benghalensis* Lower Murree hills, Y. J. Nasir

109. *Fagonia indica* a, b: Islamabad, S. A. Sultan

111. *Tribulus terrestris* Karachi, Y. J. Nasir

112. *Zygophyllum propinquum* Panjgur, Makran, Y. J. Nasir

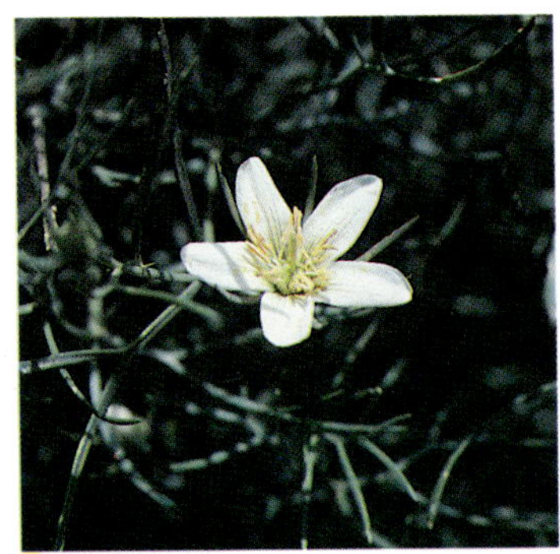

110. *Peganum harmala* Islamabad, S. A. Sultan

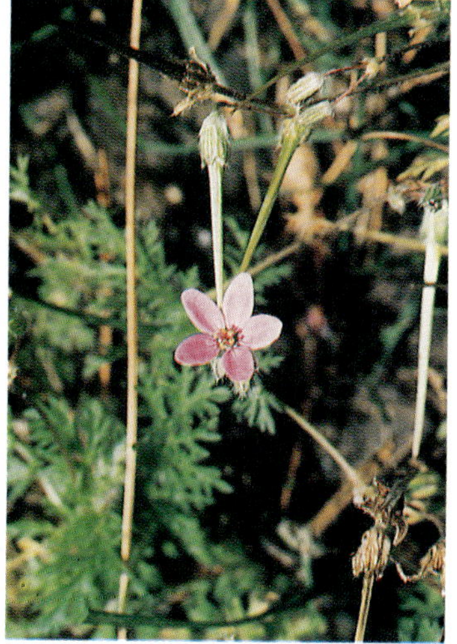

119. *Erodium cicutarium* Mokshpuri, Nathiagali, Y. J. Nasir

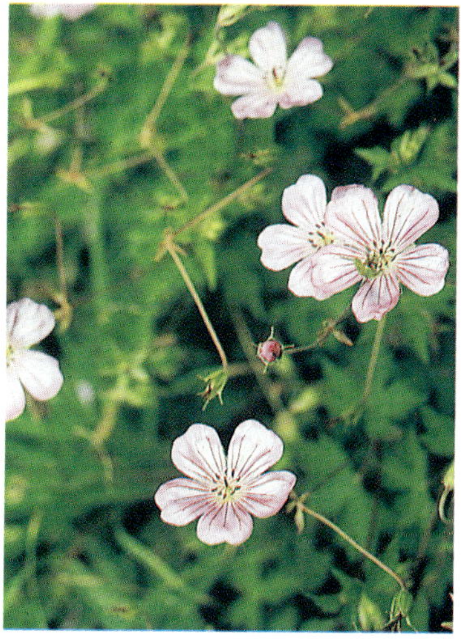

117. *Geranium collinum* Lowari pass, Y. J. Nasir

118. *Erodium malacoides* Taxila, S. A. Sultan

116. *Geranium wallichianum* Changla Gali, S. A. Sultan

114. *Geranium ocellatum* Margalla hills, S. A. Sultan

115. *Geranium rotundifolium* Margalla hills, S. A. Sultan

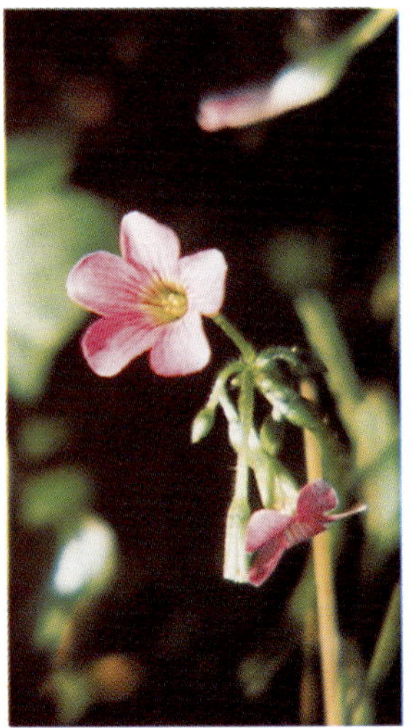

123. *Oxalis corymbosa* Islamabad S. A. Sultan

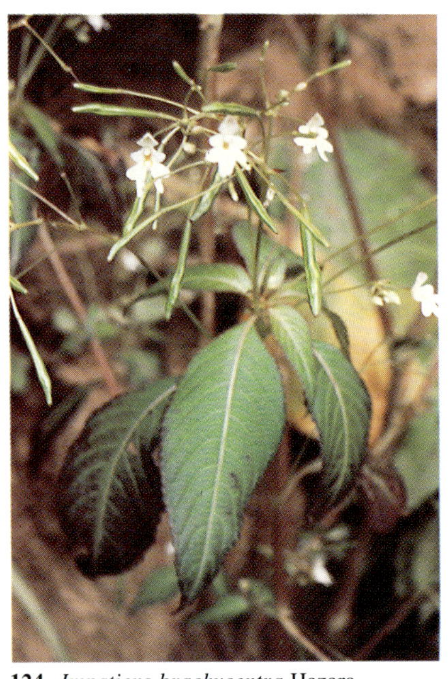

124. *Impatiens brachycentra* Hazara, Y. J. Nasir

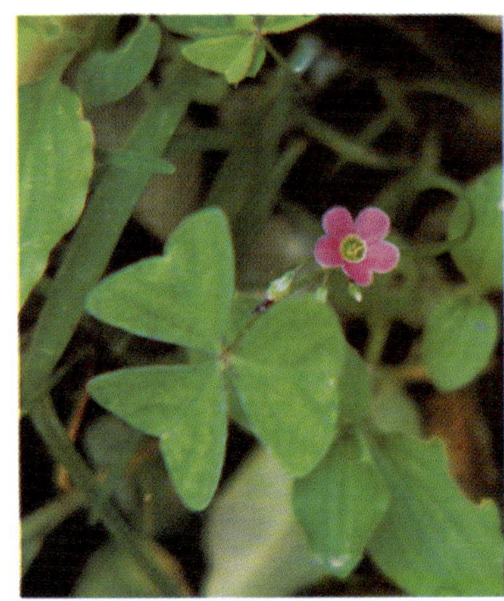

122. *Oxalis latifolia* Abbottabad, S. A. Sultan

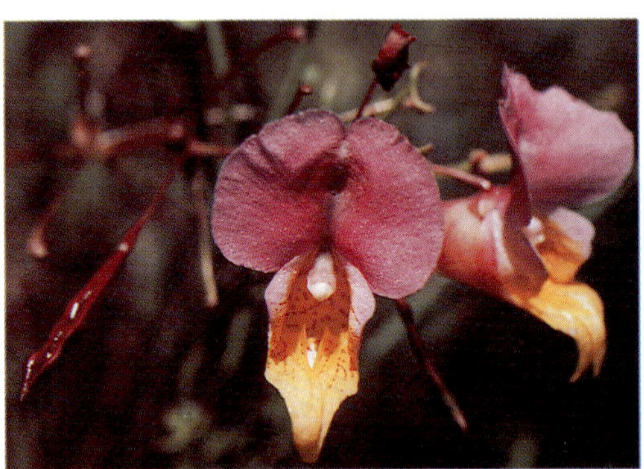

126. *Impatiens bicolor* Siran valley, Hazara, Rubina A. Rafiq

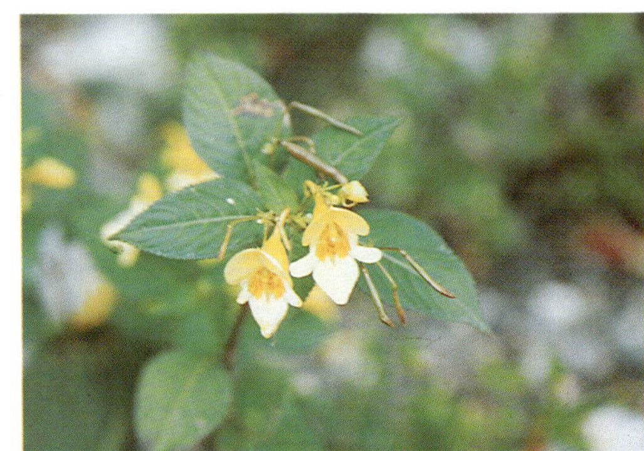

125. *Impatiens edgeworthii* Murree hills, Y. J. Nasir

120. *Oxalis corniculata* Islamabad, S. A. Sultan

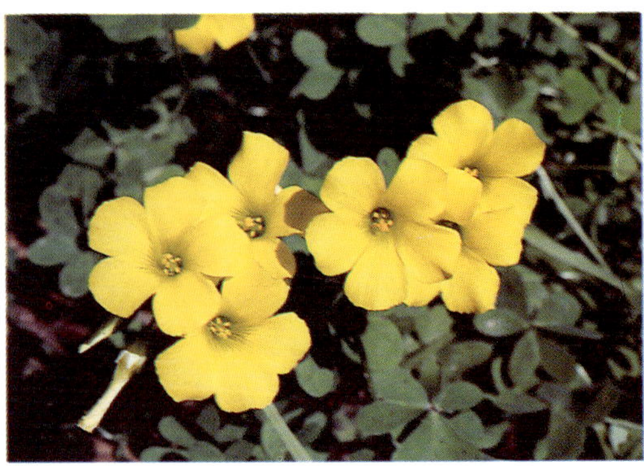

121. *Oxalis pes-caprae* Haripur, Y. J. Nasir

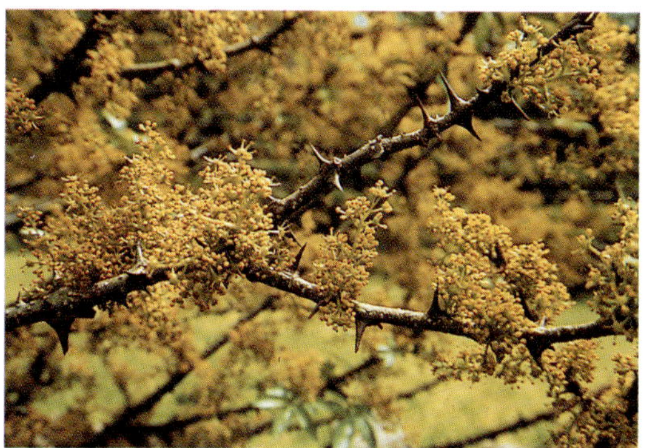

130. *Zanthoxylum armatum* Lower Swat, Fritz Berger

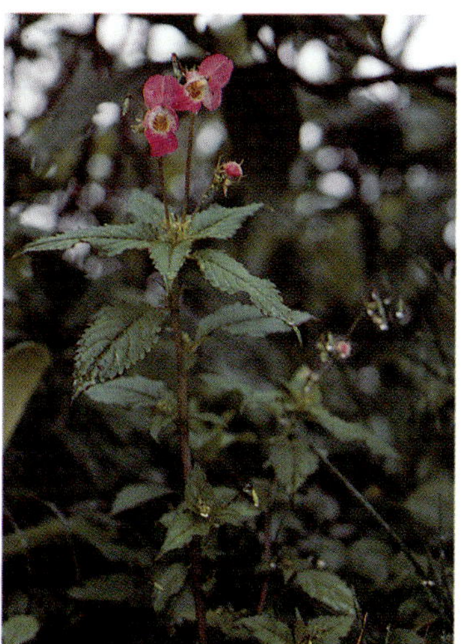

128. *Impatiens thomsonii* Sari, Kaghan, Y. J. Nasir

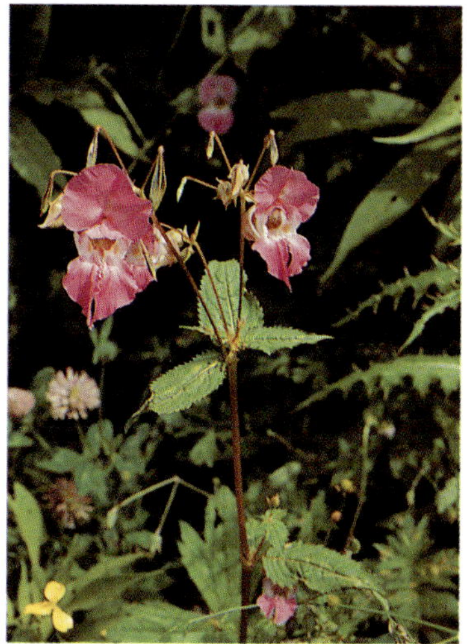

127. *Impatiens glandulifera* Kaghan valley, Y. J. Nasir

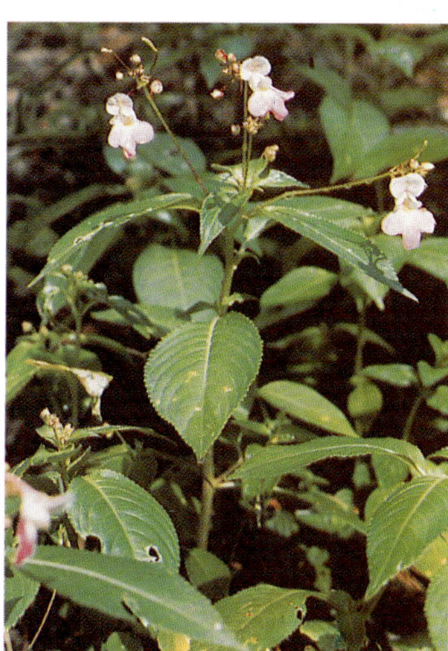

129. *Impatiens flemingi* Thandiani, Y. J. Nasir

131. *Haplophyllum erythreum* Chiltan hills, Rubina. A. Rafiq

132. *Hasplophyllum tuberculatum* Chiltan hills, Rubina A. Rafiq

134. *Maytenus royleanus* Margalla hills, S. A. Sultan

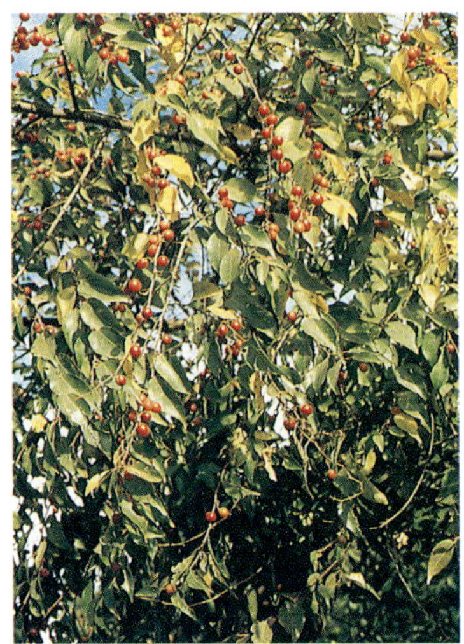

136. *Zizyphus oxyphylla* Margalla hills, Rubina A. Rafiq

135. *Helinus lanceolatus* Margalla hills, Y. J. Nasir

140. *Stocksia brahuica* Pishin, T. J. Roberts

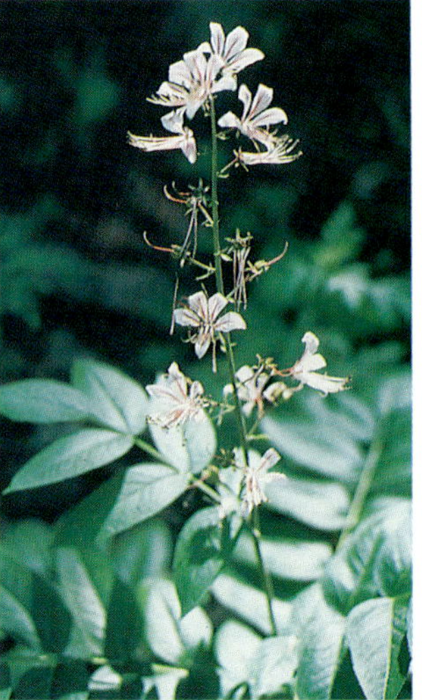

133. *Dictamnus albus* Kalam, Fritz Berger

138. *Tetrastigma serrulatum* Margalla hills, Rubina A. Rafiq

139. *Dodonaea viscosa* Islamabad, Y. J. Nasir

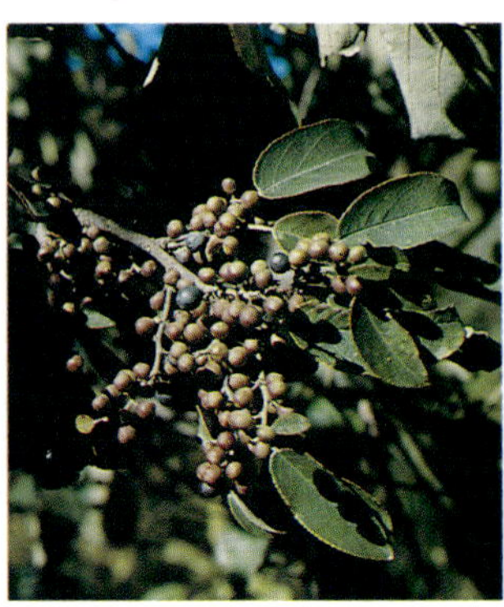

137. *Rhamuns triquetra* Margalla hills, Rubina A. Rafiq

146. *Pistacia chinensis ssp. integerrima* a, b: Margalla hills, Y. J. Nasir

144. *Acer caesium* Kaghan, Rubina A. Rafiq

145. *Acer pentapomicum* Upper Swat, Y. J. Nasir

143. *Acer oblongum* Peshawar, Rubina A. Rafiq

142. *Acer cappadocicum* Palas, Kohistan, Rubina A. Rafiq

141. *Aesculus indica* a: Swat Kohistan, Fritz Berger b: Jhica Gali, Y. J. Nasir

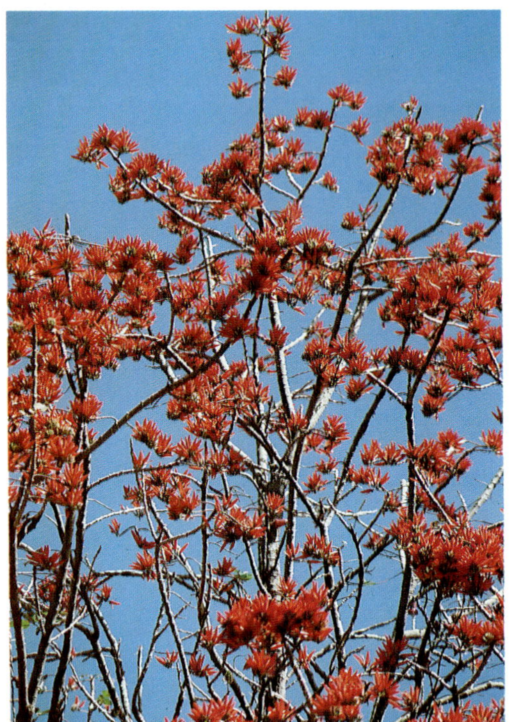

150. *Erythrina suberosa* Islamabad, S. A. Sultan

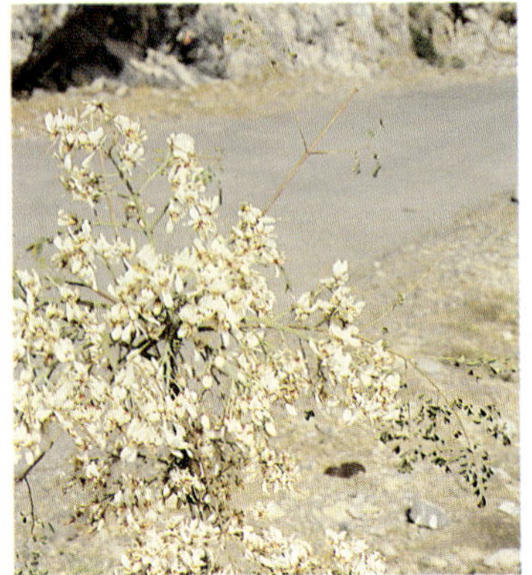

149. *Moringa oleifera* Margalla hills, Y. J. Nasir

◀a b▼

151. *Butea monosperma* Salt Range, S. A. Sultan

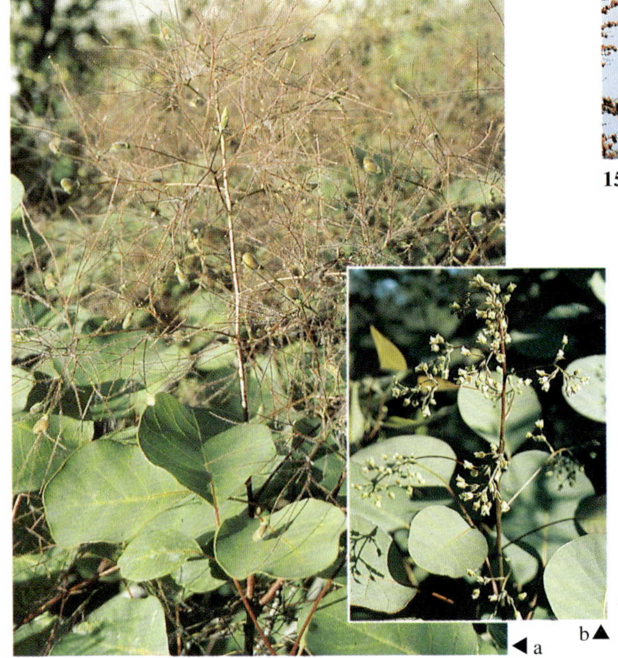

b▲ ◀a

147. *Cotinus coggyria* a, b: Margalla hills, Y. J. Nasir

148. *Lannea coromandelica* Margalla hills, Y. J. Nasir

154. *Astragalus graveolens* Lower Swat, S. A. Sultan

152. *Pueraria tuberosa* a, b: Lower Margalla hills, Y. J. Nasir

155. *Astragalus psilocentros* Salt Range, Y. J. Nasir

153. *Astragalus pyrrhotrichus* Margalla hills, S. A. Sultan

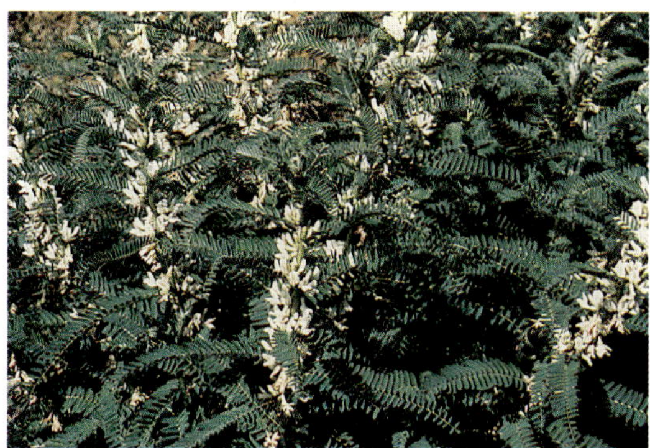

157. *Astragalus retamocarpus* Margalla hills, S. A. Sultan

156. *Astragalus stocksii* Balochistan, Y. J. Nasir

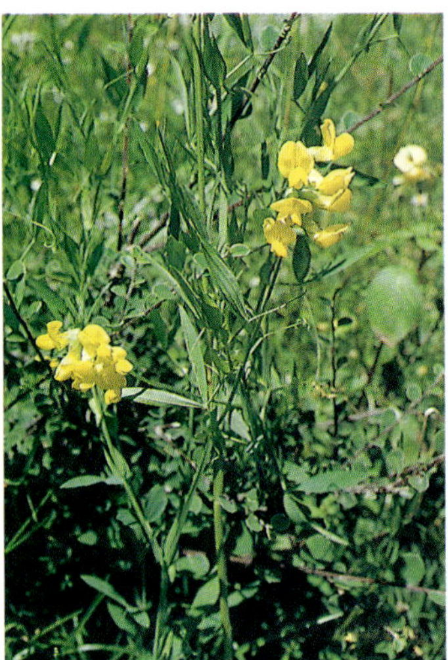

161. *Lathyrus pratensis* Gabral, Swat, Y. J. Nasir

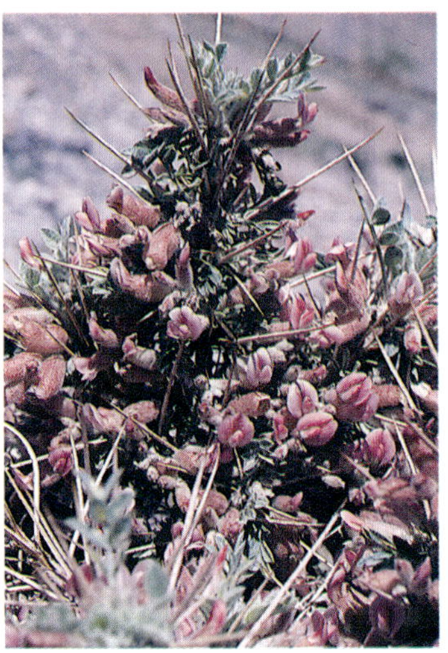

159. *Astragalus anisacanthus* Chiltan hills, Rubina A. Rafiq

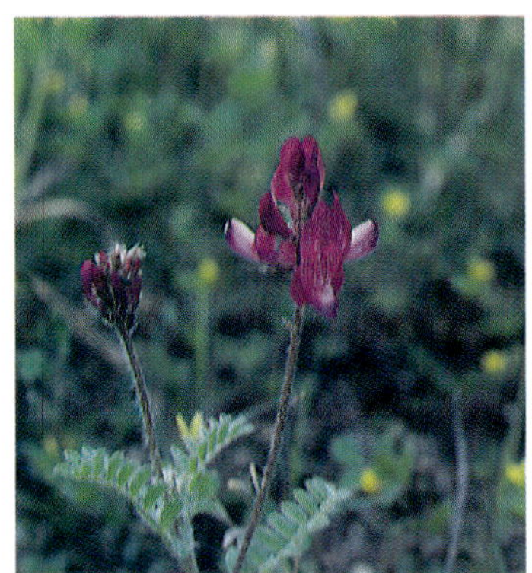

158. *Astragalus scorpiurus* Kala Chita hills, S. A. Sultan

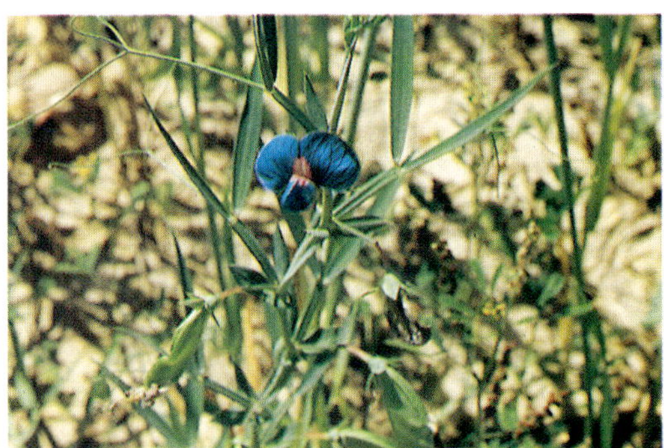

162. *Lathyrus sativus* Lahore, Y. J. Nasir

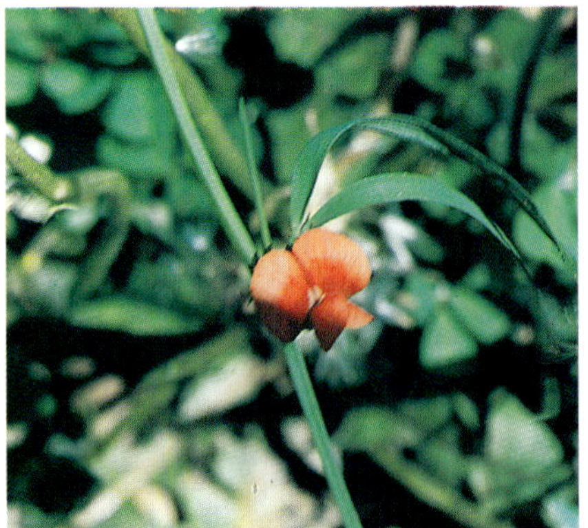

163. *Lathyrus sphaericus* Margalla hills, S. A. Sultan

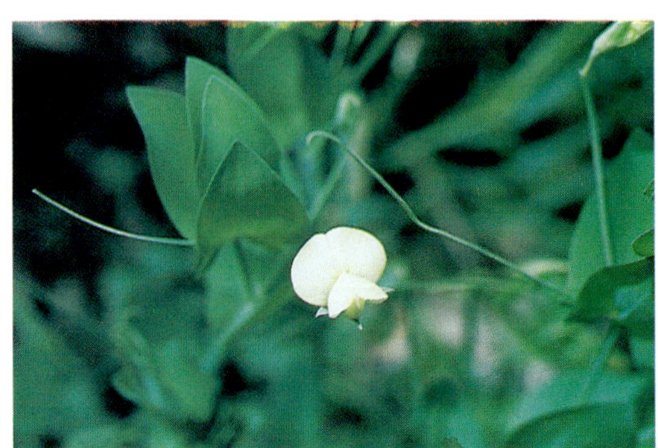

160. *Lathyrus aphaca* Islamabad, S. A. Sultan

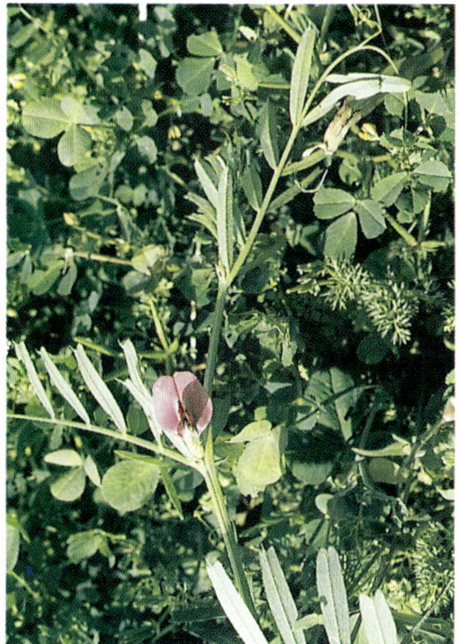

165. *Vicia sativa* Siran valley, Hazara, Y. J. Nasir

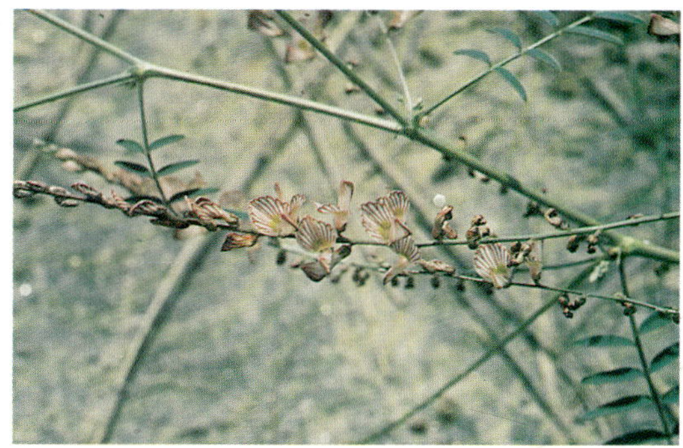

167. *Onobrychus stewartii* Attock, S. A. Sultan

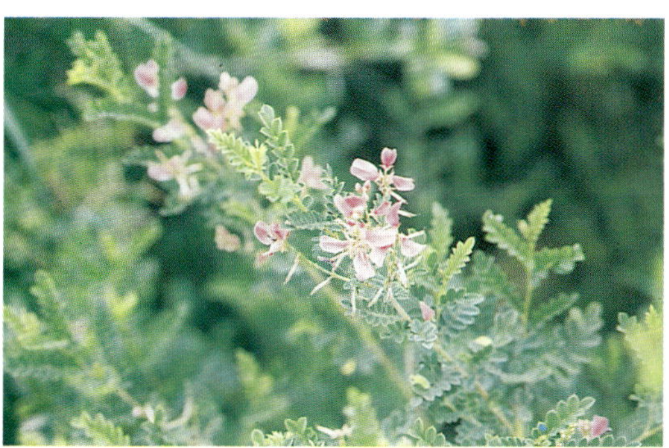

169. *Indigofera heterantha* Ghora Gali, S. A. Sultan

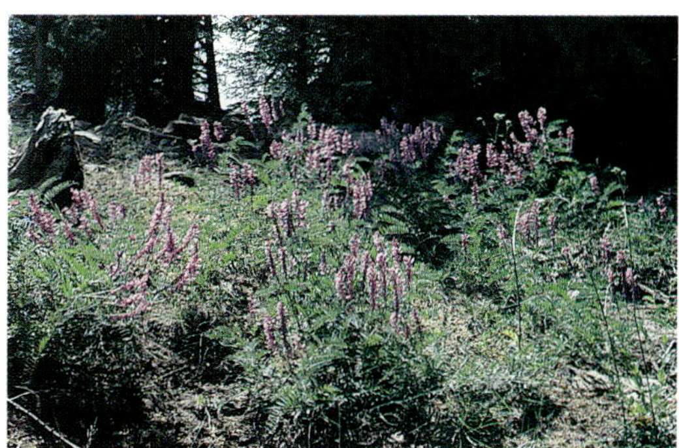

164. *Vicia bakeri* Kalam, Swat, T. J. Roberts

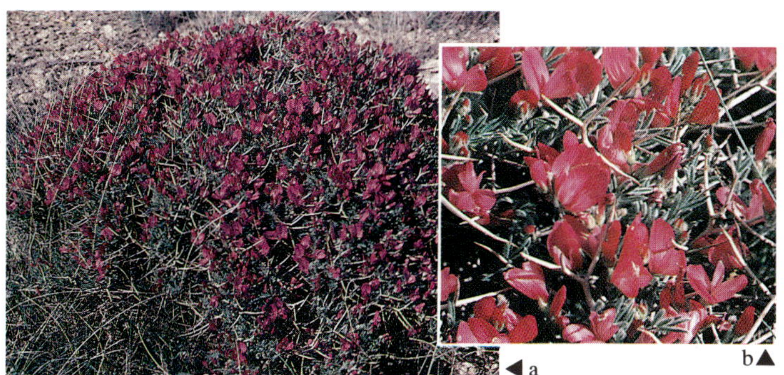

166. *Onobrychus cornuta* a, b: Ziarat, Rubina A. Rafiq

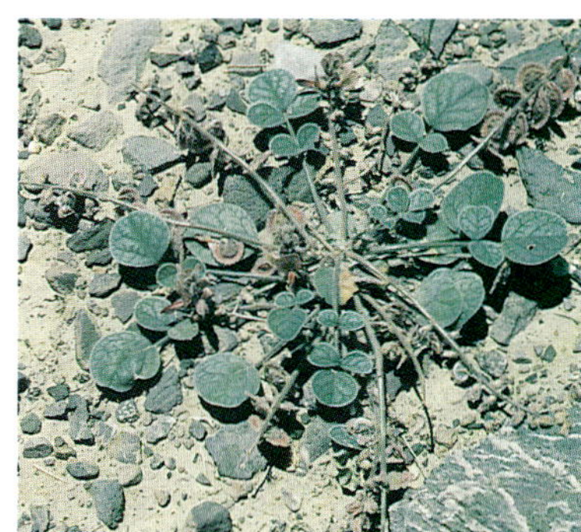

168. *Onobrychus tavernierifolia* Chiltan hills, Rubina A. Rafiq

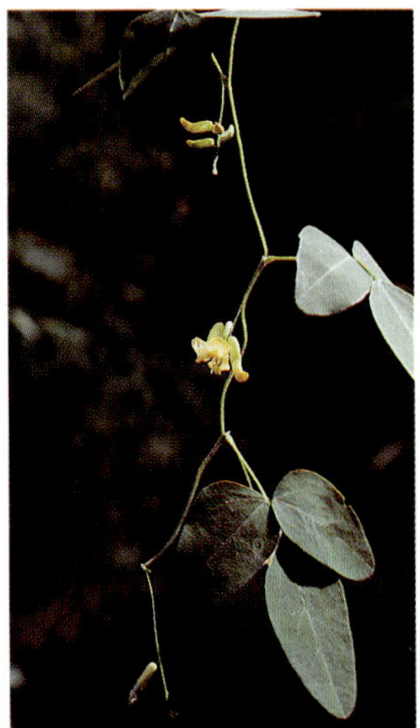

175. *Shuteria involucrata* Ghora Gali, Y. J. Nasir

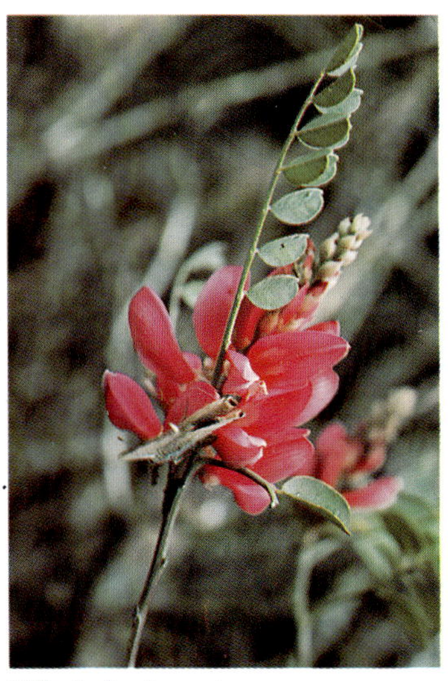

170. *Indigofera atropurpurea* Margalla hills, Y. J. Nasir

172. *Indigofera argentea* Sonmiani, Y. J. Nasir

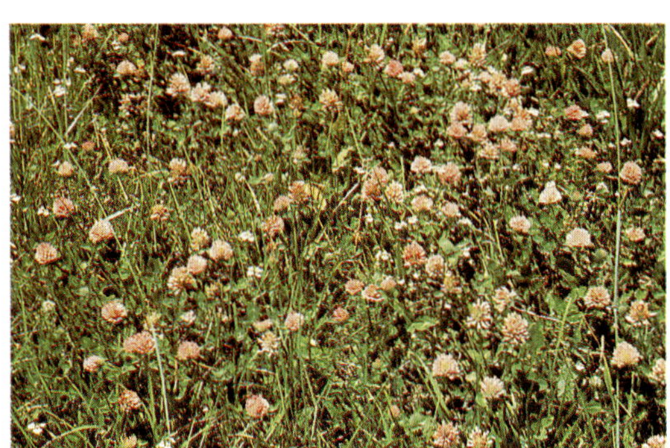

176. *Trifolium repens* Kaghan, Y. J. Nasir

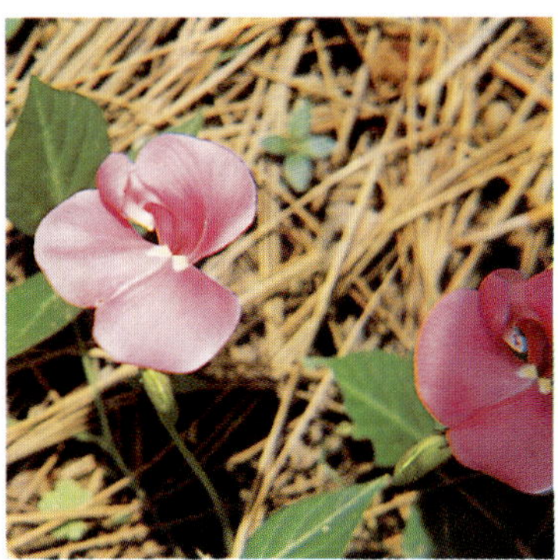

174. *Vigna vexillata* Margalla hills, S. A. Sultan

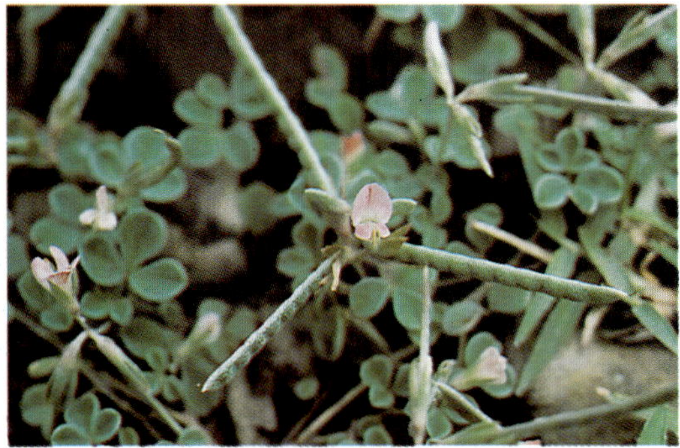

173. *Argyrolobium roseum* Margalla hills, S. A. Sultan

171. *Indigofera linifolia* Islamabad, S. A. Sultan

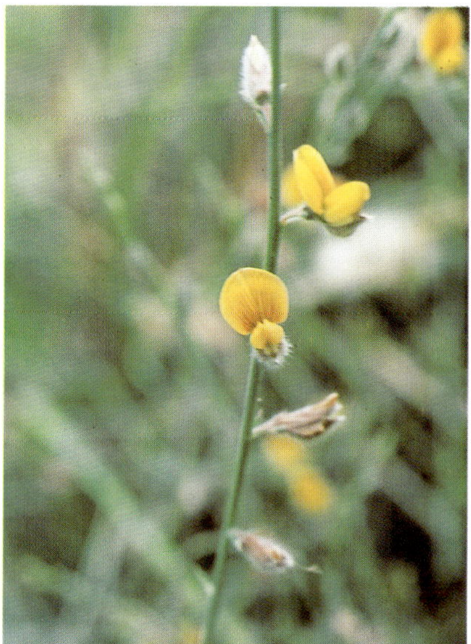

178. *Crotolaria burhia* Attock hills, S. A. Sultan

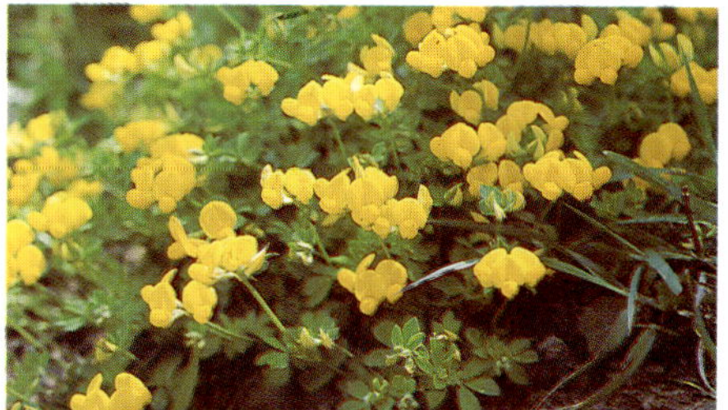

180. *Lotus corniculatus* Murree hills, Rubina A. Rafiq

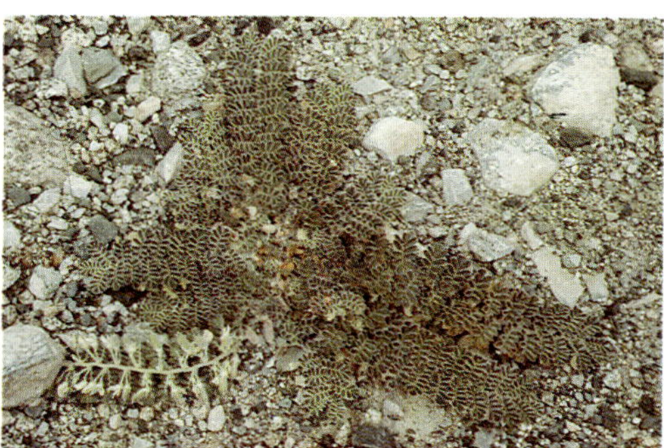

182. *Chesneya depressa* Naltar valley, Mark Mallalieu

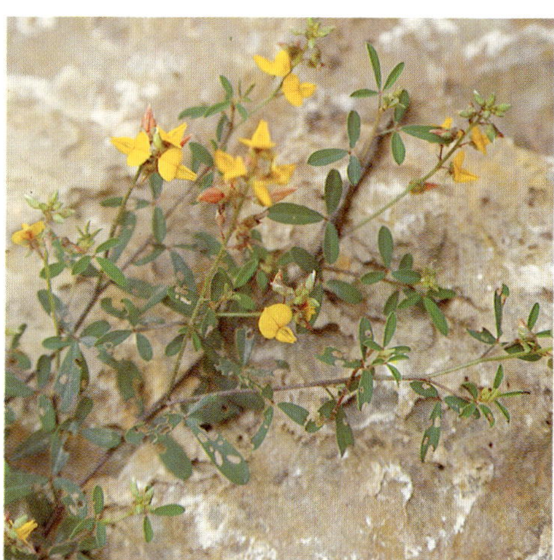

179. *Crotolaria medicaginea* Bhurban, S. A. Sultan

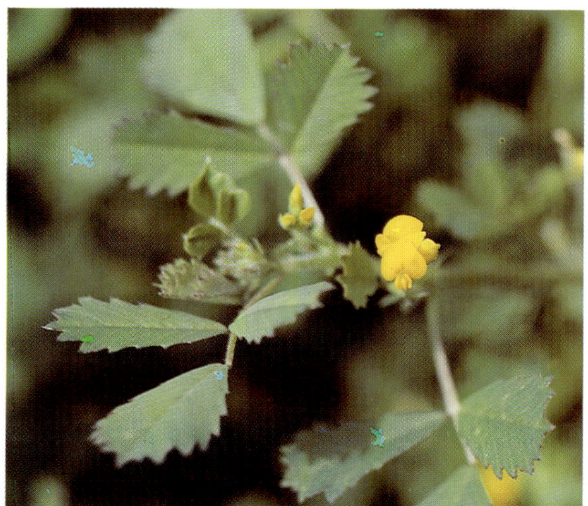

177. *Trigonella monantha subsp. incisa* Attock hills, S. A. Sultan

181. *Chesneya parviflora* Makran, Y. J. Nasir

185. *Caragana ambigua* Ziarat, Rubina A. Rafiq

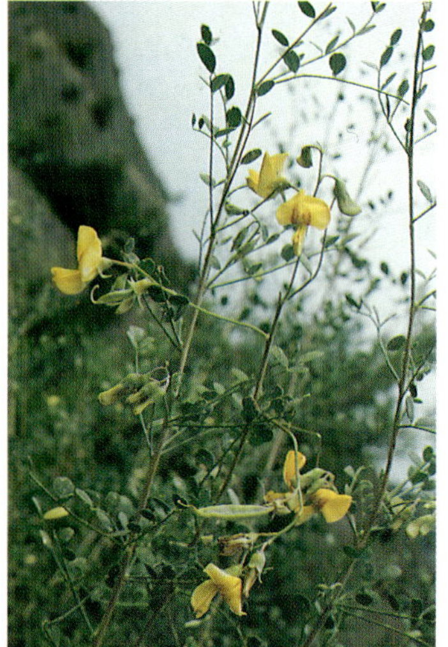

184. *Colutea paulsenii* Gilgit,
Y. J. Nasir

187. *Sophora mollis subsp. griffithii*
Makran, Y. J. Nasir

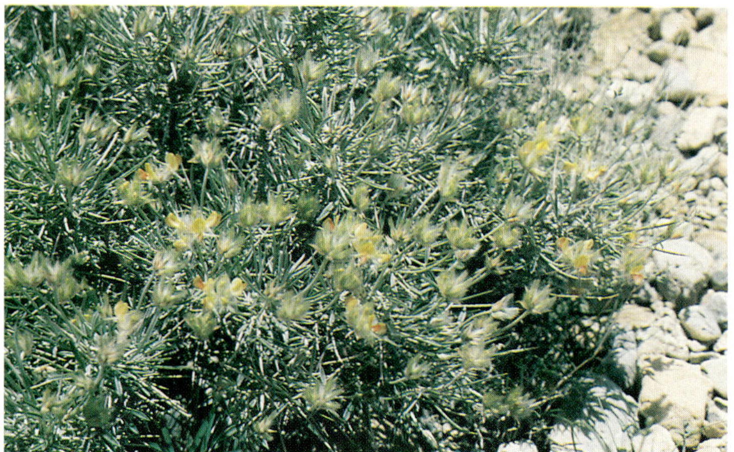

186. *Ebenus stellata* Chiltan hills, Rubina A. Rafiq

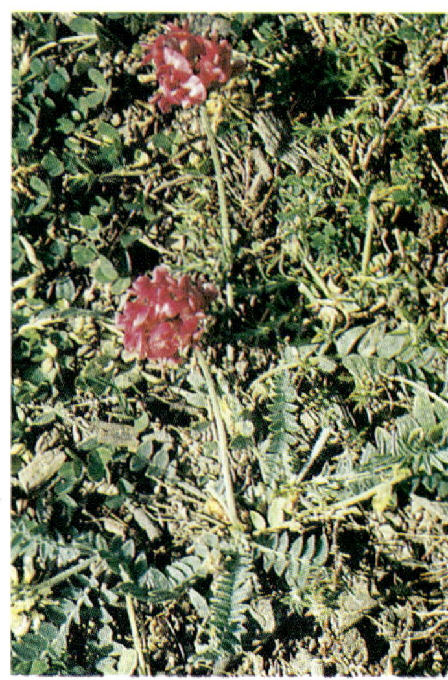

183. *Oxytropis cachemiriana* Kaghan,
Y. J. Nasir

193. *Acacia catechu* Margalla hills, Y. J. Nasir

188. *Sophora alopecuroides* Quetta, Rubina A. Rafiq

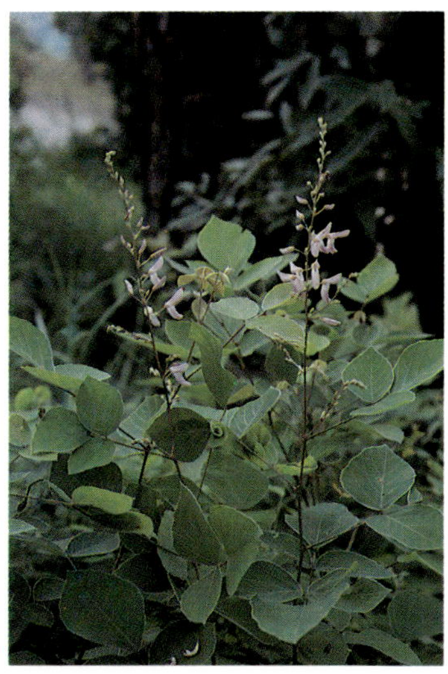

192. *Desmodium elegans* Ghora Gali, Y. J. Nasir

189. *Taverniera spartea* Makran, Y. J. Nasir

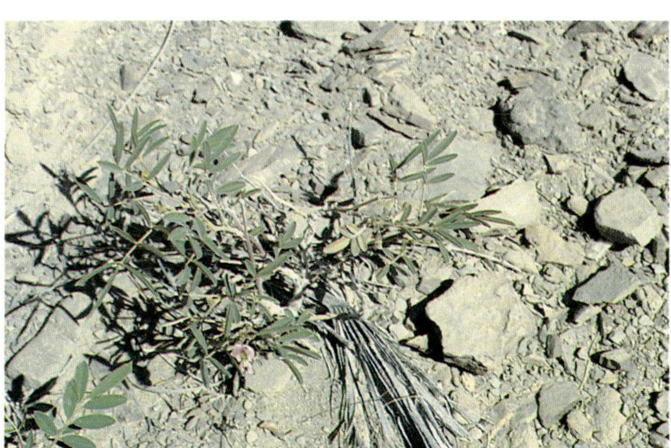

191. *Tephrosia uniflora* Hunza, Y. J. Nasir

190. *Taverniera cuneifolia* Islamabad S. A. Sultan

199. *Albizzia lebbek* a, b: Islamabad, Rubina A. Rafiq

198. *Acacia farnesiana* a, b: Islamabad, S. A. Sultan

196. *Acacia nilotica* Margalla hills, S. A. Sultan

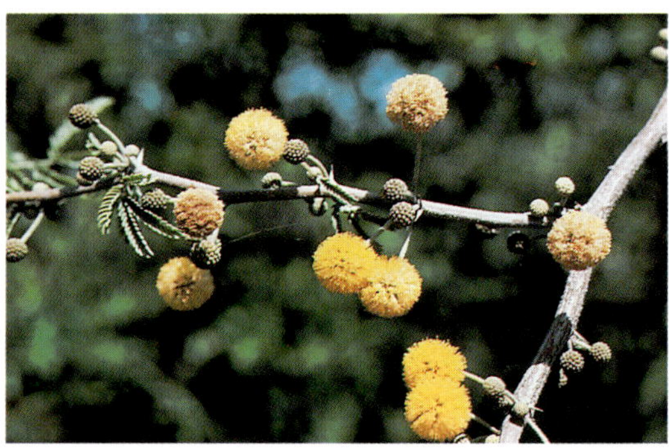

197. *Acacia hydaspica* Taxila, S. A. Sultan

194. *Acacia modesta* Margalla hills, S. A. Sultan

195. *Acacia jacquemontii* Makran, Y. J. Nasir

2 per node. Flowering spike up to 7 cm long. Flowers creamish, fragrant. Pod flat, 5-6 mm long.

Distribution: Afghanistan, Pakistan, India.

A very common small tree typical of dryer foothill tracts throughout N.W.F.P. and Punjab. Often found in association with the olive *(Olea ferruginea)* from the plains up to 900 m. Useful as a hedge plant and as fire wood. A gum extract is used in medicine. Also a source of excellent honey. May.

195. **A. jacquemontii** Benth. *Babul* Pl. 32

Branched very spiny shrub up to 3 m tall with reddish-brown bark and yellow fragrant flowers borne on pedunculate heads. Pod 5-7 cm long, glabrous.

Distribution: Pakistan, India.

In dry hot areas especially in Cholistan and Thar deserts. The bark is used in the tanning industry and leaves as fodder. February-April.

196. **A. nilotica** (L.) Delile *Kikar, Babul* Pl. 32

A very variable tree in size, habit, and degree of hairiness of the pod, on the basis of which 4 subspecies are recognized.

Distribution: Africa, Arabia, Pakistan, India.

One of the commonest trees in the plains. Forming the climax riverine forests in Sindh. Wood is used in furniture etc., and the bark in tanning industry. Gum extract is used in printing. February-March.

197. **A. hydaspica** J. R. Drummond ex Parker Pl. 32

A slender deciduous shrub up to 2 m tall with white stipular spines 10-25 mm. Flowers in globose pedunculate heads. Pods 10-15 cm long, flattened.

Distribution: Pakistan, India.

Plant parts used as fodder and wood as fuel. September-March.

198. **A. farnesiana** (L.) Willd. *Vilaiti Kikar* Pl. 32

Resembling the preceding species, but differing mainly in the pulpy turgid pods.

Distribution: Native of America. Cultivated or naturalized elsewhere.

Flower heads fragrant. Flowers are used to extract perfume (cassie oil). Plant yields a valuable gum. The bark and seeds used in tanning. Useful as a hedge plant. November-March.

2. **Albizzia** c. 4 spp.

A genus of tropical and sub-tropical trees or shrubs with compound leaves. Flowers bisexual, on globose pedunculate heads. Pods generally large, flattened, dehiscent.

199. **A. lebbek** (L.) Benth. *Siris* Pl. 32

Tree up to 18 m or more tall. Pinnae 1-4 pairs, 5-15 cm long; leaflets 3-11 pairs, often with tumid glands between them. Flowers creamish, in globose heads, fragrant, borne on peduncles up to 12 cm long. Pod 12-26 x 3-5 cm, thin, pale yellow-brown.

Distribution: Tropical Asia, Africa.

Widely cultivated as a roadside tree and naturalized. Fast growing but timber of little economic value. April-May.

3. **Mimosa** 2 spp.

Generally thorny trees, shrubs or herbs. Leaves bipinnate. Flowers small, in globose heads or spikes. Stamens twice the number of petals, exserted. Pods compressed, dehiscing by 2 valves.

200. **M. rubicaulis** Lam. *Ral* Pl. 33
subsp. **himalayana** (Gamble) Ohashi

Prickly shrubs with branches ascending from the base. Leaves 12-24 cm long, rachis prickly. Pinnae up to 14 pairs. Leaflets 5-17 pairs. Flowers pink, in panicles of globose heads 10-12 mm broad. Pods strap-like, 5-10 cm long.

Distribution: Afghanistan, Pakistan, Kashmir to Bhutan, India.

Found in the sub-Himalayan tracts especially the Margalla hills and adjoining plains from 550-1200 m. A useful hedge plant. July-August.

4. **Prosopis** 4 spp. *Mesquite*

Armed or unarmed trees or shrubs. Leaves bipinnate. Pinnae 1-2 pairs. Flowers 5-merous, in axillary cylindric spikes or globose heads. Stamens 10, free. Pod elongated, indehiscent.

1.	+	Branches thorny throughout.	201. **P. cineraria**
	–	Branches with stipular spines only.	2
2.	+	Leaflets 5-16 mm long, oblong, spaced 2.5 to mm apart.	202. **P. juliflora**

– Leaflets 15-35 mm long, broadly linear, spaced 7-18 mm apart. 203. **P. glandulosa**

201. **P. cineraria** (L.) Druce *Jand, Jandi* Pl. 33
Syn.: *P. spicigera* L.

A medium sized thorny tree up to 9 m tall. Thorns curved, flattened. Leaves up to 16 cm long, with 1-2 pinnae. Leaflets 7-14 pairs. Flowers creamish, in pedunculate spikes up to 13 cm long. Pod 12-20 x 0.6-0.9 cm, cylindric.

Distribution: Arabia, south Iran, Afghanistan, hotter parts of Pakistan and India.

A xerophytic tree found in dry places in the Indus plain and a dominant tree in the Thar, Thal and Cholistan deserts. Wood used as fuel, and leaves for fodder. December-March.

202. **P. juliflora** (Swartz) DC. *Afghani Kikar* Pl. 33

An introduced large shrub or small tree with greenish-yellow flowers and spiny stipules up to 10 mm long. Pod 14-20 x 1-1.5 cm, yellow.

Distribution: Native to West Indies and Mexico. Widely naturalized elsewhere.

Useful for reforestation in arid regions. Wood used as fuel and makes good charcoal. March-May.

203. **P. glandulosa** Torr. Pl. 33

Similar to preceding species but a tree with larger narrower leaflets that are more widely spaced.

Distribution: Native of Mexico and southern United States. Naturalized elsewhere.

Commonly planted and naturalized in Sindh, Balochistan and Punjab. It has become a pest in irrigated forest plantation especially in lower Punjab and Sindh. March-September.

CAESALPINACEAE

Cassia Family

A family of trees, shrubs or climbers, characterised by irregular bisexual flowers. Stamens 10 in number or fewer and free filaments.

1. + Leaves simple, generally bilobed. **1. Bauhinia**
   – Leaves compound. 2
2. + Leaves unipinnate. Plant unarmed. **2. Cassia**
   – Leaves bipinnate. Plants prickly. **3. Caesalpinia**

1. **Bauhinia** 3 spp.

Flowers showy in terminal or axillary panicles or racemes. Petals 5. Stamens 10 or 3 or 5 by reduction. Pod linear.

204. **B. variegata** L. *Kachnar* Pl. 33

Medium sized tree with leaves that are shallowly bilobed. Flowers large and pretty in racemes, pale purple or white. Pods 12-26 cm long, flat, dehiscent.

Distribution: Pakistan, Bhutan, Burma, India, China.

Both cultivated and wild up to 1000 m in foothill Himalaya. The floriferous trees produce a colorful display in spring. The flower buds are often cooked and eaten. February-April.

2. **Cassia** c. 6 spp.

Flowers in axillary or terminal racemes. Petals 5. Stamens 10 or less by abortion. Pod cylindrical or compressed, dehiscent or indehiscent.

About 18 species are cultivated or introduced.

205. **C. fistula** L. *Amaltas, Indian Laburnum* Pl. 34

A medium sized tree with paripinnate leaves and large drooping panicles of yellow flowers. Pod cylindric, 30-55 cm long, indehiscent.

Distribution: Pakistan, India, Burma, Sri Lanka.

A beautiful tree when in flower. Wild in the sub-Himalayan tracts, conspicuous in Murree hills. Also cultivated for its big panicles of golden yellow flowers. The pods are laxative. April-May.

3. **Caesalpinia** 1 sp.

Trees, shrubs or lianas with large bipinnate leaves. Flowers yellow or red, racemose. Stamens 10, free.

206. **C. decapetala** (Roth) Alston *Kanderi* Pl. 34

A woody straggler with prickly parts. Leaves 10-40 cm long. Pinnae 5-12 pairs; leaflets 6-14 pairs. Flowers yellow, in erect racemes 30-45 cm long. Pods 6-12 cm long, woody.

Distribution: Pakistan, eastward to Bhutan, India, Sri Lanka, Malay Isls., China, Japan.

Both cultivated and wild from Dir eastward up to 1100 m. A useful hedge plant. March-April.

ROSACEAE *Rose Family*

Herbs, shrubs or trees with stipulate, alternate, simple or compound leaves. Flowers often showy, 5-merous. Stamens many. Ovary superior, inferior or semi-inferior. Carpels one to many, united or free. Fruit dry or fleshy and very variable in appearance.

1.	+	Trees or shrubs.	2
	–	Herbs, sometimes with a more or less woody base.	7
2.	+	Leaves simple.	3
	–	Leaves compound.	5
3.	+	Carpel solitary. Fruit a drupe.	4
	–	Carpels many. Fruit a pome.	**3. Pyrus**
4.	+	Drupe tomentose.	**2. Amygdalus**
	–	Drupe glabrous.	**1. Prunus**
5.	+	Unarmed shrubs or undershrubs.	**4. Sorbaria**
	–	Generally prickly shrubs.	6
6.	+	Ovary not adnate to the calyx tube. Fruit a collection of small juicy drupes.	**5. Rubus**
	–	Ovary adnate to the calyx tube. Fruit fleshy, flask-like.	**6. Rosa**
7.	+	Fruitlets 2, enclosed in a sac or pitcher-like hypanthium.	**7. Agrimonia**
	–	Fruitlets more than 2, borne on a flat to slightly concave torus.	8
8.	+	Carpels at fruiting time hooked.	**8. Geum**
	–	Carpels at fruiting time not hooked.	9
9.	+	Receptacle dry at ripening, not separating from hypanthium.	10
	–	Receptacle fleshy at ripening, separating from hypanthium.	11

10. + Stamens and carpels up to 12 in number or less.
Flowers small, inconspicuous. **9. Sibbaldia**

– Stamens and carpels 20 or more in number.
Flowers generally quite large and showy. **10. Potentilla**

11. + Flowers yellow, solitary. **11. Duchesnea**

– Flowers white, several in corymbs. **12. Fragaria**

1. Prunus c. 10-11 spp. *Plum*

An economically important fruit bearing genus of unarmed trees or shrubs, generally flowering at an early stage. Leaves often with tumid glands. Sepals and petals 5. Stamens many. Fruit a glabrous drupe.

+ Flowers white, in pendulous racemes. 207. **P. cornuta**

– Flowers pale pink, in axillary corymbs. 208 **P. cerasioides**

207. **P. cornuta** (Wall. ex Royle) Steud. *Kalakat* Pl. 34

Tree up to 18 m tall. Leaves lanceolate or broadly so, margin minutely toothed. Drooping racemes up to 18 cm long. Fruit globose, 6-8 mm broad, purplish-black on maturity.

Distribution: Afghanistan, Pakistan, Kashmir to Burma, south west China.

Found in mixed coniferous and deciduous forests from 1500-3100 m; widespread in Hazara district, Murree hills and Azad Kashmir. The round fruit which are much favoured by birds, are often diseased and horn-like due to an insect infestation. April-May.

208. **P. cerasioides** D. Don Pl. 34

A medium sized tree from 3-10 m tall, with pretty pink clusters of flowers. Leaves elliptic to elliptic-oblong, toothed. Drupes red.

Distribution: Pakistan, Kashmir to Nepal, Bhutan, south Tibet, Assam, Burma, west China.

Uncommon. Found in the sub-Himalayan tracts from 900-1800 m. November.

2. Amygdalus 1 sp. *Almond & Peach*

Resembling *Prunus* but differing in the tomentose drupes.

209. **A. brahuica** Boiss. *Zargah* Pl. 34
subsp. **afghanica** (Pachom) Browicz
Syn.: *Prunus brahuica* (Boiss.) Aitch. and Hemsl.

A small tree or shrub with pretty pink flowers, that bloom in spring. Leaves elliptic to elliptic-lanceolate or obovate, margin toothed. Drupe 12-15 x 10-12 mm, tomentose.

Distribution: Afghanistan, Pakistan (endemic).

Locally common in juniper tract from 1900-3000 m in dry stony areas in Balochistan and south Waziristan. March-April.

3. Pyrus 1 sp. *Pear*

Deciduous trees with white or lilac flowers in corymbs. Petals clawed. Stamens more than 20. Fruit a succulent pome.

210. **P. pashia** Buch.-Ham. ex D. Don *Batangi, Medlar* Pl. 34

Medium sized tree with white flowers that appear along with the young leaves. Leaves ovate-lanceolate, toothed. Pome globose, 20-30 mm broad, brownish when mature.

Distribution: Afghanistan, Pakistan, Kashmir to south west China, Burma.

Both cultivated and wild especially conspicuous in the Murree hills tracts in spring. March-April.

4. Sorbaria 2 spp.

Unarmed shrubs or undershrub. Leaves stipulate, pinnate. Flowers in showy terminal panicles. Stamens many.

211. **S. tomentosa** (Lindl.) Rehder Pl. 35
Syn.: *Spiraea lindleyana* Wall.

A deciduous woody perennial 1.5-2 m tall with spreading reddish branches often forming thickets. Flowers creamy white, in large attractive panicles up to 30-40 cm long.

Distribution: Central Asia, Afghanistan, Pakistan, Kashmir to central Nepal, north China.

A common gregarious shrub usually along river banks, moist open slopes in the Himalaya from 1200-3300 m. July-August.

5. **Rubus** c. 13 spp. *Blackberry & Raspberry*

Prickly shrubs with usually compound leaves. Flowers bisexual, in simple or compound racemes. Sepals and petals 5. Stamens many. Fruit a collection of small juicy drupes.

+ Flowers white. Fruit orange-yellow. 212. **R. ellipticus**

– Flowers pink. Fruit red-black. 213. **R. sanctus**

212. **R. ellipticus** Smith Pl. 35

A robust prickly shrub with bristly tomentose stems, branches ascending and 3-foliolate leaves. Flowers white, 12-15 mm broad. Leaflets white-tomentose on undersurface, toothed. Fruit globose, yellow, in clusters.

Distribution: Pakistan, Kashmir to south west China, south India, Sri Lanka, Philippines.

Common in sub-Himalayan tracts especially in *Pinus roxburghii* zone from 900-1800 m. Fruit is edible and excellent. February-April.

213. **R. sanctus** Schreber Pl. 35
Syn.: *R. ulmifolius* L.

A shrub with pink flowers and fruit that are red-black. Leaflets grey or white on the undersurface.

Distribution: West and central Europe, Mediterranean, Crimea, south west Asia to west Himalaya.

Variable in size of leaves and flowers. Found up to 2500 m. April-May.

6. **Rosa** 11-12 spp. *Gulab, Rose*

Prickly or spiny shrubs with imparipinnate stipulate leaves. Flowers solitary or in clusters. Calyx persistent in fruit. Fruit fleshy and flask-like.

1. + Climbers. Flowers white. 214. **R. brunonii**

   – Generally shrubs. Flowers pink or red. 2

2. + Stem and branches with few prickles. 215. **R. macrophylla**

   – Stems and branches densely prickly. 216. **R. webbiana**

214. **R. brunonii** Lindley *Tarni* Pl. 35

A rampant thorny climber with panicles of white flowers. Leaflets 5-9, elliptic-lanceolate to elliptic-oblong, toothed. Fruit subglobose, brown.

Distribution: Pakistan, Kashmir to Bhutan, Assam, Burma, west China.

This floriferous climber decorates the whole Murree hill range in spring from 900-2300 m. April-June.

215. **R. macrophylla** Lindley *Himalayan Musk Rose* Pl. 35

A shrub 3-4 m tall, with red, prickly stems. Flowers pink, 5-7 cm broad. Leaflets (5-) 7-13, elliptic-ovate, finely toothed. Fruit 4-5 cm long, red, flask-like.

Distribution: Afghanistan, Pakistan, Kashmir to Nepal, south west China.

In forests and forest glades, valleys from 1800-3300 m in Himalaya. June-July.

216. **R. webbiana** Wall. ex Royle Pl. 35

Closely resembling the preceding species (especially the large leaved forms) from which it differs in the more prickly nature, coarser toothed and fewer veined leaves and stouter peduncles.

Distribution: Pakistan, Kashmir to west Nepal.

Usually in dry rocky places from 1400-4000 m, especially Gilgit and Baltistan. June-August.

7. Agrimonia 1 sp.

Perennial herbs with imparipinnate leaves. Flowers spicate, generally yellow. Sepals and petals 5. Fruit prickly.

217. **A. eupatoria** L. *Agrimony* Pl. 36

Herbs up to 80 cm tall. Stem leafy. Leaflets 5-13 in number, elliptic to ovate, toothed, densely white hairy on undersurface. Flowering spikes up to 27 cm long bearing many small flowers. Fruit drooping, bristly.

Distribution: Europe, Russia, Mediterranean, Pakistan, Kashmir.

In forest clearings, meadows, field borders from 1500-3100 m. May-August.

8. Geum 2 spp. *Avens*

Perennial herbs with a rosette of lyrate-pinnate leaves. Flowers solitary or in clusters. Sepals and petals 5. Styles hood-like or not at fruiting time.

218. **G. elatum** Wall. ex G. Don Pl. 36

Basal leaves 15-28 cm long, pinnately lobed; the terminal the largest. Stem leaves smaller. Flowers 3-4 cm broad, yellow. Fruit globose, hairy.

Distribution: Pakistan, Kashmir to Sikkim, south Tibet.

Common in alpine and sub-alpine areas from 2800-4200 m. June-August.

9. Sibbaldia c. 2 spp.

Prostrate perennials spreading by stolons. Leaves ternate. Flowers small, 5-merous, yellow, in clusters. Stamens and carpels up to 12 in number or less. Fruit ovoid, smooth.

219. **S. cuneata** O. Kuntze Pl. 36

A prostrate perennial spreading extensively by stolons. Flowers c. 4 mm broad, yellow, on stems up to 4-7 cm long. Leaflets wedge-shaped, apex toothed.

Distribution: Afghanistan, Pakistan, Kashmir to central Nepal, south west China.

Alpine slopes and meadows from 2900-4300 m. Widespread from Safed Koh, Chitral east to Deosai plateau. June-August.

10. Potentilla c. 35 spp.

Perennial herbs with a woody base. Leaves ternate or pinnate. Flowers solitary or in corymbs or umbel-like clusters. Sepals and petals 5. Stamens and carpels 20 or more in number. Fruitlets many, nut-like, glabrous.

A genus quite richly represented in northern areas of Pakistan.

1.	+	Flowers white.	220. **P. salesoviana**
	–	Flowers yellow or red.	2
2.	+	Flowers red or shades of it.	221. **P. nepalensis**
	–	Flowers yellow.	3
3.	+	Lower stem leaves 3-foliolate.	222. **P. atrosanguinea**
	–	Lower stem leaves pinnate.	4

4. + Petals smaller than the calyx. Annual. 223. **P. supina**

– Petals larger than the calyx. Perennials. 5

5. + Leaflets 3-5 in number, each further pinnatifid to pinnatisect. Peduncles short, not prominent. 224. **P. pamiro-alaica**

– Leaflets 11 or more in number, toothed. Peduncles long, exserted above foliage. 225. **P. peduncularis**

220. **P. salesoviana** Steph. Pl. 36

A perennial herb, with a suffruticose base. Stems many, covered with brown peeling bark. Leaflets 3-4 pairs more or less oblong, coarsely toothed, whitish on undersurface. Calyx hairy. Petals white, obovate, obtuse. Fruit hairy.

Distribution: Central Asia, Altai, Tibet, Pakistan, Kashmir.

On scree slopes, river banks etc. from 3200-4200 m in the Himalaya. June-August.

221. **P. nepalensis** Hook. f. Pl. 36

A hairy perennial herb with erect to spreading stems and long stalked 5-palmate leaves and rose or crimson coloured flowers.

Distribution: Pakistan, Kashmir to central Nepal.

Forests clearings and alpine slopes from 2000-2700 m. Very conspicuous in Hazara and Murree hills during monsoon season. June-August.

222. **P. atrosanguinea** Lodd. Pl. 37
Syn.: *P. argyrophylla* Wall. ex Lehm.

A semi-erect perennial herb with hairy parts. Leaflets 3, elliptic-ovate, 2-4 cm long, toothed, undersurface silky white. Flowers yellow, but sometimes dark red.

Distribution: Afghanistan, Pakistan, Kashmir to Nepal, Sikkim.

Alpine and sub-alpine meadows from 2400-3300 m. A variable species. June-August.

223. **P. supina** L. Pl. 37

A yellow flowered annual, with ascending to spreading branches. Leaflets 3-9, toothed. Flowers 5 mm broad. Petals shorter than calyx.

Distribution: North America, Europe, north Africa, west Asia, Mongolia, Pakistan, India.

Damp places and ditches in the plains to 2400 m throughout the country. March-May.

224. **P. pamiro-alaica** Juz. Pl. 37

Perennial with short thick branched ascending stem. Leaflets 3-5 pairs, each deeply divided into oblong lobes, under-surface especially dense appressed hairy. Flowers few, short-pedicelled.

Distribution: Central Asia, north Pakistan.

Alpine meadows from 3100-4200 m especially Hunza. June-August.

225. **P. peduncularis** D. Don Pl. 37

Leaflets many in number. Flowers 15-23 mm broad, borne on long peduncles. Achenes rather large (more or less 2 mm long).

Distribution: Pakistan, Kashmir to Nepal, China.

Alpine meadows from 3000-4100 m. June-August.

11. Duchesnea 1 sp.

Perennial herbs with runners, related to the strawberry *(Fragaria)* and *Potentilla*. From the former it differs in the solitary yellow flowers and from the latter in having a fleshy receptacle (not dry) that separates from the hypanthium.

226. **D. indica** (Andr.) Focke *Yellow-Flowered Strawberry* Pl. 37

A yellow-flowered perennial, spreading by reddish runners. Leaves 3-foliolate. Fruit subglobose, c. 1 cm broad, red.

Distribution: Afghanistan, Pakistan, Kashmir to Nepal, China, Japan and Malaysia.

Common in damp shady places in the hilly tracts up to 2400 m. The attractive red fruit are almost tasteless. March-May.

12. Fragaria 1 sp. *Strawberry*

Leaves radical, long petioled, ternate. Stems spreading by runners. Flowers white, few, in corymb-like clusters.

227. **F. nubicola** Lindl. ex Lacaita *Mewa* Pl. 37

Similar to the preceding species but with white flowers and fruit that are delicious.

Distribution: Pakistan, Kashmir to Nepal, Bhutan, north Burma, west China.

Common and widespread above 2300 m in Himalayan region. Occasionally cultivated. April-June.

SAXIFRAGACEAE *Saxifrage Family*

Herbs with alternate, simple or compound leaves. Flowers generally more than one, in cymes or racemes. Sepals and petals 5. Stamens twice as many. Carpels and styles 2. Fruit capsular.

+ Mature leaves 5-18 cm long; leaf base sheathing. Ovary superior. **1. Bergenia**

– Leaves up to 4 cm long; leaf base not sheathing. Ovary superior to half inferior or inferior. **2. Saxifraga**

1. Bergenia 2 spp.

Rhizomatous perennial herbs with stems covered by persistent remains of leaf bases. Leaves basal, large, rounded, leathery. Flowers racemose, borne on thick scapes. Capsule dehiscing by 2-3 valves.

+ Leaves more or less orbicular, margin entire to denticulate. Flowers pink or pinkish-white. 228. **B. ciliata**

– Leaves obovate-cuneate, margin distinctly toothed. Flowers white. 229. **B. stracheyi**

228. **B. ciliata** (Haw.) Sternb. *Zakham-e-Hayat* Pl. 38

Leaves 4-18 x 4-4 cm, glabrous to hairy, dark green. Flowers pink or pinkish-white. Capsule 11-13 x 5-6 mm.

Distribution: Afghanistan, Pakistan, Kashmir to Bhutan, south Tibet, Assam.

An early flowering species carpetting rocky slope crevices from 900-3000 m. March-May.

229. **B. stracheyi** (Hook. f. & Thoms.) Engl. Pl. 38

Resembling the preceding species but differing in the leaf shape and flowers that are white and pinkish stamens. Because of the colour of stamens, flowers appear pinkish-white.

Distribution: Afghanistan, Pakistan and Kashmir.

Alpine. Gregarious on rocky slopes from 2400-4300 m especially upper Kaghan valley and Swat Kohistan. June-August.

2. **Saxifraga** 19 spp. *Saxifrage*

Erect, decumbent or low growing herbs, sometimes cushion-like, with simple leaves. Styles 2, persistent. Capsule more or less globose.

230. **S. hirculus** L. var. **alpina** Engl. Pl. 38

A stoloniferous erect herb. Basal leaves petiolate. Stem and pedicels villous brown. Flowers 1-4 in number, yellow. Petals 2-3 times the length of sepals.

Distribution: Pakistan and Kashmir.

From 3900-4500 m. Alpine meadows and moist places especially Hunza and Baltistan. June-August.

PARNASSIACEAE *Grass of Parnassus Family*

A north temperate family of perennial scapigerous herbs, closely related to the *Saxifragaceae*, from which it differs mainly by an additional whorl of 5 staminodes opposite the petals.

1. **Parnassia** 4 spp.

Leaves simple. Scape bearing a solitary flower. Flowers regular, bisexual, 5-merous. Stamens 5, alternating with petals. Carpels 3-4, syncarpous. Fruit a membranous capsule dehiscing by 3-valves.

231. **P. nubicola** Wall. ex Royle Pl. 38

Glabrous herb up to 27 cm tall. Leaves 12-30 x 5-18 mm, elliptic-ovate to oblong. Scapes arising from a single leaf. Flowers 15-20 mm broad, white. Staminodes 3-lobed. Capsule obovoid, 8-10 mm long.

Distribution: Afghanistan, Pakistan, Kashmir to southeast Tibet.

Moist or open places from 2400-3200 m from Chitral to Murree hills. July-August.

GROSSULARIACEAE *Gooseberry Family*

Deciduous armed or unarmed shrubs with simple lobed leaves. Flowers solitary or in racemes, 4-5 merous. Petals smaller than sepals. Stamens as many as petals and alternating with them. Fruit a berry.

Ribes 8 spp. *Gooseberry, Currant*

Characters as those of the family.

1. + Plants prickly. Flowers solitary. 232. **R. alpestre**

 – Plants without prickles. Flowers in racemes. 2

2. + Berries red or orange in colour. Flowers unisexual. 233. **R. orientale**

 – Berries black. Flowers bisexual. 234. **R. himalense**

232. **R. alpestre** Decne. Pl. 38

Prickly shrubs upto 2 m tall. Leaves 3-5-lobed, orbicular, hairy. Flowers bisexual, solitary on short stalks, white. Berries red, sour, not edible.

Distribution: Afghanistan, Pakistan, India to south west China.

Common on open slopes up to about 2400 m in the Himalaya. May-June.

233. **R. orientale** Desf. Pl. 38

Shrubs 1-2.5 m tall with young branches glandular sticky. Leaves 10-22 x 15-50 mm, suborbicular to reniform, 3-5-lobed, lobes toothed to obtuse. Racemes up to 6 cm long. Female racemes 10-23 mm long.

Distribution: Southeast Europe, west and central Asia, Altai, Kashmir to Bhutan, Tibet, Mongolia, west China.

In dry stony places from 2300-3900 m from Balochistan, N.W.F.P. up to Chitral and Gilgit. April-June.

234. **R. himalense** Royle ex Decne. Pl. 39

Glabrous shrub with pendulous racemes of greenish-yellow flowers and 3-5-lobed leaves. Berries black, 6-8 mm long.

Distribution: Pakistan, Kashmir to Bhutan, west and central China.

From 2400-4200 m in Himalaya. The fruit is used for making jam. April-June.

CRASSULACEAE *Stone Crop Family*

Herbaceous succulent plants, with thick fleshy leaves, generally reproducing by asexual means (bulbils, vegetative parts). Flowers disposed in compound panicles on scapigerous stalks.

Fruit a collection of follicles.

1. + Stamens twice as many as petals. Corolla tube longer than lobes, much smaller than the basal ones. **3. Rosularia**

   – Stamens less than twice the petal number. Corolla tube shorter than the lobes. 2

2. + Rhizomes if present, slender and without scale leaves. Petals 5 or more. Flowers bisexual. **1. Sedum**

   – Rhizomes present with scaly leaves.Petals 4. Flowers unisexual. **2. Rhodiola**

1. Sedum c. 13 spp. *Stone Crop*

Annual or perennial herbs found mostly in stony and rocky places. Leaf rosettes generally absent. Petals spreading.

+ Low growing plants with white or light pink flowers. Leaves acute. 236. **S. hispanicum**

– Plants 20-40 cm tall. Flowers rose-pink to pink-purple. Leaves rounded. 235. **S. ewersii**

235. **S. ewersii** Ledeb. Pl. 39

A many stemmed perennial glabrous herb 20-40 cm tall. Leaves fleshy, rounded. Flowers rose-pink to pink-purple, in terminal flat topped clusters. Petals 5.

Distribution: Afghanistan, Pakistan in Chitral, Swat and Hazara district, Kashmir to Uttar Pradesh (India).

Common in rock crevices from 1800-4500 m. June-July.

236. **S. hispanicum** L. Fig. facing p. 102

A small annual growing in clumps with white to pale pink flowers. Leaves fleshy.

Distribution: Southeast Europe, Mediterranean, Asia Minor.

Found in lower hills from Chitral to Hazara area up to 1200 m. March-April.

2. Rhodiola c. 7 spp.

Related to *Sedum*, but differing in the unisexual flowers and presence of rhizomes with scaly leaves.

1.	+	Flowers yellow.	237. **R. heterodonta**
	–	Flowers pink or red.	2
2.	+	Leaves linear to linear-lanceo-late up to 1.2 mm broad. Flower clusters 2-3 cm broad.	238. **R. quadrifida**
	–	Leaves oblanceolate to elliptic-ovate, 2-6 mm broad. Flower clusters 3-4.5 cm broad.	239. **R. himalensis**

237. **R. heterodonta** (Hook. f. & Thoms.) Boriss. Fig. facing p. 102

Leaves toothed. Flowers yellow, in terminal clusters, 2-3 cm broad.

Distribution: Afghanistan, Pakistan, Kashmir to central Nepal.

Alpine screes, gravel areas from 3300-4700 m. June-August.

238. **R. quadrifida** (Pall.) Fisch. & Mey. Pl. 39
Syn.: *Sedum quadrifidum* Pall.

A low growing cushion-like plant 5-6 cm tall, with narrow leaves and reddish clusters of up to 5 flowers.

Distribution: Pakistan, Kashmir to Sikkim, Tibet, north west China, Mongolia, Siberia, Eurasia.

Alpine rocky places from 3200-4300 m especially Chitral, Astor, Baltistan. June-August.

239. **R. himalensis** (D. Don) S. H. Fu Pl. 39

Resembling the previous species, but with larger leaves and clusters of flowers, 30-45 mm broad.

Distribution: Pakistan, Kashmir to south west China.

Alpine rocky places. 3300-4600 m. June-July.

3. Rosularia c. 3 spp.

Perennials differing from *Sedum* or *Rhodiola,* in the corolla tube which is longer than the lobes, the stem leaves which are generally much smaller than the basal leaves, and stamens twice as many as the petals.

\+ Radical leaves minutely hairy, obtuse. Plants of the plains and hills up to 2400 m. 240. **R. adenotricha**

– Radical leaves glabrous, acute. Plants of mountainous areas from 2400-4300 m. 241. **R. alpestris**

240. **R. adenotricha** (Wall. ex Edgew.) Jansson Pl. 40
Syn.: *Sedum adenotrichum* Wall.

Perennial glandulose plants with a rosette of radical leaves. Flowers white or tinged pink, in loose few-flowered clusters. Petals acute.

Distribution: Afghanistan, Pakistan, Kashmir to west Nepal.

Common and widespread from Balochistan, N.W.F.P., Chitral, Gilgit and Murree hills in rocky places. March-April.

241. **R. alpestris** (Kar. & Kir.) A. Bor. Pl. 40

Perennial with underground tubers. Leaves in rosettes, fleshy, oblanceolate, 12-20 mm long. Stems leafy, up to 10 cm tall. Flowers few, pinkish or whitish-pink.

Distribution: Central Asia, Afghanistan to Himachal Pradesh (India).

Stony places up to 3300 m. July-August.

HAMAMELIDACEAE

Witch Hazel Family

Shrubs or trees with simple alternate leaves. Flowers borne in heads or spikes. Petals 4-5, sometimes absent. Fruit capsular.

Parrotiopsis 1 sp.

242. **P. jacquemontiana** (Decne.) Rehder *Pishor, Paser* Pl. 40

A branched shrub up to 3 m tall. Leaves 3-7 cm long, suborbicular, crenate-serrate. Bracts 4-5 in number, petal like, creamy-white, 2-3 cm long. Flowers yellow, in globose heads 1-1.5 cm broad. Capsule 4-valved.

Distribution: Afghanistan, Pakistan, Kashmir to Himachal Pradesh (India).

Forest slopes, especially in Chitral and Swat. Gregarious. 1900-2200 m. April-May.

LYTHRACEAE *Crepe Myrtle Family*

A tropical or sub-tropical family of trees, shrubs or herbs with opposite, entire leaves. Flowers bracteate, bisexual. Petals often crinkly. Fruit capsular.

Woodfordia 1 sp.

Shrubs with flowers in axillary cymes. Flowers 6-merous with narrow petals. Stamens 12, curved. Capsule 2-valved.

243. **W. fruticosa** (L.) S. Kurz *Dhawi* Pl. 40

A shrub with spreading branches and ovate-lanceolate or ovate leaves. Flowers orange-red to bright red, in clusters of 3-17. Capsule 6-8 x 2-4 mm, more or less elliptic.

Distribution: Tropical Africa, Asia, Sri Lanka east to China.

Common in sub-Himalayan tracts. Often on cliffs. Flowers and leaves medicinal. Leaves and twigs yield a dye used in printing. Birds are especially attracted to the nectar of the flowers. Found from 600-1200 m. April-May.

PUNICACEAE *Pomegranate Family*

Sometimes included in the *Lythraceae*, but differing in the inferior ovary and ebracteate flowers.

Punica 1 sp.

Small trees or shrubs with opposite, simple, exstipulate leaves. Flowers solitary, showy. Stamens many, persistent. Ovary inferior. Fruit subglobose with a leathery skin, crowned with the persistent calyx lobes and withered stamens.

244. **P. granatum** L. *Anar, Pomegranate* Pl. 40

A shrub or small tree up to 4 m tall. Leaves oblanceolate to elliptic-obtuse, 15-30 x 5-12 mm. Flowers orange-red, showy. Fruit 3-9 cm broad, globose, pale red to scarlet, with a leathery skin. Seeds juicy, red or pinkish.

Distribution: South Europe, central and west Asia.

Widely cultivated for its edible fruit, also used as ornamental bush for its brightly coloured beautiful flowers. Wild in the sub-Himalayan tracts and hills up to 2100 m. Bark of the root and wood is used as a vermifuge and in stomach complaints. September-December.

ONAGRACEAE

Willow-Herb Family

Perennial or annual herbs with simple leaves. Flowers regular or irregular, bisexual, solitary or in racemes. Sepals and petals 2-7. Stamens as many as or twice the number of sepals. Ovary inferior. Fruit a capsule, berry or nut.

+ Corolla tube up to 3 mm long.
  Capsules 1-2.5 mm broad, obscurely ribbed. **1. Epilobium**

– Corolla tube 4-120 mm long.
  Capsule 3-12 mm broad, distinctly ribbed. **2. Oenothera**

1. Epilobium 18 spp.

Willow Herb

Flowers in axillary racemes. Sepals and petals 4. Stamens 8, longer than the others. Capsule linear, pubescent or globose.

1. + Leaves opposite or verticellate. Flowers regular.
   Floral tube up to 3 mm long. 2

   – Leaves alternate. Flowers showy more or
   less irregular. Floral tube absent. 247. **E. angustifolium**

2. + Plants up to 40 cm tall, glabrous.
   Leaves ovate. 245. **E. laxum**

   – Plants up to 2 m tall, hairy.
   Leaves lanceolate. 246. **E. hirsutum**

245. **E. laxum** Royle Pl. 40

Perennial herb with simple or branched stems, 8-40 cm tall, glabrous. Leaves ovate, 2.5-7 x 8-30 mm, serrate. Flowers showy, pink. Capsule 4-7 cm long.

Distribution: Afghanistan, Pakistan, Kashmir to Sikkim, Tibet, Assam, west China.

Sub-alpine and alpine meadows and forest clearings from 2600-4300 m in the Himalaya. July-August.

246. **E. hirsutum** L. Pl. 41

Robust perennial hairy herb up to 2 m tall. Leaves lanceolate or oblanceolate, 2-10 x 0.5-1 cm. Flowers rose-pink, white in centre. Capsule 3-10 cm long.

Distribution: Temperate Eurasia and Africa.

In marshy places up to 2000 m in Balochistan, N.W.F.P., Punjab extending to Gilgit. June-October.

247. **E. angustifolium** L. Pl. 41

Stout perennial growing in clumps, with racemose flowers which are pink to magenta. Leaves lanceolate, 3-18 x 0.5-3.4 cm. Capsule 4-7 cm long.

Distribution: Circumboreal.

Alpine and sub-alpine especially in Hazara district and Murree hills. Rocky ledges, slopes etc. July-August.

2. Oenothera 4 spp.

A native of the new world. Mostly weedy annual or perennial herbs with alternate, exstipulate leaves. Flowers regular, 4-merous. Floral tube well-developed. Stigma 4-lobed. Capsule elongated, more or less 4-angled.

+	Flowers rose-pink. Perennial.	248. **O. rosea**
-	Flowers yellow. Annuals.	249. **O. affinis**

248. **O. rosea** L'Herit ex Ait. Pl. 41
Syn.: *Hartmannia rosea* (L'Herit. ex Ait.) G. Don

Branched herb up to 40 cm tall. Leaves elliptic to narrow ovate, 20-40 x 10-20 mm. Flowers closing in cloudy weather. Capsule club-like, 12-26 mm long.

Distribution: Native of central and south Texas, USA. Naturalized elsewhere.

Common from 600-2700 m from Swat, Hazara and Murree hills. Near damp places or in and around cultivation. April-September.

249. **O. affinis** Camb. Pl. 41

Erect hairy annual, usually branched. Flowers orange yellow, 8-13 cm long. Capsule 2-4 cm long.

Distribution: Native of south America. Naturalized elsewhere.

Found in and around cultivated areas from 900-1500 m especially in Hazara district. July-August.

CACTACEAE *Cactus Family*

A large family native to drier parts of tropical America, with stem and leaves modified to a xeric habit. Flowers solitary, showy. Fruit a berry which is usually edible.

1. **Opuntia** 1-2 spp.

A genus characterised by the small caducous leaves which are more or less cylindrical (not flat) and sessile rotate flowers. Stems flat and leaf-like, spinose.

250. **O. monacantha** (Willd.) Ham. *Nagphana, Prickly Pear* Pl. 41

Prickly branched shrub to 2 m tall, with leaf-like grass green flattened stems up to 60 cm long. Flowers yellow, 6-7 cm broad. Berry purplish-red when ripe.

Distribution: Native of Brazil and Argentine. Naturalized elsewhere.

Both cultivated as a hedge row especially in the Salt Range and naturalized in the plains to 1200 m. Fruit edible. April-May.

CUCURBITACEAE *Melon Family*

A tropical and sub-tropical family of annual or perennial herbs, prostrate or climbing by tendrils. Leaves entire or much dissected. Flowers solitary, often showy, unisexual, solitary or not. Sepals and petals 5. Carpels 3, united. Ovary inferior or semi-inferior. Fruit a berry or pepo.

1.	+	Corolla lobes long fimbriate.	**1. Trichosanthes**
	–	Corolla lobes not so.	2
2.	+	Flowers 4-5 mm long.	**2. Cucumis**
	–	Flowers more than 5 mm long.	3
3.	+	Flowers white. Leaves glabrous.	**4. Coccinea**
	–	Flowers yellow. Leaves hairy.	**3. Momordica**

1. **Trichosanthes** 2 spp.

Perennial or annual herbs climbing by means of tendrils. Leaves palmately 3-7-lobed with cordate base. Flowers unisexual, male in racemes and female solitary. Stamens 3. Fruit indehiscent.

251. **T. cucumerina** L. *Jangli Chachinda* Pl. 41

Annual climber with branched tendril. Leaves orbicular-reniform, denticulate. Petals 2-8 cm long, white. Fruit ovoid, 4.5-7 x 3-4 cm, yellowish on maturity.

Distribution: Pakistan, India, Sri Lanka, Bangladesh, Burma, Malaysia, Philippines, north Australia.

Lower hills and plains up to 700 m. July-October.

2. Cucumis 2 spp. *Cucumber*

Monoecious annual or perennial trailing or climbing herbs. Leaves 3-7-lobed. Tendrils unbranched. Male flowers in clusters, female solitary. Fruit smooth or variously ornamented.

252. **C. prophetarum** L. *Kharchvit, Wild Cucumber* Pl. 41

A trailer with hispid branched stems. Leaves 3-lobed; lobes dentate or lobulate, scabrid on both surfaces. Male flowers 2-3. Corolla 3-5 mm long, hairy. Fruit ovoid or more or less subglobose, longitudinally stripped green and white, spiny.

Distribution: North Africa, Arabia, Pakistan, India.

Sandy arid places especially in Sindh and Las Bela. February-April.

3. Momordica 1 sp.

Annual or perennial trailers or climbers. Leaves hairy, 3-9-lobed or sometimes simple. Tendrils simple or branched. Flowers yellow, usually bracteate; female flowers solitary, male in clusters. Fruit dehiscent or indehiscent.

253. **M. balsamina** L. *Jangli Karela* Pl. 42

Monoecious annual with simple tendrils. Leaves 3-5-lobed, orbicular in outline. Flowers borne on peduncles, 1.5-7 cm long. Fruit ovoid-globose, orange-red on maturity, surface tuberculate.

Distribution: South and tropical Africa and Asia, Arabia, Australia.

In sandy desert places in Sindh, N.W.F.P. and Punjab up to 300 m. Young fruit are cooked and pickled. Plant considered as a tonic and in stomach complaints. August-November.

4. Coccinea 1 sp.

Dioecious climbers and trailers with simple tendrils. Leaves 5-lobed, glabrous. Female flowers solitary, male usually in clusters. Fruit indehiscent.

254. **C. grandis** (L.) Voigt. *Kanduri* Pl. 42

An extensive branched climber with bright green glabrous leaves 4-11 cm long, cordate at the base. Flowers solitary, showy, white. Fruit elongated, deep red when ripe.

Distribution: Tropical Africa, India, Pakistan, Malaya.

Waste ground or near cultivation. December-May.

BEGONIACEAE *Begonia Family*

Monoecious herbs or undershrubs, often rhizomatous or with tubers. Leaves simple, alternate. male flowers with 2 outer and 2 inner perianth segments. Female flowers with ovary 3-locular; ovary often winged or angled. Fruit capsular or berry.

Begonia 2 spp.

Fruit capsular, angled or winged. Leaf base often oblique, margin entire or serrate.

255. **B. picta** Sm. Pl. 42

A delicate rhizomatous herb with long petioled cordate, doubly serrate leaves and 1-2 flowering scapes. Male flowers with 2 outer segments larger than the inner two, pale pink. Stamens many. Female flowers with 5 segments. Ovary and fruit winged, pubescent.

Distribution: Pakistan, Kashmir to Bhutan, Assam.

Rare. Damp shady rocks or slopes from 600-2000 m, confined to Murree hills area and Kashmir. August-September.

UMBELLIFERAE *Carrot Family*

A large and economically important family, predominantly herbs with hollow stems and generally compound leaves, with sheathing leaf base. Flowers small, many 5-merous, in simple or compound umbels. Ovary inferior, styles 2. Fruit splitting into 2 one-seeded mericarps.

1. + Plants thistle-like, spiny.
   Flowers deep blue. **1. Eryngium**

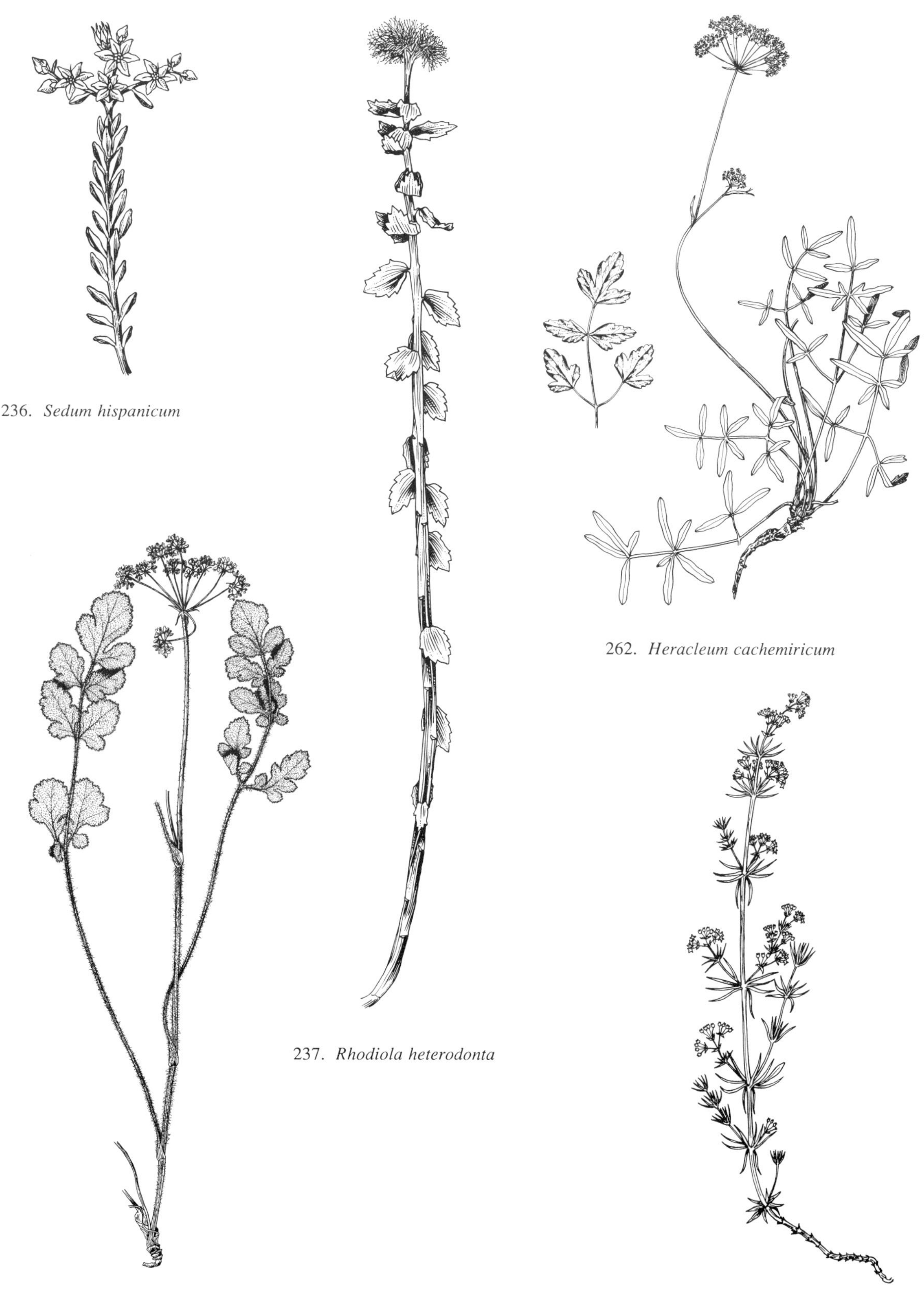

236. *Sedum hispanicum*

237. *Rhodiola heterodonta*

262. *Heracleum cachemiricum*

263. *Heracleum canescens*

277. *Galium verum*

353. *Amberboa moschata*

342. *Echinops echinatus*

414. *Heliotropium crispum*

294. *Launaea resedifolia*

412. *Heliotropium subulatum*

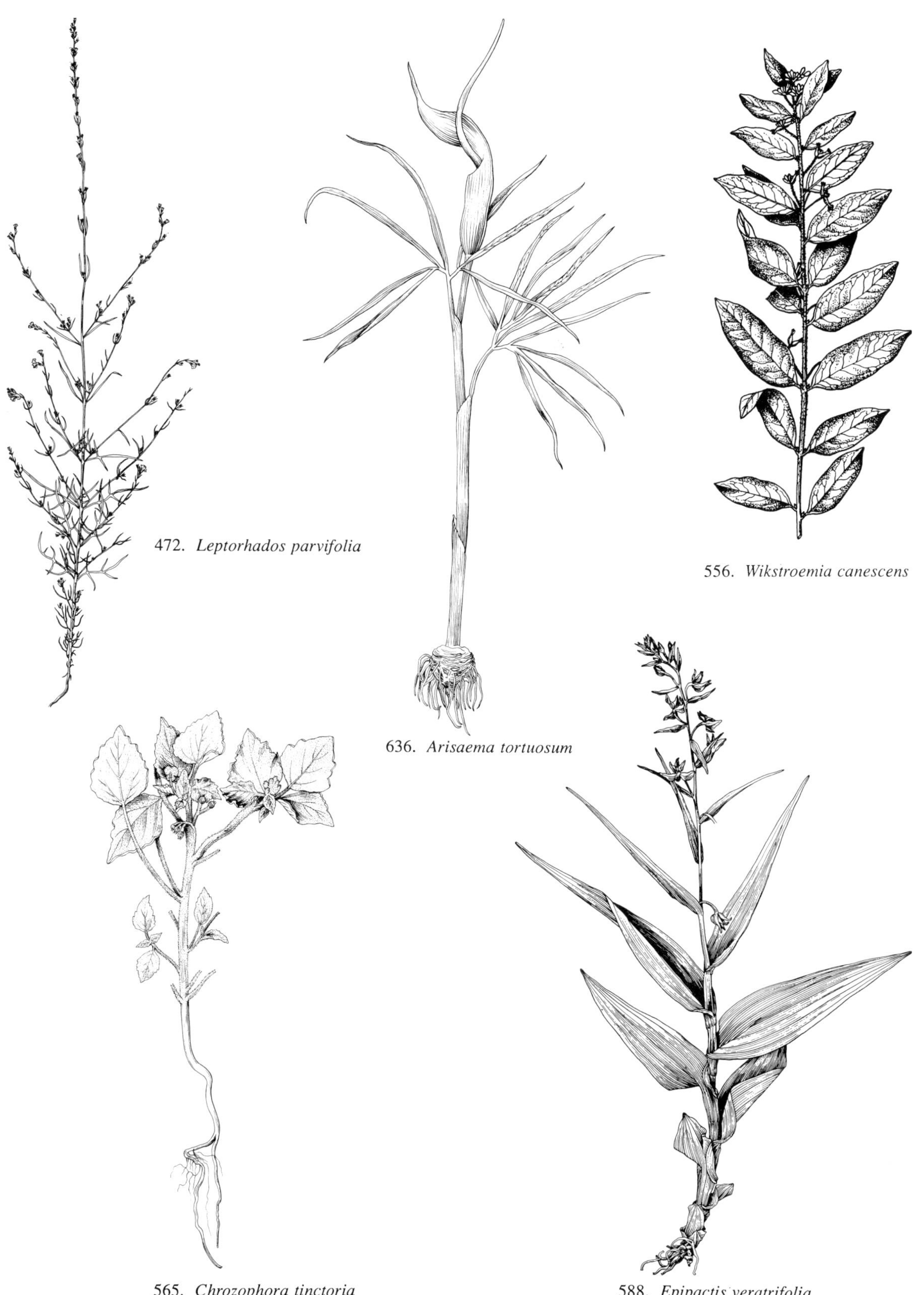

472. *Leptorhados parvifolia*

636. *Arisaema tortuosum*

556. *Wikstroemia canescens*

565. *Chrozophora tinctoria*

588. *Epipactis veratrifolia*

	–	Plants not as above. Flowers pink, yellow or white.	2
2.	+	Flowers yellow.	3
	–	Flowers white to pinkish-white or pink.	5
3.	+	Leaves entire.	**2. Bupleurum**
	–	Leaves compound.	4
4.	+	Fruit wall spongy, wavy.	**3. Prangos**
	–	Fruit wall not spongy, glabrous.	**4. Ferula**
5.	+	Fruit compressed dorsally, ridges winged.	6
	–	Fruit slightly compressed to round in outline, ridges not winged but sometimes with wavy margin.	**7. Scaligera** **8. Torilis** **9. Conium**
6.	+	Fruit vittae not extending to the base of fruit; petals always radiant.	**5. Heracleum**
	–	Fruit vittae extending to the base of fruit. Petals radiant or not.	**6. Psammogeton** **10. Cortia**

1. Eryngium c. 3 spp. *Eryngo*

Leaves toothed or spiny, like the involucral bracts. Flowers sessile, bracteate. Fruit elliptic to cylindrical, ridges not prominent.

256. **E. biebersteinianum** Nevski ex Bobrov Pl. 42
Syn.: *E. coeruleum* M. Bieb.

Plants 30-60 cm tall. Flowers in numerous subglabrous heads, blue-purple. Fruit ridges not prominent.

Distribution: Pakistan, Kashmir, west and central Asia.

Fairly common in the lower hills from 1200-1500 m from Balochistan, N.W.F.P. up to Swat. May.

2. Bupleurum 25 spp.

Leaves simple, entire. Flowers usually yellow, in compound umbels. Fruit elliptic, oblong or ovoid, ridges prominent.

257. **B. longicaule** Wall. ex DC. Pl. 42

Perennial up to 30 cm tall, few branched from the base. Leaves linear-lanceolate, upper stem clasping. Flowers black, in globose heads.

Distribution: Pakistan, Kashmir to Bhutan.

In open places from 3000-4000 m especially sub-alpine zone, Kaghan and Galis. June.

3. Prangos 2 spp.

Perennial herbs with 4-6 pinnate leaves. Fruit corky, ridges winged, vittae many.

258. **P. pabularia** Lindl. Pl. 42

Plants up to 1 m tall with delicately divided pale green leaves and heads of yellow flowers. Fruit oblong, 10-15 mm long with many transverse folds.

Distribution: Afghanistan, Pakistan, Russia.

Not common, but locally abundant especially lower Swat, Swabi. April.

4. Ferula 2 spp.

Tall perennial herbs, often with woody base. Leaves pinnately divided, with a prominent sheathing base. Petals yellow. Fruit compressed dorsally, elliptic to oblong, ridges winged.

+ Plant up to 1 m tall. Leaf segments 2-3 cm long, lanceolate or broadly so. 259. **F. jaeschkeana**

– Plants 2-2.5 m tall. Leaf segments up to 1 cm long, linear. 260. **F. oopoda**

259. **F. jaeschkeana** Vatke Pl. 43

Leaves up to 45 cm long, 2-3-pinnate. Leaf base oblong, 10-12 cm long. Fruit 12-16 x 10-13 mm, brownish-red.

Distribution: Afghanistan, Pakistan.

Fairly common on hillsides from 900-3000 m especially sub-Himalayan tracts from Chitral to Kashmir. April-May.

260. **F. oopoda** (Boiss. & Buhse) Boiss. Pl. 43

A more robust and taller plant than the previous species growing up to 182 cm tall with finely dissected leaves. Has a spectacular inflorescence of globular yellow umbels.

Distribution: Iran, Russia, Pakistan.

Open dry places from 1500-2400 m especially Balochistan. April-May.

5. Heracleum 6 spp.

Perennial herbs with 1-3-pinnate leaves. Flowers white. Fruit compressed laterally, elliptic, ridges prominent, not winged.

1.	+	Robust plants upto 2 m tall. Leaf segments more than 10 cm long.	261. **H. candicans**
	–	Slender plants upto 1 m tall. Leaf segments less than 10 cm long.	2
2.	+	Leaf segments linear to lanceolate. Plants pubescent.	262. **H.cachemiricum**
	–	Leaf segments broader, oval to ovate. Plants villous.	263. **H. canescens**

261. **H. candicans** Wall. ex DC. Pl. 43

Perennial herbs upto 2 m tall. The handsome leaves are large, pinnately lobed, white below. Leaf stalk enlarged into boat-shaped sheath at the base. Flowers large, outer petals bilobed, white.

Distribution: Pakistan, India to south west China.

A widespread and common species on open slopes from 2000 to 4000 m in the Himalaya. May-August.

262. **H. cachemiricum** C. B. Clarke Fig. facing p. 102

Herb 30-90 cm tall. The handsome leaves are largely basal, segments narrow, margin entire to toothed. Fruit 6-11 mm long, oblong, vittae very narrow, ridges broadly winged.

Distribution: Pakistan and Kashmir.

Found in forests and open slopes flowering before the monsoon, from 1000-2500 m in Chitral, Swat and Hazara district. Variable in leaf shape and size. June-July.

263. **H. canescens** Lindl. Fig. facing p. 102

Plants 25-60 cm tall with villous parts. Leaf pinnae ovate, toothed. Fruit 6-8 x 5-6 mm, bilobed at the apex. Vittae very narrow.

Distribution: Pakistan, Kashmir to Uttar Pradesh (India).

Found from 1000-2100 m. Open places, slopes. July-August.

6. Psammogeton 4 spp.

Annual herbs with 1-2-pinnate leaves. Bracts membranous, white. Petals white to pinkish, notched. Fruit ovoid, hairy to hispid.

264. **P. canescens** (DC.) Vatke Pl. 43

Pubescent plant 10-40 cm tall. Leaf segments 2-3-lobed. Bracts 5. Fruit densely hairy, with hairs up to 1.5 mm long.

Distribution: Iran, Afghanistan, Pakistan.

A species of dry regions from 900-1800 m in Balochistan, N.W.F.P. and northern Punjab. March-April.

7. Scaligera 5 spp.

Perennial plants with tuberous roots. Leaves 2-3-pinnate. Bracts white margined. Fruit ovoid, ridges thin. Vittae very small.

265. **S. stewartiana** (Nasir) Nasir Pl. 43

Leaves finely dissected. Leaf segments up to 5 cm long. Bractlets white-margined. Fruit ridges very thin. Vittae obscure.

Distribution: Pakistan.

Foothill Himalaya and adjacent plains from 500-1000 m. March-April.

8. Torilis 3 spp.

Hispid plants with leaves 1-2-pinnate; segments narrow. Flowers white to light purple. Fruit 4-6 mm long, prickly.

266. **T. leptophylla** (L.) Reichb. Pl. 43

Erect plant 25-50 cm tall with bipinnate leaves. Flowers umbellate. Fruit more or less ovoid, 4-5 mm long, bristly.

Distribution: Africa, Europe, central and south Asia. Introduced to USA.

Common weed in plains and lower hills up to 1400 m. March-April.

9. Conium 1 sp. *Morkach, Hemlock*

Leaves 2-4-pinnate. Petioles sheathing. Inflorescence terminal and axillary. Flowers white. Fruot ovoid.

267. **C. maculatum** L. Pl. 44

Large stout plants 1-2 m tall. Leaves 2-pinnate; segments oval, serrate. Pedicels 2-3 times as long as the flowers. Fruit 2-3 mm long.

Distribution: North and south America, Europe, central Asia, Afghanistan, Pakistan, India, China.

Found in and around cultivated areas, slopes, waste places mainly in Murree hills. Poisonous. Plant parts used in spasmodic diseases. Fruit yields the alkaloid conine. June-August.

10. Cortia 1 sp.

Perennial plants with reduced stems. Leaves 2-3-pinnate. Flowers white to pinkish, umbellate. Fruit with lateral ridges winged.

268. **C. depressa** (D. Don) Norman Pl. 44

Small plant up to 18 cm tall with a fibrous base. Bracts and bractlets divided. Leaf segments narrow. Fruit 4-5 mm long.

Distribution: Pakistan, Kashmir.

Alpine areas from 3800-5000 m in Gilgit, Hunza and Baltistan. June-August.

ARALIACEAE *Ivy Family*

Trees or shrubs, often climbers. Leaves simple or compound, alternate. Flowers umbellate, generally arranged in racemes or panicles. Petals 5 or more. Ovary inferior. Fruit a drupe or berry.

+ Generally climbers with simple exstipulate leaves. Fruit a berry. **1. Hedera**

– Shrubs or herbs with compound stipulate leaves. Fruit a drupe. **2. Aralia**

1. Hedera 1 sp.

Shrubby climbers with clinging adventitious roots from the stems. Flowers disposed in a panicle of umbels, 5-merous. Berry globose.

269. **H. nepalensis** K. Koch *Kurie, Himalayan Ivy* Pl. 44

Extensive climbers on trees or cliffs. Leaves simple, variously lobed to subentire. Flowers yellow. Berry subglobose, 6-10 mm broad orange at maturity.

Distribution: Afghanistan, Pakistan, Kashmir to Bhutan, Assam, south west China, Burma.

In forests from 800-3000 m. Common from Chitral to Kashmir. July-August.

2. Aralia 1 sp.

Shrubs with 1-2-pinnate leaves. Flowers 5-merous.

270. **A. cachemirica** Decne. Pl. 44

Perennial herb up to 2 m tall with imparipinnate leaves. Leaflets 3-7 in number, ovate, margin toothed. Flowers in axillary or terminal paniculate umbels, yellow-green. Drupe subglobose, black on maturity.

Distribution: Afghanistan, Pakistan, Kashmir to Bhutan, south east Tibet.

Widespread in forests from 2200-3500 m. June-August.

CAPRIFOLIACEAE *Honeysuckle Family*

Shrubs with simple, opposite and entire leaves. Flowers bisexual, 5-merous, in cymes or panicles, regular or irregular. Corolla 2-lipped or rotate, tubular or salverform. Stamens epipetalous. Ovary inferior, 2-5-locular. Fruit a drupe, berry or capsular.

+ Flowers regular. Style very short. Stigma 3-lobed. **1. Viburnum**

– Flowers irregular. Style long and slender. Stigma capitate. **2. Lonicera**

1. Viburnum 6 spp.

Shrubs or small trees. Flowers in terminal or axillary panicled cymes or corymbs. Corolla lobes spreading. Stamens 5, attached at the base of corolla tube. Fruit a drupe.

+	Leaves ovate to suborbicular, crenate, undersurface tomentose.	271. **V. cotinifolium**
–	Leaves elliptic-oblong, toothed, undersurface glabrous, sparsely hairy on the nerves.	272. **V. grandiflorum**

271. **V. cotinifolium** D. Don *Taliana* Pl. 44

A shrub up to 3 m tall with roundish leaves that are white tomentose on undersurface. Flowers white, appearing before leaves. Drupe 8-9 mm, oblong, reddish-black.

Distribution: Afghanistan, Pakistan (Swat, Hazara, Gilgit, Murree hills and Sulaiman Range), Kashmir to Bhutan.

In forests from 900-3300 m preferring warmer dryer slopes than *V. grandiflorum*. Fruit edible. March-April.

272. **V. grandiflorum** Wall. ex DC. *Guch* Pl. 44
Syn.: *V. nervosum* auct. non D. Don

A more gregarious shrub than the preceding species, with clusters of pink flowers, that appear just when the snow melts. Drupe about 10 mm long, ellipsoid, black when ripe

Distribution: Pakistan, Kashmir to Bhutan, south Tibet.

One of the commonest shrubs of Himalaya between 1500-3200 m often forming dominant undershrub in forest clearings. Flowers fragrant and the fruit is edible. November-May.

2. Lonicera 18 spp. *Honeysuckle*

Deciduous or evergreen shrubs with irregular flowers generally in axillary pairs and bracteate. Bracteoles 2. Corolla tubular, 2-lipped, sac-like or not at the base. Fruit a berry.

1.	+	Erect shrubs. Leaves 2-11 cm long. Corolla 2-lipped.	2
	–	Prostrate shrubs. Leaves 5-20 mm long. Corolla limb unequally 5-lobed.	273. **L. myrtillus**
2.	+	Stem and branches whitish, hollow. Leaves pubescent-villous. Flowers white. Berries white.	274. **L. quiquelocularis**

– Stem and branches darker coloured, solid. Leaves glabrous. Flowers yellowish-white, suffused with pink. Berries red. 275. **L. heterophylla**

273. **L. myrtillus** Hook. f. & Thoms. Pl. 45

A more or less prostrate shrub with small leaves. Flowers 7-8 mm long, yellowish-white, suffused with some pink. Corolla tube cylindrical, 5-lobed. Berries red.

Distribution: Pakistan, Kashmir to Nepal, south west China.

Forest and alpine areas from 2300-4000 m. Variable in shape and size of leaves and bracts. June-July.

274. **L. quinquelocularis** Hardwicke *Phut* Pl. 45

A branched shrub up to 3 m tall, with hollow stems. Leaves 2-4 x 1-3 cm broad, broadly elliptic-ovate, pubescent to villous. Flowers paired, creamy white. Berries white.

Distribution: Afghanistan, Pakistan, Kashmir to south west China.

Our largest and commonest honeysuckle from 900-2700 m, especially in Murree hills, Azad Kashmir. Forest shrubberies. April-May.

275. **L. heterophylla** Decne. Pl. 45

Small more or less erect shrub up to 1 m. Leaves 3-4 x 1.2-2 cm, broadly lanceolate to lanceolate-ovate, glabrous. Flowers 8-10 mm long, pale yellow, suffused red. Berries 2, free, globose, red.

Distribution: North west Himalaya, west Tibet.

Alpine from 3300-4200 m. Gregarious on slopes. June-July.

SAMBUCACEAE *Elder Family*

Trees or shrubs with opposite, imparipinnate leaves. Flowers regular, bisexual in terminal corymbs or cymes. Corolla 3-5-lobed. Stamens 5. Fruit a drupe.

A monotypic family, sometimes included in the *Caprifoliaceae* from which it differs in the compound stipulate leaves, the valvate corolla lobes and anthers which dehisce extrorsely.

1. Sambucus 1 sp.

Characters as those of the family.

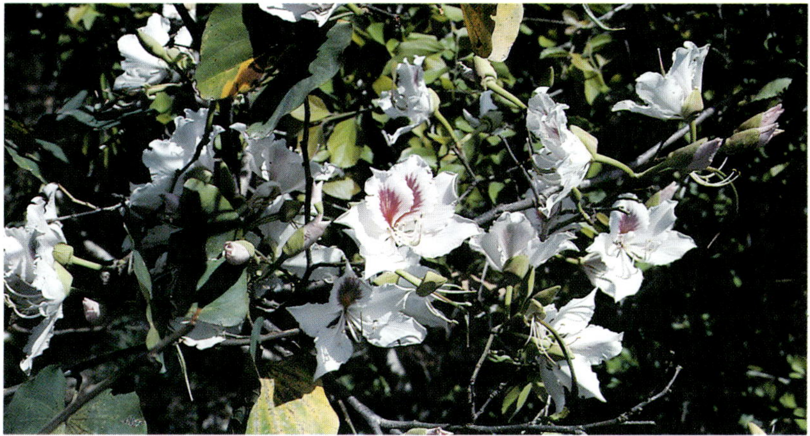

204. *Bauhinia variegata* Margalla hills, S. A. Sultan

201. *Prosopis cineraria* Hoshab, Makran, Y. J. Nasir

202. *Prosopis juliflora* Karachi, Y. J. Nasir

203. *Prosopis glandulosa* Salt Range, Y. J. Nasir

200. *Mimosa rubicaulis subsp. himalayana* Margalla hills, Y. J. Nasir

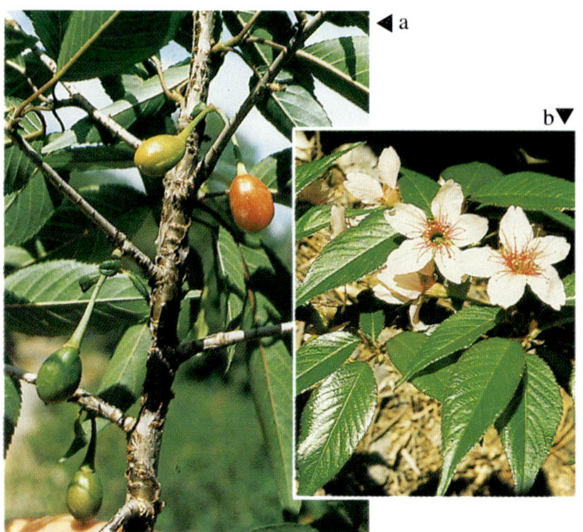

208. *Prunus cerastioides* a, b: Margalla hills, Y. J. Nasir

207. *Prunus cornuta* Shogran, Kaghan, Y. J. Nasir

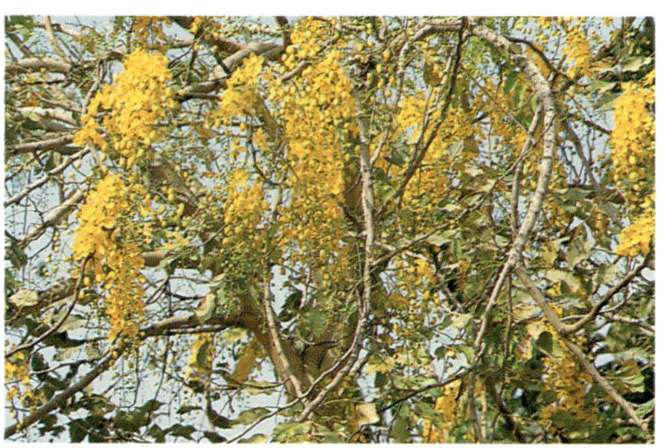

205. *Cassia fistula* Islamabad, S. A. Sultan

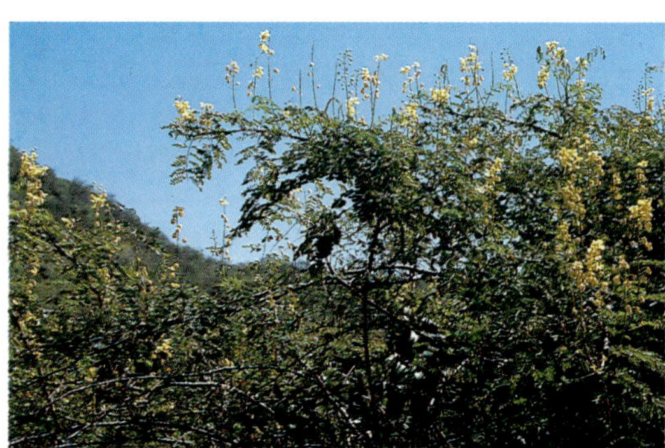

206. *Caesalpinia decapetala* Margalla hills, Y. J. Nasir

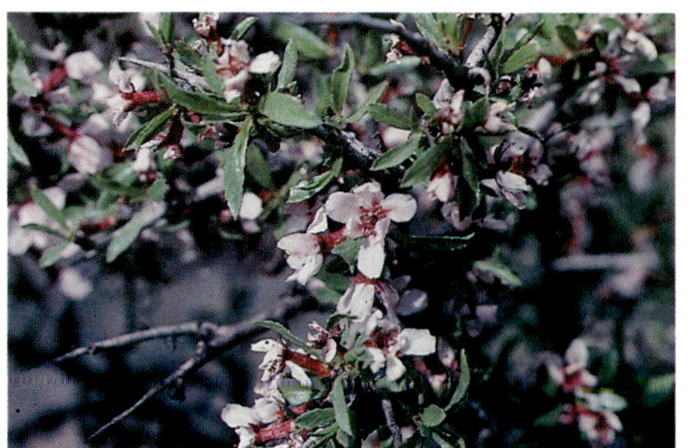

209. *Amygdalus brahuica subsp. afghanica* Ziarat, Rubina A. Rafiq

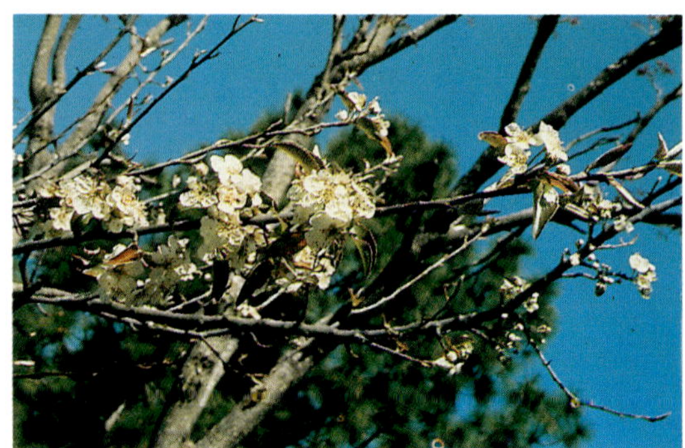

210. *Pyrus pashia* Margalla hills, Y. J. Nasir

215. *Rosa macrophylla* Chitral, Maj Gen M. Shuaib Quraishi

214. *Rosa brunonii* Margalla hills, S. A. Sultan

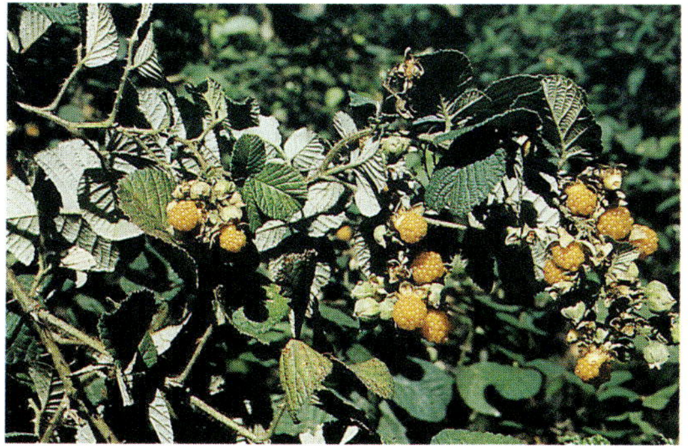

212. *Rubus ellipticus* Margalla hills, Y. J. Nasir

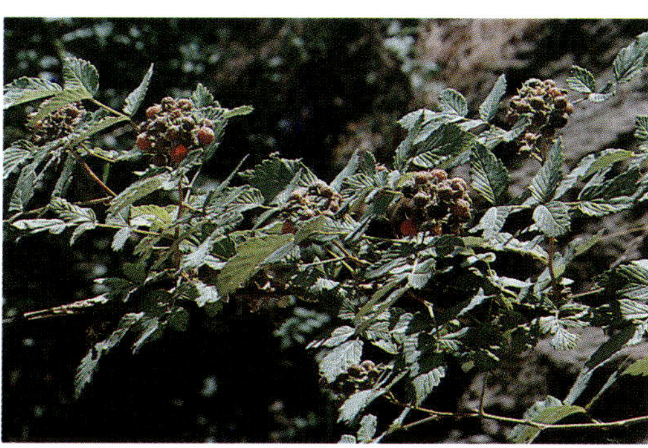

213. *Rubus sanctus* Ghora Gali, Y. J. Nasir

211. *Sorbaria tomentosa* Babusar valley, Chilas T. J. Roberts

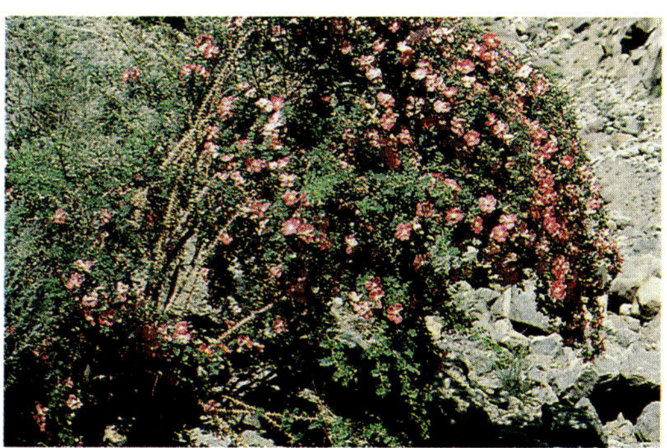

216. *Rosa webbiana* Naltar, Gilgit, T. J. Roberts

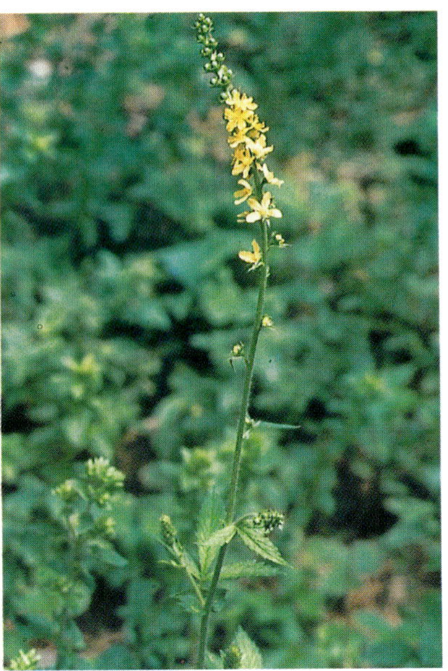

217. *Agrimonia eupatoria* Thandiani, Y. J. Nasir

218. *Geum elatum* Thandiani, Y. J. Nasir

221. *Potentilla nepalensis* Mokshpuri, T. J. Roberts

219. *Sibbaldea cuneata* Makran, Kaghan, Y. J. Nasir

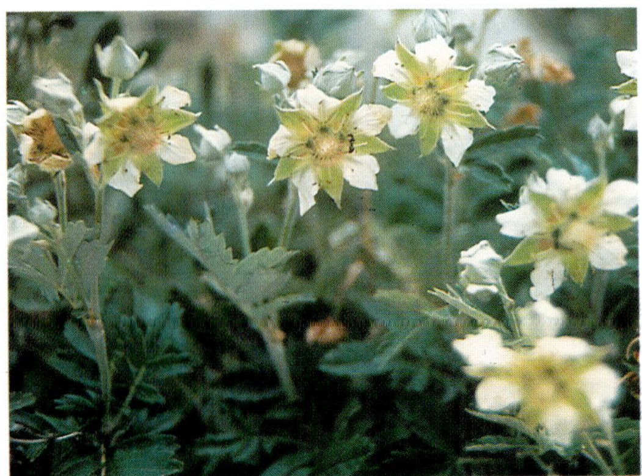

220. *Potentilla salesoviana* Deosai, Fritz Berger

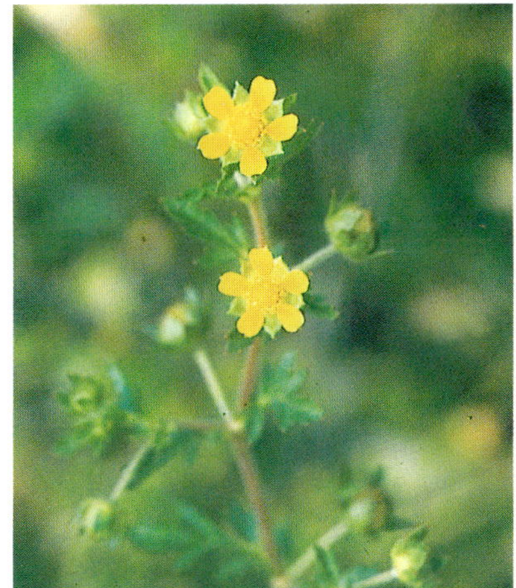

223. *Potentilla supina* Islamabad, S. A. Sultan

227. *Fragaria nubicola* Murree hills, Y. J. Nasir

224. *Potentilla pamiro-alaica* Khunjerab, Mark Mallalieu

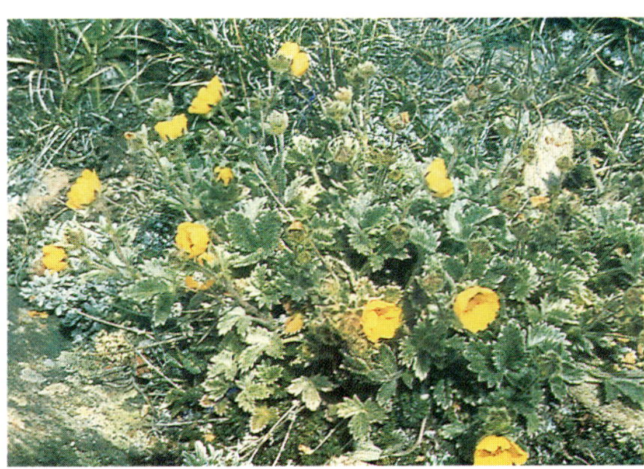

222. *Potentilla atrosanguinea* Babusar, T. J. Roberts

226. *Duchesnea indica* Islamabad, Y. J. Nasir

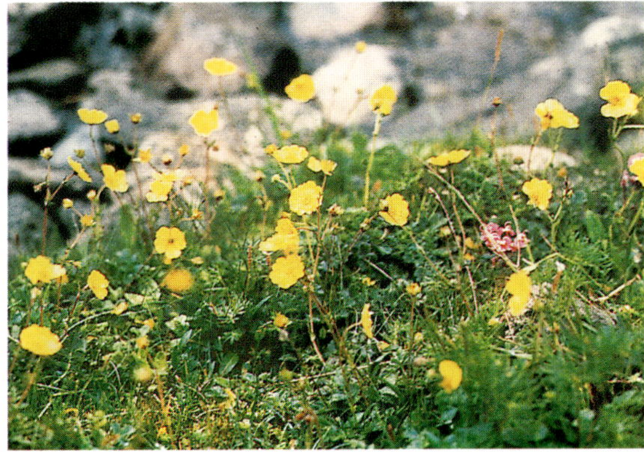

225. *Potentilla peduncularis* Kalam, Fritz Berger

232. *Ribes alpestre* Hunza, Y. J. Nasir

228. *Bergenia ciliata* a. Nathia Gali, T. J. Roberts
b. Murree Hills, Y. J. Nasir

229. *Bergenia stracheyi* Kaghan, T. J. Roberts

233. *Ribes orientale* Skardu, Fritz Berger

230. *Saxifraga hirculus* Khunjerab, Y. J. Nasir

231. *Parnassia nubicola* Dunga Gali, T. J. Roberts

238. *Rhodiola quadrifida* Babusar Pass, Fritz Berger

239. *Rhodiola himalensis* Hunza, Mark Mallalieu

234. *Ribes himalense* Shogran, Y. J. Nasir

235. *Sedum ewersii* Gabral, Kalam, Fritz Berger

243. *Woodfordia fruticosa* Margalla hills, Y. J. Nasir

245. *Epilobium laxum* Kaghan, Y. J. Nasir

244. *Punica granatum* Margalla hills, Y. J. Nasir

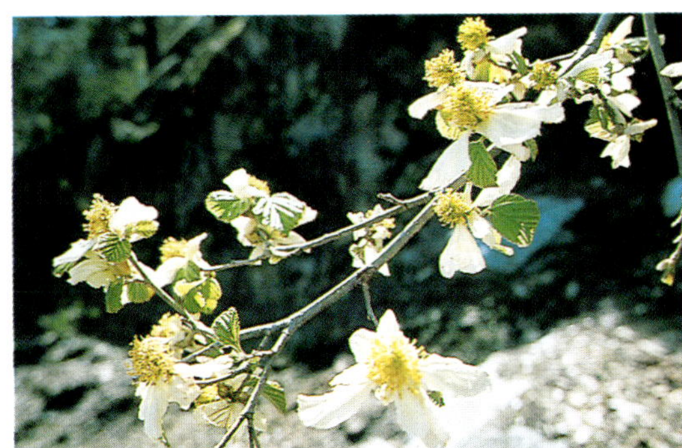

242. *Parrotiopsis jacquemontiana* Swat, Ushu, Fritz Berger

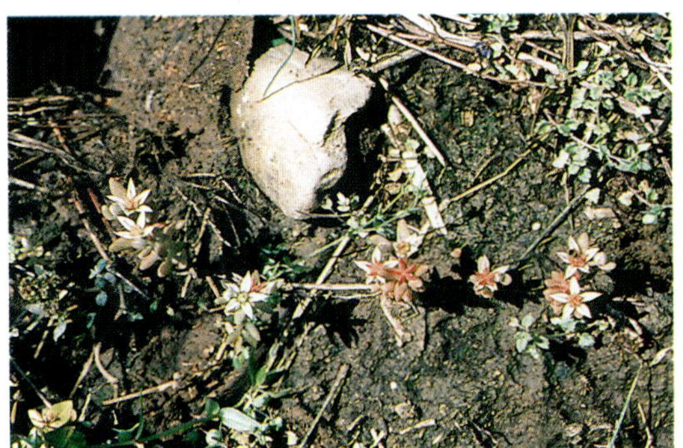

240. *Rosularia adenotricha* Tarbela, Y. J. Nasir

241. *Rosularia alpestris* Kalam, Y. J. Nasir

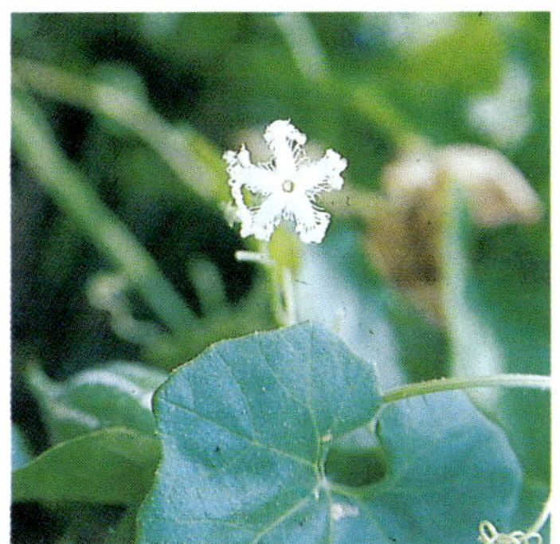

251. *Trichosanthes cucumerina* Margalla hills, S. A. Sultan

252. *Cucumis prophetarum* Karachi, Y. J. Nasir

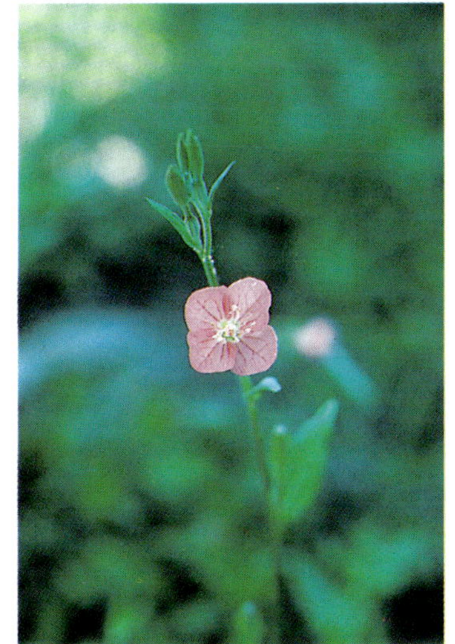

248. *Oenathera rosea* Murree foothills, Y. J. Nasir

250. *Opuntia monacantha* Islamabad, S. A. Sultan

249. *Oenathera affinis* Battagram, Hazara, T. J. Roberts

247. *Epilobium angustifolium* Kalam, Y. J. Nasir

246. *Epilobium hirsutum* Islamabad, Y. J. Nasir

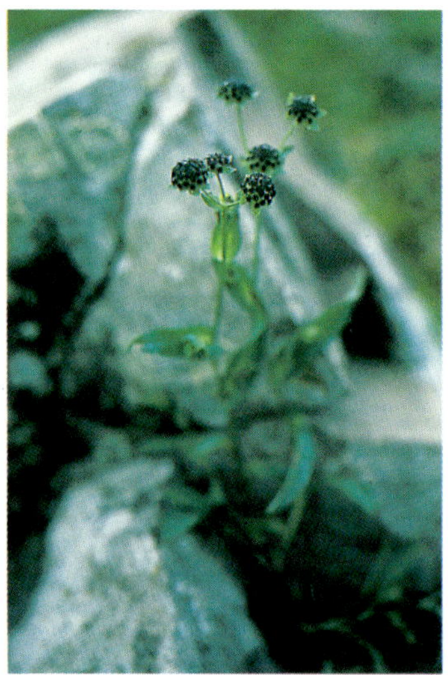

257. *Bupleurum longicaule* Sari, Kaghan, Y. J. Nasir

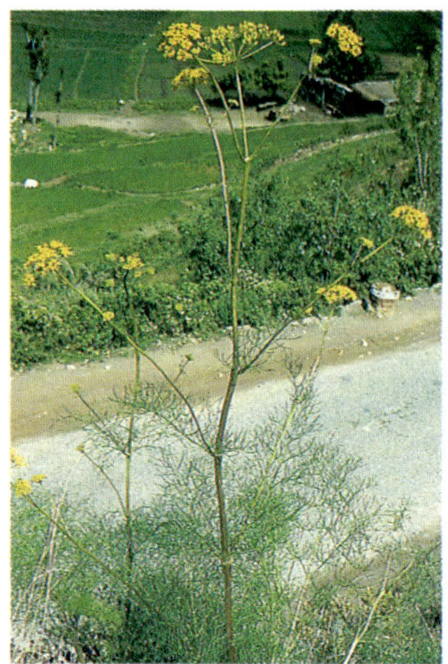

258. *Prangos pabularia* Swat, Fritz Berger

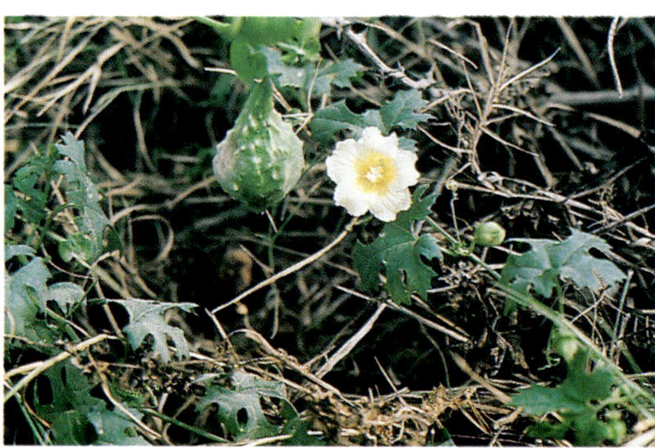

253. *Momordica balsamina* Karachi, Y. J. Nasir

255. *Begonia picta* Murree hills, Y. J. Nasir

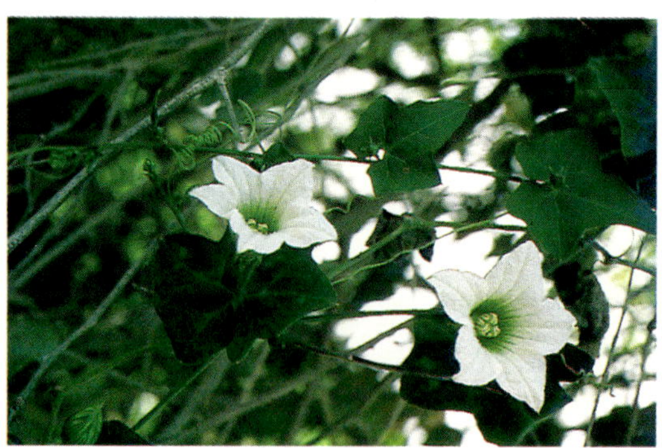

254. *Coccinea grandis* Sindh, Y. J. Nasir

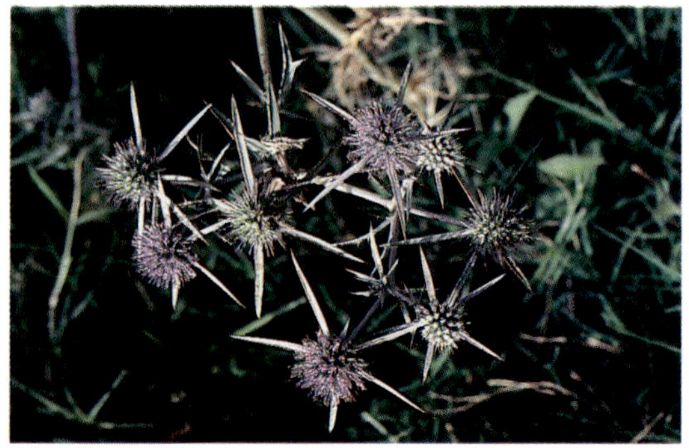

256. *Eryngium biebersteinianum* Abbottabad, Rubina A. Rafiq

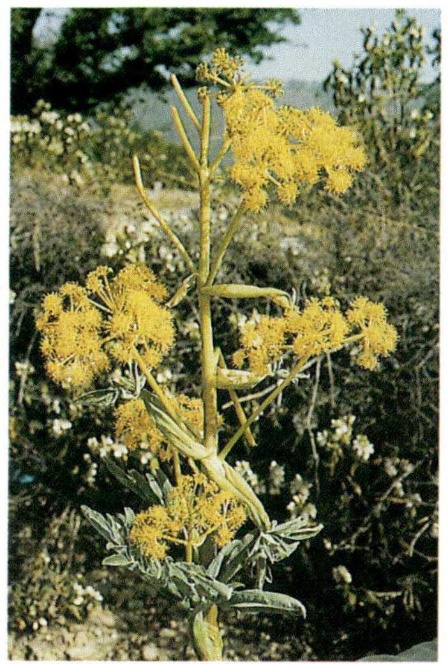

259. *Ferula jaeschkeana* Margalla hills, S. A. Sultan

260. *Ferula oopoda* Chiltan hills, T. J. Roberts

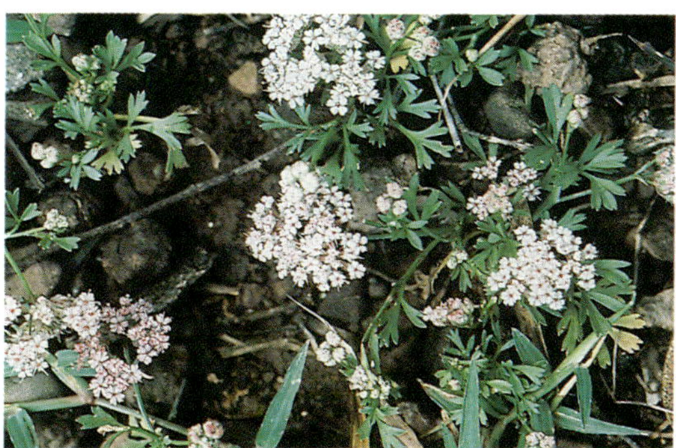

264. *Psammogeton canescens* Islamabad, S. A. Sultan

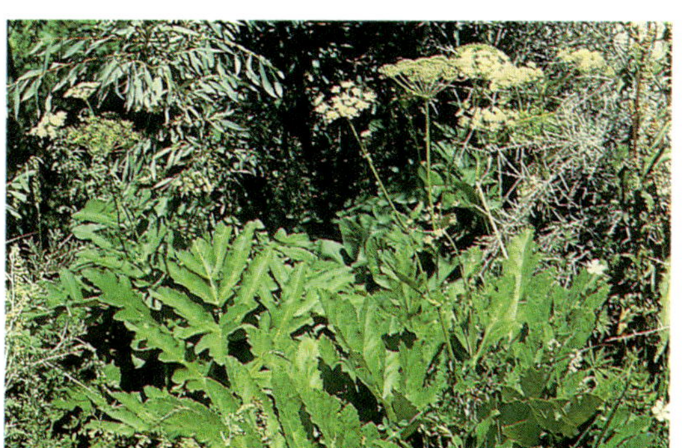

261. *Heracleum candicans* Mastuj Valley, Chitral, T. J. Roberts

266. *Torilis leptophylla* Haripur, S. A. Sultan

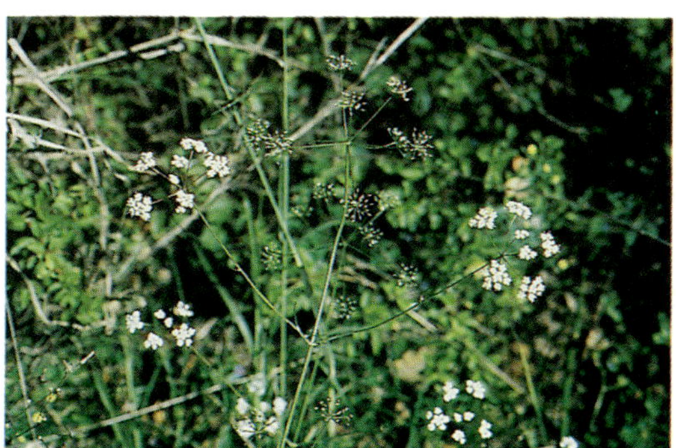

265. *Scaligera stewartiana* Islamabad, S. A. Sultan

272. *Viburnum grandiflorum* Kaghan, Rubina A. Rafiq

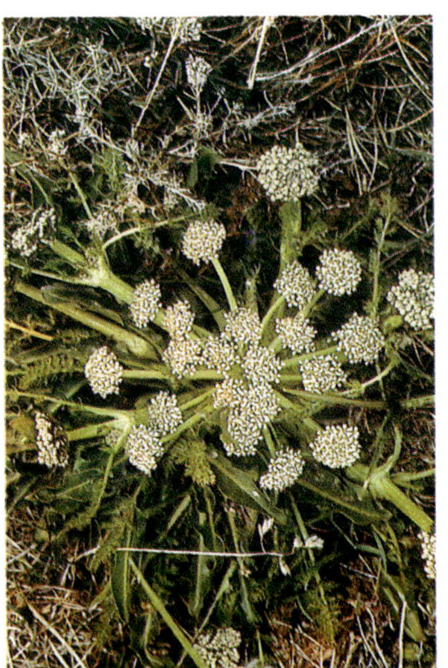

268. *Cortia depressa* Khunjerab, Mark Mallalieu

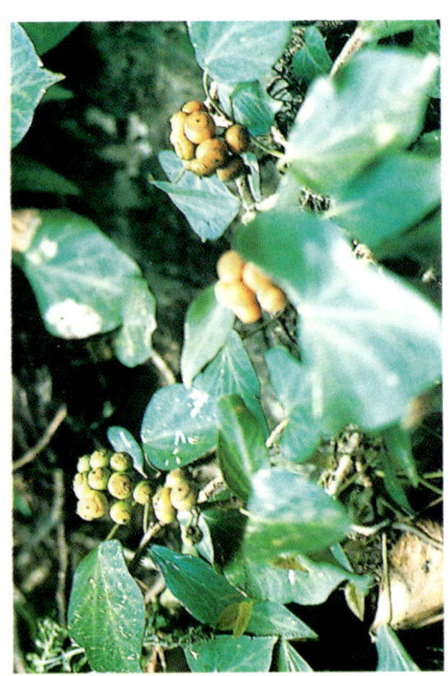

269. *Hedera nepalensis* Swat, Fritz Berger

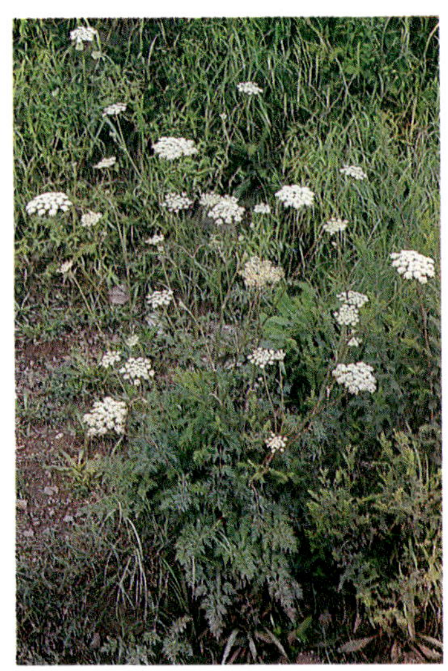

267. *Conium maculatum* Murree hills, S. A. Sultan

271. *Viburnum cotinifolium* Mohodand, Swat, Fritz Berger

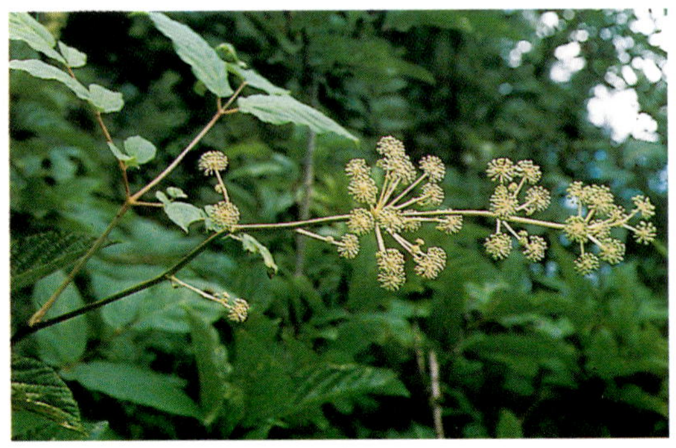

270. *Aralia cachemirica* Kaghan, Rubina A. Rafiq

276. *Sambucus wightiana* a. Kaghan valley, T. J. Roberts
b. Mahodand, Fritz Berger

274. *Lonicera quinquelocularis*
Changla Gali, Rubina A. Rafiq

275. *Lonicera heterophylla* Lowari pass, Y. J. Nasir

280. *Leptodermis virgata* Margalla hills, Maj Gen M. Shuaib Quraishi

278. *Galium boreale* Kaghan, Y. J. Nasir

273. *Lonicera myrtillus* Kaghan, Rubina A. Rafiq

279. *Gaillonia macrantha* Kalat, Rubina A. Rafiq

281. *Spermadictyon suaveolens* Margalla hills, Rubina A. Rafiq

283. *Valeriana jatamansi* Ghora Gali, Rubina A. Rafiq

282. *Valeriana pyrolifolia* Kaghan, Y. J. Nasir

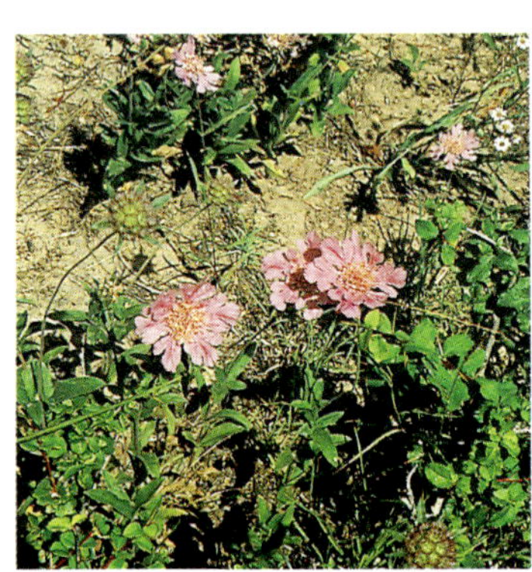

285. *Scabiosa speciosa* Kaghan, Rubina A. Rafiq

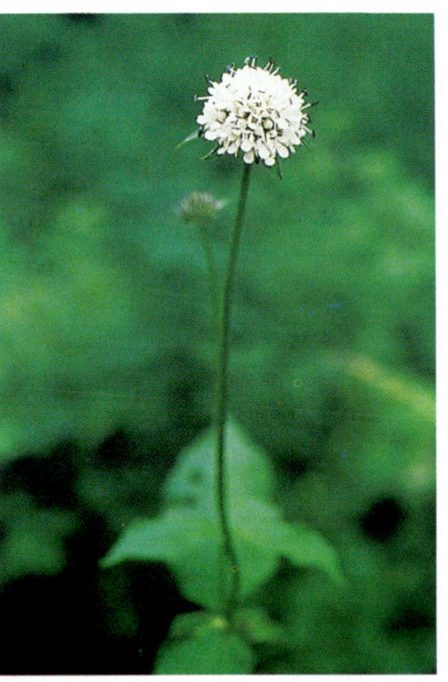

287. *Dipsacus inermis* Kaghan, Y. J. Nasir

284. *Scabiosa olivieri* Islamabad, S. A. Sultan

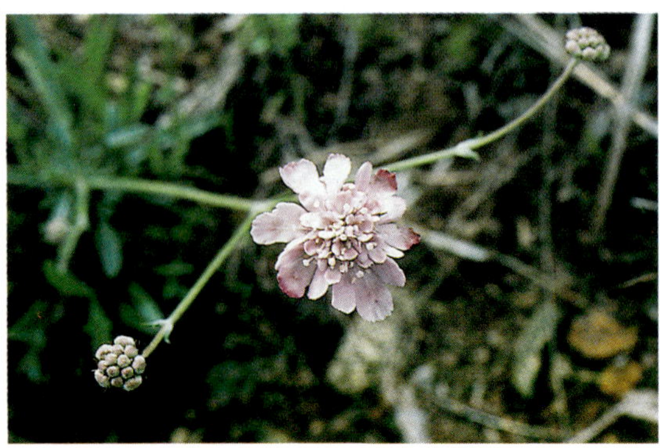

286. *Scabiosa candollei* Islamabad, S. A. Sultan

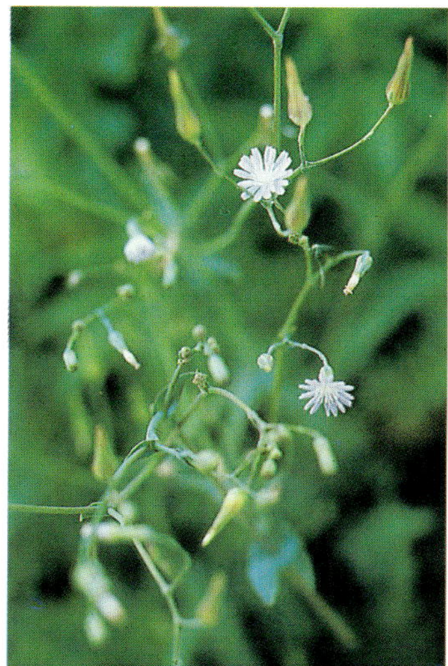

290. *Lactuca dissecta* Islamabad, Y. J. Nasir

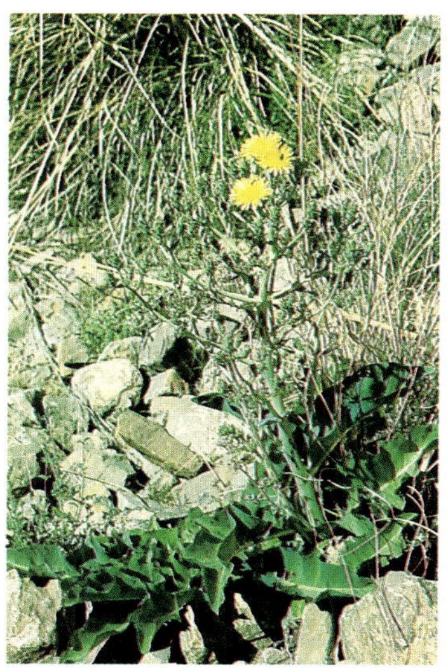

291. *Streptorhampus persicus* Chiltan hills, Rubina A. Rafiq

292. *Taraxacum officinale* Swat, Y. J. Nasir

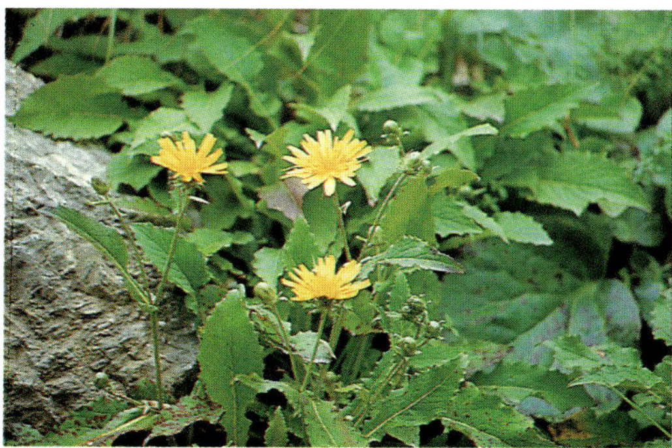

293. *Dubyea oligocephala* Changla Gali, Rubina A. Rafiq

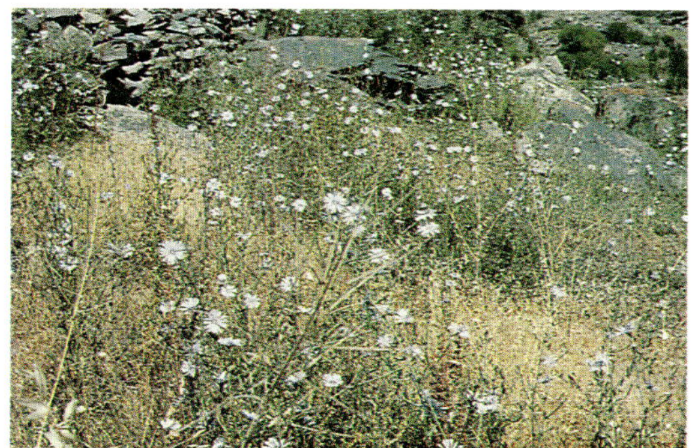

289. *Cichorium intybus* Abbottabad, S. A. Sultan

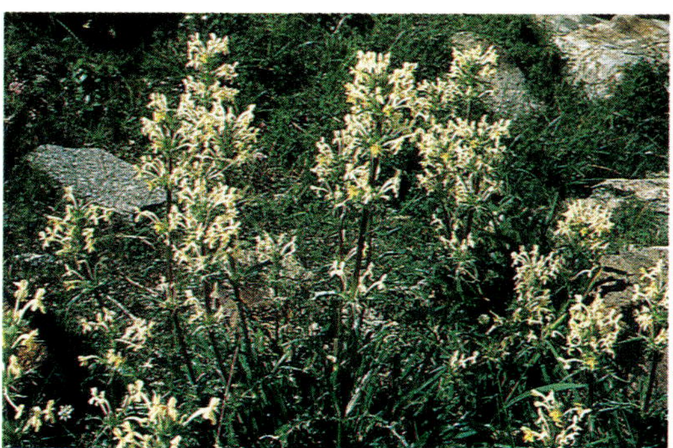

288. *Morina coulteriana* Kaghan, Y. J. Nasir

299. *Tragopogon gracilis* Lower Swat, Y. J. Nasir

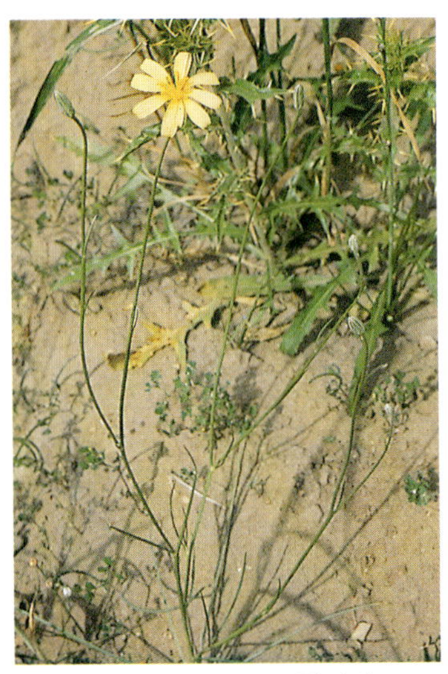

300. *Scorzonera virgata* Thal desert, Y. J. Nasir

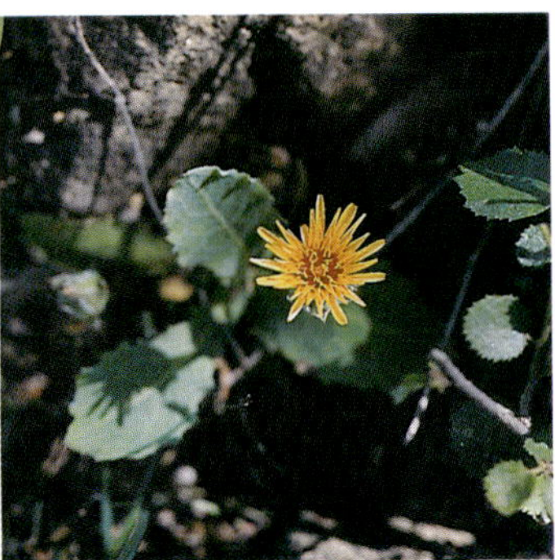

298. *Reichardia orientalis* Salt Range, Y. J. Nasir

295. *Launaea procumbens* Thal desert, Y. J. Nasir

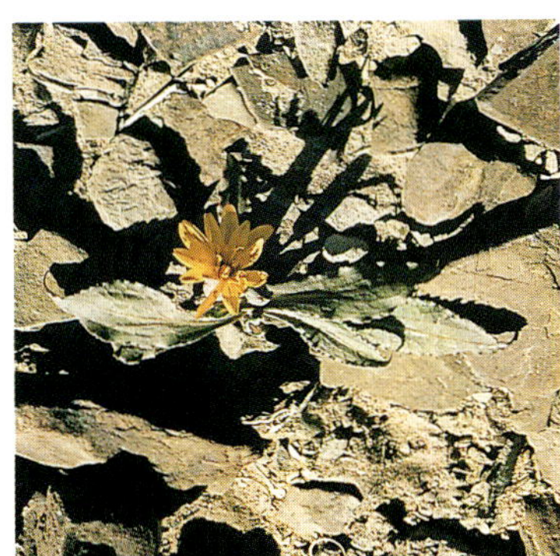

301. *Scorzonera sp.* Makran, Y. J. Nasir

296. *Youngia japonica* Changla Gali, S. A. Sultan

297. *Sonchus asper* Islamabad, S. A. Sultan.

276. **S. wightiana** Wall. ex Wight & Arn. *Gandala, Dwarf Elder* Pl. 45
Syn.: *S. ebulus* auct. non L.

Shrubs up to 1.7 m tall with leaves up to 30 cm long. Leaflets 5-9 in number, lanceolate, toothed. Flowers white, small, in corymbose clusters. Drupes globose, 4-5 mm broad, orange, turning blackish on maturity.

Distribution: Afghanistan, Pakistan, Kashmir to Uttar Pradesh (India).

Gregarious on open slopes from 2000-3000 m. Inner drier Himalayan valleys, where it is often the dominant undershrub. Plant parts medicinal. June-July.

RUBIACEAE *Madder or Coffee Family*

Trees or shrubs with opposite or whorled simple leaves. Stipules interpetiolar. Flowers regular, arranged in cluster or heads. Corolla 4-5-lobed, tubular. Stamens as many as corolla lobes. Ovary inferior. Fruit a berry, drupe, capsule or nutlet.

1.	+	Herbs.	**1. Galium**
	–	Shrubs or undershrubs.	2
2.	+	Fruit capsular, dehiscent, calyx not enlarged in fruit.	3
	–	Fruit indehiscent, calyx enlarged in fruit.	**2. Gaillonia**
3.	+	Bracteoles free. Capsule shallowly 5-valved.	**4. Spermadictyon**
	–	Bracteoles united. Capsule 5-valved to the base.	**3. Leptodermis**

1. Galium c. 24 spp.

Annual or perennial herbs with angled stems. Flowers in paniculate or corymbose cymes. Corolla rotate. Fruit of 2 mericarps.

+	Flowers yellow.	277. **G. verum**
–	Flowers white.	278. **G. boreale**

277. **G. verum** L. *Ladies Bedstraw* Fig. facing p. 102

Slender plants up to 30 cm tall with heads of small yellow flowers in terminal cluster. Leaves 8-10 per node.

Distribution: Eurasia, north Africa, Iran, Afghanistan, Pakistan, Kashmir to Uttar Pradesh (India).

Alpine and sub-alpine stony places from 1600-3200 m. June-August.

278. **G. boreale** L. Pl. 45

Similar but with white flowers. Leaves lanceolate, 3-4 per node, upper surface glabrous and 3-nerved. Fruit hairy.

Distribution: Eurasia, Iran, Afhanistan, Pakistan, Kashmir, Korea, China, Japan, north America.

Fairly common in cultivated areas, and alpine zone in open places from 2000-3300 m in Himalaya. June-July.

2. Gaillonia 4 spp.

Perennial prickly shrubs with opposite linear leaves. Flowers funnel-form, pink or reddish. Stamens 5.

279. **G. macrantha** Blatt. & Hallb. *Kharbat* Pl. 45

A shrub up to 40 cm tall, with pubescent parts. Older branches whitish. Leaves 5-18 x 1-2 mm. Corolla 16-18 mm long, hispid, lobes obtuse. Stamens included in corolla tube. Fruit 6-7 mm long, densely villous.

Distribution: Afghanistan, Pakistan.

Fairly common in drier areas especially Balochistan from 1500-2200 m. May-August.

3. Leptodermis c. 3 spp.

Shrubs with opposite leaves. Flowers purplish or white, in head-like clusters. Corolla funnel shaped, 5-lobed.

280. **L. virgata** Edgew. Pl. 45

A small bushy undershrub with greyish-white stems, leaves linear and pretty pale purple flowers. Capsule c. 4-5 mm long.

Distribution: Pakistan, Kashmir, Himachal Pradesh (India).

Usually dry places in the sub-Himalayan tracts and Salt Range from 600-1900 m. July-September.

4. **Spermadictyon** 1 sp.

Erect branched shrubs with opposite leaves. Leaves with prominent lateral nerves. Flowers in panicles or cymes. Corolla funnel-form, 5-lobed. Stamens 5.

281. **S. suaveolens** Roxb. Pl. 46
Syn.: *Hamiltonia suaveolens* Roxb.

Branched shrub 1.5-2 m tall. Leaves elliptic-ovate, 5-18 x 3-8 cm. Flowers 10-12 mm long, pale bluish-lilac, fragrant, in head-like clusters in terminal panicles. Capsule ovoid.

Distribution: Pakistan, Kashmir, Nepal, south east Tibet.

Common on steep banks in chir pine zone up to 1400 m in sub-Himalayan tracts and Salt Range. October-February.

VALERIANACEAE *Valerian Family*

Annual or perennial herbs. Flowers small, irregular, usually branched, bisexual. Corolla 3-5-lobed, tube gibbous at the base. Stamens 1-4, epipetalous. Carpels 3, syncarpous. Fruit an achene.

1. **Valeriana** 9 spp.

Perennial herbs with aromatic underground parts. Leaves entire or divided. Flowers white or pink, in corymbose cymes, panicles or heads. Stamens 3. Achenes crowned by the persistent pappose calyx.

+	Radical and stem leaves suborbicular to reniform. Inflorescence more or less compact. Achenes glabrous.	282. **V. pyrolifolia**
–	Radical leaves cordate, stem leaves divided. Inflorescence spreading. Achenes tomentose.	283. **V. jatamansi**

282. **V. pyrolifolia** Decne. Pl. 46

Plants up to 30 cm tall, more or less stout. Stem solitary or 2-3. Radical leaves 2.5-10 x 1.4-8 cm, crenate. Flowers in compact corymbose heads or cymes. Achenes glabrous.

Distribution: Pakistan, Kashmir to Uttar Pradesh (India).

Shade tolerant-species growing in rock crevices in alpine areas from 2300-4500 m. Replaces *V. jatamansi* at higher elevations. April-June.

283. **V. jatamansi** Jones *Mushkbala* Pl. 46

More slender plants than in the previous species, with a lighter coloured foliage, different leaf shape, open and spreading inflorescence and tomentose achenes. Flowers pale pink.

Distribution: Afghanistan, Pakistan, Kashmir to Bhutan, Assam, Tibet Burma, west & central China.

A common species in forested areas from 900-3200 m. Widespread in N.W.F.P., Swat, Hazara and Murree hills. The rhizome yields an aromatic oil which is used in the preparation of tranquilizers, in suppression of urine and in perfumed powders. March-May.

DIPSACACEAE *Teasel Family*

Herbs with exstipulate, whorled or opposite leaves. Flowers bisexual, in heads. Corolla 4-5-lobed. Stamens 4, epipetalous. Ovary inferior. Fruit an achene.

+ Corolla 5-lobed. Bracts of inflorescence herbaceous. **1. Scabiosa**

– Corolla 4-lobed. Bracts of inflorescence spinose. **2. Dipsacus**

1. Scabiosa 4 spp. *Scabious*

Perennial or annual herbs with opposite leaves. Calyx limb of 5 stiff setae. Corolla tubular, 5-lobed. Achene crowned by the persistent calyx.

1. + Plant annual. 284. **S. olivieri**

– Plants perennial. 2

2. + Flower heads 2-4 cm broad. Petals pink. Involucral bracts-lanceolate. 285. **S. speciosa**

– Flower heads 8-12 mm broad. Petals white or lilac. Involucral bracts ovate to elliptic-ovate. 286. **S. candollei**

284. **S. olivieri** Coult. Pl. 46

Hairy annual upto 40 cm tall. Leaves pinnatifid or pinnatisect. Flowers mauve, in smaller few-flowered heads. Calyx setae c.1 cm long, light brown.

Distribution: Caucasus, central Asia, Iran, Afghanistan, Pakistan.

Short lived annual species. Common in dry areas especially Balochistan, N.W.F.P., Swat and northern Punjab in spring in lower hills and adjacent plains. April-May.

285. **S. speciosa** Royle Pl. 46

Plants 30-60 cm tall. Leaves lanceolate or broadly so, 3-6 x 0.4-1.2 cm, toothed. Flowers in radiant heads. The ray flowers pink to mauve, and larger than the disc flowers. Calyx setae black.

Distribution: Pakistan, Kashmir to Himachal Pradesh (India).

Open alpine and sub-alpine slopes from 2600-4300 m in the Himalaya. July-August.

286. **S. candollei** DC. Pl. 46

A less attractive and slender perennial with flowering heads 8-12 mm broad and white flowers. Calyx setae brown.

Distribution: Afghanistan, Pakistan, Kashmir to Uttar Pradesh (India).

Dry foothills to 1500 m in the Himalaya and Sulaiman Range. April-May.

2. Dipsacus 2 spp.

Herbs with opposite and receptacular spinose bracts. Flowers infundibuliform, corolla 4-lobed. Stamens 4, of which only 2 are fertile.

287. **D. inermis** Wall. *Upalhak, Teasel* Pl. 46

Spectacular perennial plants 1.5-2 m tall, with attractive globose heads, 20-45 cm broad, of creamish-white to white flowers. Receptacular bracts spiny, 4-6 mm long.

Distribution: Afghanistan, Pakistan, Kashmir to south west China, Burma.

Forest slopes, open places from Chitral, Swat Kohistan, Kaghan, Hazara and Murree hills from 1600-4000 m. June-August.

MORINACEAE *Morina Family*

A monogeneric family of perennial herbs with spiny leaves, related to *Dipsacaceae*, but differing in the whorled flowers, 2-lobed calyx, 2-lipped and ringent corolla and terminal stigma.

Morina 2 spp.

Corolla 2-lipped, 5-lobed, infundibuliform. Stamens 2 or 4, attached at the corolla throat. Ovary inferior. Stigma capitate. Fruit a thick walled achene.

288. **M. coulteriana** Royle Pl. 47

Herbs 30-75 cm tall spiny margined leaves and pale yellow flowers. Achene 4-5 mm long, whitish.

Distribution: Pakistan to Uttar Pradesh (India), Tibet, central Asia.

A pretty sight when growing gregariously on open slopes and forest clearings Chitral, Swat, Kohistan and Hazara from 2400-4200 m. Flowering at onset of monsoon. July-August.

COMPOSITAE

Sunflower Family

A cosmopolitan family, which in terms of size is the largest in Pakistan, comprising of c. 650 species, including those in cultivation.

Perennial, biennial or annual herbs with opposite or alternate simple leaves. Inflorescence a capitulum, in which the flowering heads are borne on a receptacle and are of two types: The disc florets which are tubular and generally occupy the central portion of the receptacle and the ray florets which are petal-like or strap shaped and occupy the rim of receptacle. Receptacle naked or with bristles etc. Involucral bracts 1-many seriate, spiny or not. Stamens 5, anthers fused together, filaments free. Ovary inferior. Fruit an achene usually beaked or not, pappose or plumose.

1. + Flowers all ligulate. Plants with a milky juice. 2

– Flowers, at least some, tubular.
Plants without milky juice. 4

2. + Flowers blue, bluish-purple or pink-violet.
(Sometimes yellow in **Lactuca**). **1. Cichorium**
12. Epilasia

– Flowers yellow. 3

3. + Milky juice white. **3. Steptorrhampus**
4. Taraxacum
5. Dubyaea
7. Youngia
8. Sonchus
9. Reichardia
10. Tragopogon
11. Scorzonera

– Milky juice yellow. **6. Launaea**

4.	+	Flower heads radiate, with both ray and disc flowers (except in **Leontopodium, Anaphalis, Cyathocline).**	5
	–	Flower heads with usually tubular flowers.	9
5.	+	Disc flowers regular.	6
	–	Disc flowers irregular, 2-lipped.	**39. Ainsliaea** **40. Gerbera**
6.	+	Anther bases long appendaged. Styles without any appendages.	7
	–	Anther bases not long appendaged. Styles with minute appendages at tip.	8
7.	+	Pappus of scales or absent.	**13. Tridax** **14. Bidens** **15. Inula** **16. Francoueria** **17. Leontopodium** **18. Anaphalis** **19. Pentanema** **29. Anthemis** **30. Achillea** **31. Leucanthemun** **32. Calendula** **35. Tanacetum**
	–	Pappus with long soft hairs.	**25. Tussilago** **26. Senecio** **27. Ligularia** **28. Hertia**
8.	+	Flower heads discoid.	**20. Cyathocline**
	–	Flower heads radiate.	**21. Solidago** **22. Aster** **23. Erigeron** **24. Bellis**
9.	+	Leaves and involucral bracts spiny.	10
	–	Leaves and involucral bracts not spiny.	13
10.	+	Flowers in dense globose heads.	**41. Echinops**

	–	Flowers not so.	11
11.	+	Leaves variegated green and white.	**42. Silybum**
	–	Leaves not so.	12
12.	+	Stems and branches distinctly winged.	**44. Onopordum**
	–	Stems and branches not so.	**43. Cousinea** **45. Cnicus** **46. Cirsium** **47. Carthamus** **48. Centaurea** **49. Tricholepis** **50. Amberboa** **51. Serratula** **52. Saussurea**
13.	+	Styles distinctly exserted from corolla. Leaves opposite.	**36. Adenostemma** **37. Ageratum**
	–	Styles barely exserted. Leaves usually alternate.	**34. Artemisia** **33. Matricaria** **38. Vernonia**

1. Cichorium 1 sp.

Flowers blue, blue-purple or pinkish. Pappus absent or scaly. Involucral bracts 2-seriate, the outer shorter.

289. **C. intybus** L. *Kasni, Chicory* Pl. 47

Perennial suberect to spreading herb. Leaves pinnatifid, toothed. Flowers in terminal or axillary heads, 25-33 mm broad.

Distribution: Europe, west Asia, widely introduced.

Common weed of open places, roadsides, cultivated areas from 1300-2500 m. It has spread rapidly into northern mountain areas, wherever there is cultivation. June-September.

2. Lactuca c. 21 spp.

Annual or perennial herbs with yellow, blue or purplish flowers with pinnately lobed or pinnatisect leaves. Involucral bracts in 3-4 rows. Achenes flat, 4-20-ribbed, beaked.

290. **L. dissecta** D. Don Pl. 47

Branched annual herb up to 30 cm with pinnately lobed leaves. Lobes dentate. Ray florets strap-shaped, purplish-blue. Achenes 2-2.5 mm long.

Distribution: Central Asia, Afghanistan, Pakistan, Kashmir to Nepal, Tibet.

Common spring annual in waste places, roadsides etc. from 400-2700 m. March-May.

3. Steptorhamphus 2 spp.

A genus closely related to *Lactuca*, but differing mainly in the fruit with a 2-seriate pappus, in which the outer ring is composed of very short hairs and an inner ring of dense long hairs.

291. **S. persicus** (Boiss.) O. & B. Fedtsch. Pl. 47
Syn.: *Lactuca persica* Boiss.

A robust perennial 30-70 cm tall. Basal leaves large, 12-28 x 5-9 cm, oblong-lanceolate, pinnatipartite, toothed. Cauline ones smaller, auriculate. Flowers yellow, in stout panicles. Achenes 4-5 mm long.

Distribution: Iran, Turcomania, Afghanistan, Pakistan, Caucasus.

In dry places from 1700-2200 m, mainly in Balochistan and Chitral. April.

4. Taraxacum c. 90 spp. *Dandelion*

Glabrous perennial with rosette of basal leaves which are pinnately lobed. Flowers yellow, all ligulate, borne on scapigerous heads. Achenes oblong, terete, 10-ribbed, bearing the persistent pappus.

A taxonomically problematic genus with about 40 species recorded from Pakistan. Species delimitations are ill-defined due to polyploidy and breeding behaviour. Many are apomicts.

292. **T. officinale** Weber Pl. 47

A common herb with yellow flowers and runcinate-lyrate leaves. Plant parts exude a milky juice. The majority of our species are misidentified and go under the name of this species.

Distribution: Cosmopolitan.

Very polymorphic and widespread in forest clearings, meadows etc. Found from sea level to 3100 m throughout the country. February-April.

5. Dubyaea 1 sp.

293. **D. oligocephala** (Sch.-Bip.) Stebbins Pl. 47

A dandelion-like plant found on shady slopes in forests from 2300-3300 m mainly in Hazara and Murree hills. Uncommon. July-August.

Distribution: Pakistan, Kashmir.

6. Launaea c. 17 spp.

Perennial herbs with a yellowish juice and glabrous leaves. Stem leaves few, lobed or pinnatifid. Flowers yellow, rayed. Receptacle smooth. Anther base auricled. Achene ribbed, narrow. Pappus with white simple hairs, connate at base into a ring.

+	Plants more or less erect. Inflorescence erect.	294. **L. resedifolia**
–	Plants prostrate. Inflorescence more or less prostrate.	295. **L. procumbens**

294. **L. resedifolia** (L.) O. Kuntze Fig. between p. 102-103

Glabrous branched herb 30-50 cm tall. Radical leaves runcinate-pinnatifid, with numerous, unequal lobes that are irregularly toothed. Flowers in terminal heads. Achenes 5-6 mm long.

Distribution: Egypt, Arabia, Iran, Afghanistan, Pakistan.

Dry sandy or gravely places up to 1000 m throughout Sindh, Punjab, Balochistan and N.W.F.P. March-April.

295. **L. procumbens** (Roxb.) Ramayya & Rajagopal Pl. 48
Syn.: *L. nudicuale* auct. non (L.) Hook. f.

Stems few, branches prostrate, 15-50 cm long, naked or with few leaves. Radical leaves 5-20 cm long, margin (in older leaves) beset with minute whitish teeth. Flowers in heads c. 2 cm broad. Achenes 2-3 mm long.

Distribution: Egypt, Arabia, Iraq, Iran, central Asia, Afghanistan, Pakistan, Kashmir to Nepal, India.

Another common widespread weed of the plains and lower hills up to 1000 m. March-May.

7. Youngia 3 spp.

Annual or perennial herbs. Flowers yellow, ligulate, in corymbose or cymose heads, 4-17 in number. Achenes 3-5-ribbed. Pappus not plumose.

296. **Y. japonica** (L.) DC. Pl. 48
Syn.: *Crepis japonica* Benth.

Perennial herb up to 35 cm tall. Leaves basal, lyrate-pinnatifid or pinnatisect. Peduncles 8-14 mm long. Flowers numerous, small, yellow. Inner involucral bracts linear-lanceolate. Achenes c. 2-3 mm long. Pappus white.

Distribution: Afghanistan, Pakistan, Kashmir to Bhutan, India, Sri Lanka, China, Japan, Indo-China, Philippines, Malaysia, Hawaii. Widely introduced.

Plains to 2200 m. Common along roadsides etc. March-May.

8. Sonchus 4 spp.

Annual or perennial herbs with stem leaves amplexicaul, pinnatifid with a dentate margin. Flowers ligulate, paniculate. Receptacle smooth. Achenes glabrous, compressed, ribbed. Pappus in 2 rows.

297. **S. asper** (L.) Hill Pl. 48

Flowers golden yellow. Cauline leaves with rounded lobes. Achenes smooth between the ribs.

Distribution: North Africa, north and west Asia, Europe.

Cosmpolitan weed. Common in waste places or as a weed from plains to 2700 m almost throughout the country. March-June.

9. Reichardia 1 sp.

Annual or perennial herbs with entire to deeply divided leaves. Inflorescence a many-flowered capitulum, ligulate, homogamous. Flowers yellow.

298. **R. orientalis** (L.) Hoch. Pl. 48
Syn.: *R. tingitana* (L.) Roth.

A dwarf spring annual only 4-8 cm tall, unbranched. Leaves 1-4 x 0.5-2 cm, oblanceolate, margin wavy-denticulate to pinnatifid; upper leaves smaller. Flower heads 20-25 mm broad on more or less thickened peduncles.

Distribution: North Africa, Canary Isls., Spain, Portugal, Greece, Palestine, north Iraq, Arabia, Iran, Afghanistan, Pakistan, northeast India.

In dry places up to 800 m in Balochistan, N.W.F.P. and Punjab. March-April.

10. Tragopogon c. 2-3 spp.

Leaves grass-like. Flowers yellow or white, ligulate, solitary on long peduncles. Receptacle smooth. Achenes 5-10-ribbed, beaked. Pappus plumose, 2-seriate.

299. **T. gracile** D. Don Pl. 48

Glabrous perennial. Flower heads yellow, 2-4 cm broad, borne on a slender peduncle. Achenes 12-17 mm long. Pappus brownish.

Distribution: Afghanistan, Pakistan, Kashmir to east Nepal.

Open places from 1900-4000 m in Balochistan, N.W.F.P., lower Chitral, Swat, Gilgit and Hazara. April-June.

11. Scorzonera 15-16 spp.

Perennial, biennial or annual herbs often with a thick roostock or tuberous underground parts. Leaves radicle or cauline. Receptacle smooth. Flowers yellow or rarely white. Achenes cylindrical, smooth or ribbed. Pappus 3-seriate, hairs plumose.

Resembling *Tragopogon,* but differing in the achenes not beaked, pappus 3-seriate and involucral bracts 2-seriate.

+ Plant stemless, with underground tuber.
  Leaves ovate-lanceolate or elliptic-lanceolate. 301. **S. sp.**

– Stem well developed. Tubers absent.
  Leaves linear-lanceolate. 300. **S. virgata**

300. **S. virgata** DC. Pl. 48
Syn.: *S. divaricata* Hook. f.

Branched perennial with green divaricate branches. Leaves narrow, with inrolled margin. Flower heads long stalked. Involucre bracts longer than the outer ones.

Distribution: Pakistan to Himachal Pradesh (India).

Stony places from 2100-4000 m. June-July.

301. **S. sp.** Pl. 48

Stemless perennial with underground tubers. Leaves elliptic-lanceolate or ovate-lanceolate, acuminate, margin wavy. Flowers ligulate, yellow.

In rocks, c.800 m, south Makran (Balochistan). February-March.

12. Epilasia 1 sp.

Annual tomentose herbs with narrow leaves. Outer involucral bracts leafy, ligulate. Flower pinkish-violet.

302. **E. acrolasia** (Bunge) C. B. Clarke Pl. 49

Stems up to 30 cm tall, erect. Leaves lanceolate, white cottony. Outer bracts larger than the inner ones. Achenes 4-5 mm long. Pappus dark brown.

Distribution: Iran, central Asia, Afghanistan, Pakistan.

In dry places from 1100-1400 m especially Balochistan, N.W.F.P. and Swat. March-May.

13. Tridax 1 sp.

Perennial herbs with opposite and usually lobed leaves. Flower heads on long peduncles, rayed, heterogamous. Ray-florets female, strap-shaped; disc-florets bisexual, tubular and 5-lobed.

303. **T. procumbens** L. Pl. 49

A straggling hairy herb. Leaves ovate, irregularly serrate. Flower heads 10-18 mm broad, on peduncles 10-18 cm long. Ray florets yellow, disc-florets white.

Distribution: South America, widely naturalized elsewhere.

Naturalized in many places in the plains from Sindh to Punjab. October-February.

14. Bidens c. 4 spp.

Herbs with opposite leaves. Flower heads corymbose, homogamous with disc flowers only or heterogamous. Ray florets female or neuter, yellow or white. Disc-florets bisexual, fertile. Achenes angled or dorsally compressed.

304. **B. biternata** (Lour.) Merrill & Sheriff Pl. 49
Syn.: *B. pilosa* auct. non L.

Erect spreading herb up to 100 cm tall. Leaves variable in shape, 3-fid to 1-2-pinnatifid; leaflets ovate or lanceolate, serrate. Flower head long-pedunculate. Disk-florets yellow.

Distribution: Tropical Africa, Arabia, Afghanistan, Pakistan, India, Siam, Malaya, Japan, Korea, Manchuria, Indo-China.

Watery places etc., in the plains. December-February.

15. Inula c. 20 spp.

Perennial or biennial herbs. Involucral bracts many-seriate. Flowers both ligulate and tubular. Achenes angular or ribbed. Pappus persistent.

1. + Perennial herbs. 2
   – Shrubs. 3

2. + Prostrate glabrous plant with succulent parts. Flower heads c. 20 mm broad. 305. **I. grantioides**
   – Erect hairy herb with leafy stem. Flower heads 40-60 mm broad. 306. **I. grandiflora**

3. + Leaves more or less thick, densely hairy on under surface. 307. **I. cappa**
   – Leaves thin, undersurface glabrous. 308. **I. cuspidata**

305. **I. grantioides** Boiss. Pl. 49

Bushy perennial with pale green succulent leaves. Leaves lobed at the apex. Achenes 2-2.5 mm long, ribbed.

Distribution: Arabia, Pakistan.

In dry arid places and rocks. Common in Sindh, lower Balochistan, N.W.F.P. February-April.

306. **I. grandiflora** Willd. Pl. 49

Perennial herb up to 60 cm tall. Stem simple, leafy. Leaves elliptic-lanceolate or oblong, 12-26 mm long, with clasping base. Flower heads up to 6 cm broad, radiate.

Distribution: Pakistan, Kashmir to central Nepal.

Locally abundant on moist forest slope clearings in Kaghan valley from 2000-3200 m. July-August.

307. **I. cappa** (Buch.-Ham. ex D. Don) DC. Pl. 49

Branched shrub up to 2.5 m tall. Leaves more or less thick, densely hairy on undersurface. Flowers 6-8 mm broad, in corymbose heads.

Distribution: Pakistan, Kashmir to Bhutan, Assam to China, Thailand, Java.

An aromatic shrub found in the sub-Himalayan tract from 900-1800 m. August-October.

308. **I. cuspidata** (DC.) C. B. Clarke Pl. 49

Leaves thin, green on undersurface. Ray florets 2-3-toothed.

Distribution: Pakistan, Kashmir eastward to Uttar Pradesh (India).

This shrub grows on forest edges, slopes from 1200-2300 m in Himalaya. August-September.

16. Francoueria 1 sp.

Closely resembling *Pulicaria,* but differing mainly in the achenes that are not ribbed and the pappus (i.e. scales and hairs united basally to form a ring).

309. **F. undulata** (L.) Lack Pl. 50
Syn.: *Pulicaria crispa* (Forssk.) Benth. & Hook. f. ex Oliv. & Hiern.

Perennial 25-50 cm tall with cottony branches. Leaves linear to narrow oblong, with crisp margin and cottony undersurface. Flower heads 6-8 mm broad, yellow.

Distribution: North and west Africa, Arabia, Iraq, Iran, Afghanistan, Pakistan, India.

Common in the plains in waste places and cultivated areas. February-April.

17. Leontopodium c. 5-6 spp. *Edelweiss*

Perennial herbs. Flower heads small, crowded together and subtended by an involucre of spreading leaves. Receptacle smooth. Flowers all tubular. Achenes not ribbed. Pappus hairs 1-seriate.

310. **L. himalayanum** DC. Pl. 50

Tufted woolly plant, stems leafy. Leaves linear. Flower heads white, flat topped. Involucral bracts star-like, up to 4 cm long.

Distribution: Pakistan, Kashmir to south west China.

Alpine and sub-alpine open places, slopes 2400-4100 m. June-August.

18. Anaphalis c. 7 spp.

Woolly perennials with simple alternate leaves. Flower heads small, with disc florets only, subtended by papery involucral bracts. Achenes with pappus 1-seriate.

+ Involucral bracts acute, more or less spreading in flower. 311. **A. triplinervis**

– Involucral bracts more or less obtuse, erect in flower. 312. **A. margaritacea**

311. **A. triplinervis** (Sims) C. B. Clarke Pl. 50

Plant 25-50 cm tall with leafy stems. Leaves elliptic to elliptic-lanceolate, 4-8 cm long, acutish, 3-5-nerved, densely woolly on under surface. Flower heads 8-10 mm broad, white. Involucral bracts white.

Distribution: Afghanistan, Pakistan, Kashmir to south west China, south Tibet, Taiwan.

Common in open places from 1800-3000 m. July-August.

312. **A. margaritacea** (L.) Benth. Pl. 50
Syn.: *A. cinnamonea* C. B. Clarke

Similar to the preceding species but with obtuse and more or less erect involucral bracts, and fewer nerved (1-3) leaves.

Distribution: North America, Kashmir to Bhutan, Indo-China, China, Japan, east Russia.

Common in open places especially Murree hills and Kaghan from 1800-3000 m. July-August.

19. Pentanema c. 4-5 spp.

Annual herbs with entire leaves. Flowers radiate. Achenes not ribbed. Pappus uniseriate; hairs 5-20 in number.

313. **P. vestitum** (Wall. ex DC.) Ling Pl. 50
Syn.: *Inula vestita* Wall. ex DC.

Erect branched herb 10-25 cm tall with densely hairy parts. Leaves oblanceolate. Flower heads 10-15 mm broad, yellow. Achenes more or less 1 mm long, glabrous.

Distribution: Afghanistan, Pakistan, India.

Open places in plains. March-April.

20. Cyathocline 1 sp.

Annual herbs with aromatic parts and alternate pinnatisect leaves. Flower heads discoid, small. Outer floret many-seriate, female, filiform; inner florets bisexual, regular. Achenes oblong, smooth, without pappus.

314. **C. purpurea** (Buch. -Ham. ex D. Don) O. Kuntze Pl. 50
Syn.: *C. lyrata* Cass.

Branched annual. Leaves 2-11 cm long, segments toothed. Flower heads in panicled corymbs, rose-purple.

Distribution: Pakistan, Kashmir to Bhutan, Assam, India, Burma, Thailand, Indo-China, China.

In moist places. Naturalized, also cultivated in gardens. Plains and low hills to 800 m. February-March.

21. Solidago 1 sp. *Golden Rod*

Erect perennial herb with leafy stems and radiate flower heads arranged in terminal racemes or panicles. Flowers yellow. Ray florets strap-shaped, spreading, disc florets tubular. Achenes ribbed. Pappus hairs 2-seriate, free.

315. **S. virga-aurea** L. Pl. 51

Stem unbranched, leafy, ending in a terminal paniculate raceme or raceme of unpleasant smelling yellow flowers. Leaves 50-120 x 10-30 mm, acute, subglabrous; upper similar but smaller. Achenes c. 3 mm long. Pappus hairs brownish-white.

Distribution: Temperate Eurasia, Pakistan, Kashmir to central Nepal.

Forest clearings, slopes from 2200-3200 m throughout northern Pakistan. July-August.

22. Aster 8-9 spp.

Perennial or biennial herbs with leafy stems. Flower heads radiate. Receptacle flat, smooth. Ray florets in 1 row, blue, purplish or white. Disc florets yellow, tubular. Achenes compressed. Pappus persistent.

1.	+	Leaves usually with a wavy margin.	316. **A. flaccidus**
	–	Leaf margin entire or toothed.	2
2.	+	Leaves linear-lanceolate to oblanceolate, entire or toothed. Ray florets bluish-pink, spreadingat flowering.	317. **A. altaicus**

– Leaves lanceolate to oblong. Ray florets pale-pink, often recurved at flowering. 318. **A. molliusculus**

316. **A. flaccidus** Bunge Pl. 51

A rock plant with pretty mauve to pinkish-blue flowers up to 40 mm broad and leaves generally with an undulate margin.

Distribution: Pakistan, Kashmir to west Nepal, south west China, central Asia.

An alpine growing amongst rocks, screes from 3300-4700 m. July-August.

317. **A. altaicus** Willd. Pl. 51

Stems branched, slender, leafy. Flowers mauve to pinkish-blue, ray florets spreading.

Distribution: Siberia, Iran, Afghanistan, Pakistan, Kashmir to Nepal, Tibet, China, Korea.

Found at elevations between 2700-3900 m, especially in northern areas. June-August.

318. **A. molliusculus** (DC.) C. B. Clarke Pl. 51

Flower heads up to 30 mm broad. Florets pale pink, generally recurved at flowering time.

Distribution: Pakistan to Uttar Pradesh (India).

A handsome cliff plant with pale pink flowers especially in Hazara area. Found from 1800-2500 m. April-May.

23. Erigeron c. 19-20 spp.

Like *Aster,* but differing in the uniseriate ray florets and pappus hairs of the fruit which are 2-3-seriate.

+ Flower heads 10-20 mm, ray florets pale lilac. 319. **E. bellidioides**

– Flower heads 20-50 mm, rayflorets purplish-pink. 320. **E. multiradiatus**

319. **E. bellidioides** (Buch.-Ham. ex D. Don) Benth. ex Clarke Pl. 51

Stem slender. Basal leaves lanceolate, often toothed. Flower heads solitary or 2-3. Ray florets pale lilac, fading to white.

Distribution: Pakistan, Kashmir to Bhutan.

Fairly common on rocky banks from 1800-2500 m. April-May.

320. **E. multiradiatus** (Lindley ex DC.) C. B. Clarke Pl. 51

Similar to above, but with larger and darker coloured flowers.

Distribution: Pakistan, Kashmir to Bhutan, China.

Common from 2200-4200 m. Open places. June-September.

24. Bellis 1 sp.

Annual or perennial low growing herbs with a rosette of leaves. Flower heads radiate on leafless scapes. Involucral bracts 1-2-seriate. Receptacle more or less conical. Ray florets 1-seriate, white; disc florets yellow. Pappus absent.

321. **B. perennis** L. *Daisy* Pl. 51

Perennial. Leaves obovate; scapes short, 10-30 mm long. Flower head 10-12 mm broad. Ray florets white, disc florets yellow.

Distribution: Europe, Cyprus, west Syria, Azerbaijan. Naturalized elsewhere.

In shady places in forests from 1900-3000 m. April-June.

25. Tussilago 1 sp.

Perennial herbs with flowers appearing before the leaves. Flower heads solitary, radiate. Ray florets yellow, many; disc ones tubular, few. Pappus with long soft hairs.

322. **T. farfara** L. *Coltsfoot* Pl. 52

Flower shoots several, 5-20 cm long, nodding. Flower heads 2-3 cm broad. Involucral bracts purplish, often glandulose.

Distribution: North Africa, Europe, Asia, north America.

In damp places, moist banks from 2300-3200 m in Swat, Chitral, Hazara and Murree hills. March.

26. Senecio 20-21 spp.

Flower heads radiate, usually in corymbose clusters. Involucral bracts 1-seriate. Flowers yellow or orange. Achenes ribbed, without a beak.

+ Robust perennial, 70-150 cm tall.
Leaves dense white woolly on undersurface.
Flower heads numerous, 10-15 mm broad. 323. **S. chrysanthemoides**

– Slender annual up to 40 cm tall.
Leaves not hairy on undersurface.
Flower heads few, 5-8 mm broad. 324. **S. desfontanei**

323. **S. chrysanthemoides** DC. Pl. 52

Leaves deeply pinnate, upper surface dark green, lower white tomentose. Ray florets yellow, strap-shaped, with a toothed apex. Achenes ribbed. Pappus hairs white.

Distribution: Pakistan, Kashmir to south west China.

Conspicuous during monsoon season in forest clearings and slopes. Gregarious from 2300-3400 m. Poisonous to grazing livestock. July-September.

324. **S. desfontanei** Druce Pl. 52
Syn.: *S. coronopifolius* Desf.

A slender glabrous herb with spreading branches. Stem leafy. Flower heads 5-8 mm broad, long stalked. Ray florets few, yellow. Pappus hairs white.

Distribution: South Europe, Afghanistan, Pakistan, Kashmir to Himachal Pradesh (India).

Common in dry areas in northern Balochistan, Waziristan, Chitral, N.W.F.P. up to Ladakh from 2500-4600 m. June-August.

27. Ligularia 4 spp.

Resembling *Senecio,* but leaf entire and not lobed, upper ones amplexicaul.

325. **L. thomsonii** (Clarke) Kitamura Pl. 52

Perennial up to 120 cm tall. Leaves prominent, reniform-orbicular, up to 26 cm broad, toothed; upper ones smaller, amplexicaul. Flower heads in corymbose clusters, arranged on a stout scape, yellow. Ray florets few.

Distribution: Pakistan, Kashmir.

A conspicuous plant in grassy areas, forest clearings, slopes, in damp places from 2200-3800 m. July-August.

28. Hertia 1 sp.

Glabrous undershrubs with alternate, fleshy leaves. Flower heads heterogamous. Yellow. Ray florets fertile, uniseriate.

326. **H. intermedia** (Boiss.) O. Kuntze Pl. 52
Syn.: *Othonnopsis intermedia* Boiss.

Branched shrub up to 40 cm tall. Stem and branches brownish-white. Leaves elliptic-oblong, thickish. Flower heads upto 25 mm broad on long peduncles.

Distribution: Iran, Afghanistan, Pakistan.

In dry arid open places, hillocks from 1400-1900 m, especially Balochistan and N.W.F.P. Plant parts poisonous. April-May.

29. Anthemis 4-5 spp.

Leaves 1-3-pinnatisect. Flower head solitary. Achenes more or less obconical, round or angled.

+	Disc florets yellow	327. **A. odontostephana**
–	Disc florets pinkish.	328. **A. rhodocentra**

327. **A. odontostephana** Boiss. Pl. 52
Syn: *A. gayana* Boiss. var. *multiramosa* Parsa

Annual with hairy parts, up to 18 cm tall. Leaves basal pinnatisect. Receptacle subglobose of yellow disc florets. Ray florets white, ovate.

Distribution: Iraq, Iran, Afghanistan, Pakistan.

Fairly common in Balochistan. April-May.

328. **A. rhodocentra** Iranshahr Pl. 52
Syn: *A. gayana* Boiss. var. *rosea* Parsa

Annuals up to 20 cm tall. Leaves similr to the above species. Receptacle shortly conical, pinkish. Ray florets white, oblong, 2-4 mm long.

Distribution: Iran, Afghanistan, Pakistan.

Fairly common in dry places up to 2000 m in Balochistan. April-May.

30. Achillea 2 spp.

Perennial herbs. Leaves alternate, generally 3-4-pinnatipartite. Flower heads radiate. Flowers white or yellow. Ray florets lobed or 2-3-toothed at the apex. Receptacle flat or convex. Achenes smooth. Pappus absent.

+ Flower heads yellow. 329. **A. wilhelmsii**

– Flower heads white. 330. **A. millefolium**

329. **A. wilhelmsii** C. Koch. Pl. 53
Syn.: *A. santolina* auct. non L.

Leaves linear, pinnatipartite; lobes 3-5-dentate. Flower heads up to 35 in number. Ray florets 3-5, yellow; disc florets 12-20.

Distribution: Turkey, Syria, Iraq, Iran, Afghanistan, Pakistan.

Open dry places, slopes from 1500-4000 m mainly in Balochistan and Waziristan. May-June.

330. **A. millefolium** L. *Yarrow* Pl. 53

Perennial up to 70 cm tall. Leaves finely 2-3-pinnatisect. Flower heads numerous. Ray florets white, disc florets up to 20 in number.

Distribution: Europe, Caucasia, Iran, Siberia, Himalaya.

Common from Swat, Gilgit to Hazara and Murree hills in forest clearings, lower alpine zones from 2100-4000 m. June-August.

31. Leucanthemum 1 sp.

Annual or perennial herbs, with simple leaves. Flower heads radiate. Receptacle flat, smooth. Ray florets white or sometimes yellow; disc florets yellow. Achenes more or less terete, 10-ribbed.

331. **L. vulgare** Lam. *Ox-eye Daisy* Pl. 53
Syn.: *Chrysanthemum leucanthemum* L.

A perennial with obovate, crenate or toothed leaves. Ray flowers 20-25 in number, white; disc florets yellow.

Distribution: Temperate Eurasia; naturalized elsewhere.

Forest clearings, slopes from 1700-3000 m. Extremely common in Murree hills, carpeting open slopes in spring with dazzling white flowers. May-June.

32. Calendula 2 spp. *Marigold*

Annual or perennial herbs with alternate leaves. Flower heads heterogamous, terminal, yellow or orange. Ray florets female, 1-2-serrate, fertile, disc florets bisexual, tubular. Bracts 1-2-seriate, linear. Receptacle flat.

332. **C. arvensis** L. *Gul-e-Ashrafi, Field Marigold* Pl. 53

Branched leafy low growing annual. Flower heads 10-15 mm broad. Ray florets orange-yellow. Disc florets of similar colour. Achenes curved.

Distribution: South Europe, Caucasus, Iran, Afghanistan, Pakistan.

Common spring annual, gregarious in open places, road sides, near fields etc. up to 900 m. March-April.

33. Matricaria 1 sp.

Branched leafy annuals with dissected leaves. Flower heads usually discoid, solitary or corymbose. Involucral bracts 2-3-seriate. Receptacle more or less conical, smooth. Disc florets yellow.

333. **M. aurea** (L.) Schulz Pl. 53
Syn.: *Cotula aurea* L.

Glabrous low growing herb. Leaves 2-3-pinnatisect into linear segments. Flower heads solitary or 2-3.

Distribution: North Africa, south west Asia to west Himalaya.

A spring weed of the plains and lower hills up to 1200 m Chitral, Swat, Hazara and lower Murree hills and Kala Chita hills. March-April.

34. Artemisia 39 spp. *Wormwood*

Herbs or shrubs usually scented. Leaves alternate. Heads small, solitary or in leafy clusters with disc florets only. Achenes very small without pappus.

334. **A. brevifolia** Wall. ex DC. Pl. 53
Syn.: *A. maritima* L. sensu Hook. f.

Strongly aromatic shrubby perennial upto 50 cm tall. Leaves dissected into linear segments, greyish-green to almost white. Flower heads small, yellow in axillary clusters.

Distribution: Afghanistan, Pakistan, India.

One of the most important plants of the high dry areas of the country, replaces grass forming *Artemisia* steppes. Widespread in Balochistan, Chitral, Gilgit and Baltistan. Important source of anthelmintic Santonin. Plant-parts locally used in medicines. July-September.

35. Tanacetum 13-14 spp. *Tansy*

Perennial herbs with pinnately divided leaves. Flower head radiate or discoid, usually in corymbs. Involucral bracts 3-4-seriate. Receptacle flat. Ray florets white; disc florets yellow.

+ Leaf segments very slender. Involucral bracts glabrous. Flower heads c. 4 mm broad. 335. **T. gracile**

– Leaf segments broad obtuse. Involucral bracts coriaceous, pubescent. Flower heads 6-7 mm broad. 336. **T. artemisioides**

335. **T. gracile** Hook. f. & Thoms. Pl. 53

Perennial up to 55 cm tall with a woody rootstock and many slender stems. Leaves pinnately divided into obtuse-linear segments. Flower heads c. 4 mm broad, yellow. Ray florets absent.

Distribution: Pakistan, Kashmir.

Dry stony places from 2600-3400 m, especially northern areas, Hunza, Gilgit and Baltistan. June-August.

336. **T. artemisioides** Sch.-Bip. ex Hook. f. Pl. 54

Perennial with finely dissected aromatic leaves and few yellow flowered heads in 5-10 loose corymbs. Flower heads 6-7 mm broad.

Distribution: Pakistan, west Tibet.

Rocky open places in Gilgit, Hazara and Baltistan from 2300-2800 m. April-May.

36. Adentostemma 1 sp.

Annual herbs with opposite leaves. Flower heads in terminal corymbs or panicles, discoid. Florets many, tubular. Involucral bracts 2-seriate. Receptacle smooth. Pappus hairy.

337. **A. lavenia** (L.) O. Kuntze Pl. 54

Glabrous erect plants up to 1.5 m tall. Leaves 5-20 x 4-8 cm, ovate or broadly so. Involucral bracts glandular. Flowers white. Achenes minutely tubercled.

Distribution: Pantropical weed.

A weed of moist and damp places from plains to 1500 m. April.

37. Ageratum 1 sp.

Resembling *Adenostemma,* but plant aromatic, flower heads in branched corymbs and scaly pappus.

338. **A. conyzoides** L. Pl. 54

Branched annual up to 60 cm tall. Leaves 3-10 x 2-6 cm, ovate. Flowers tubular, white or purplish-pink.

Distribution: Pantropical. Widespread weed.

Occasionally cultivated and as an escape from plains to 900 m. Very common in Changa Manga forest near Lahore. February-March.

38. Vernonia 4 spp.

Herbs or shrubs with alternate leaves. Flower heads discoid, corymbose or in panicles. Involucral bracts 3 or more seriate. Receptacle smooth. Flowers tubular. Pappus hairy, persistent.

339. **V. cinerascens** Schultz-Bip. Pl. 54

Branched undershrubs with greyish-white pubescence. Leaves elliptic-lanceolate to oblanceolate, 2-5 cm long. Flower heads 4-6 mm broad. Flowers small, purplish.

Distribution: Tropical Africa, Pakistan, India.

In dry stony places in lower Sindh, lower Balochistan, Salt Range and N.W.F.P. November-February.

39. Ainsliaea 2 spp.

Erect perennial herbs with leafless stems. Leaves alternate, radical. Flowers in clusters or spikes, white or tinged pale lilac. Corolla tubular, more or less 2-lipped.

340. **A. aptera** DC. *Aarons Rod* Pl. 54

Flowers borne on leafless stalks, 30-70 cm tall, appearing before the leaves. Leaves broadly ovate to suborbiculate, pubescent, cordate, margin wavy-toothed; flowers 2-4, star-shaped, borne on interrupted spikes.

Distribution: East Afghanistan, Pakistan, Kashmir to Bhutan.

A delicate appearing plant found in moist forests from 2100-3000 m in Murree hill tract and Hazara area. April-mid May.

40. Gerbera 2 spp.

Perennial herb with radical leaves and solitary flower heads borne on long leafless scapes. Flowers white; ray florets strap-shaped, 2-3-toothed, spreading. Disc florets tubular, 2-lipped.

341. **G. gossypina** (Royle) Beauv. Pl. 54

Leaves broadly lanceolate to elliptic, 6-13 x 1.5-5 cm, margin, sinuate, minutely toothed, undersurface cottony. Flower heads 3-4 cm broad, borne on scapes up to 31 cm long. Flowers white, sometimes tinged with pink.

Distribution: Pakistan to west Nepal.

Common in pine forests in Murree hills and Hazara area on rocky places, steep banks from 1000-2500 m. May-June.

41. Echinops 6-7 spp. *Globe Thistle*

Perennial spiny herbs. Flower heads discoid, in dense globose clusters. Involucral bracts many-seriate. Flowers tubular, white or pale blue. Pappus of scale-like bristles.

\+ Flower heads 20-30 mm broad, white. Outer involucral bracts, at least some prominently spinose. 342. **E. echinatus**

– Flowers 5-7 cm broad, pale bluish. Involucral bracts not spinose or with few spines. 343. **E. cornigerus**

342. **E. echinatus** Roxb. Fig. between p. 102-103

Low growing, much branched herb with white cottony stems. Leaves pinnatifid with lobes ending in spines up to 20 mm long; undersurface white tomentose. Flower heads subtended by stout spines. Flowers tubular, lobes narrow, recurved.

Distribution: Pakistan, India.

Common weedy species of dry desert areas in the plains. April-May.

343. **E. cornigerus** DC. Pl. 55

Like *E. echinatus,* but with larger flower heads with white or pale blue flowers.

Distribution: Afghanistan, Pakistan, Kashmir to west Nepal.

Dry open places, slopes from 2300-3200 m. Fairly common in dry northern areas of Pakistan. July-August.

42. Silybum 1 sp.

Herbs with spiny leaves. Flower heads solitary, discoid. Receptacle flat. Flowers 2-lipped, pink-purple.

344. **S. marianum** (L.) Gaertner *Milk Thistle* Pl. 55

A prickly erect plant 60-100 cm or more tall with leafy stem. Leaves variegated green-white, margin spiny-dentate, upper ones amplexicaul. Flower heads 4-5 cm broad, subtended by involucral spines 10-20 mm long. Pappus white.

Distribution: Mediterranean, south and east Iraq, Iran, Afghanistan, Pakistan, India.

Gregarious in waste places, roadsides from plains to 1700 m in N.W.F.P. and north Punjab. April-May.

43. Cousinea c. 30 spp.

Branched spiny herbs generally with tomentose hairy parts. Leaves variously lobed to pinnatipartite. Involucral bracts spinose. Flowers yellow or pink. Achenes glabrous with deciduous pappus.

+	Stem branched. Flower heads yellow, 12-14 mm broad.	345. **C. prolifera**
–	Stem simple. Flower heads pink-purple, 40-50 mm broad.	346. **C. thomsonii**

345. **C. prolifera** Jaub. & Spach. Pl. 55

Annual herbs with whitish stem dichotomously branched. Leaves hoary white, margin subentire, sparsely spiny. Flower heads yellow.

Distribution: Iran, Afghanistan, Pakistan, Kashmir, central Asia.

Xerophytic weed from plains to 700 m. March-June.

346. **C. thomsonii** C. B. Clarke Pl. 55

An erect perennial up to 50 cm tall with simple white hoary stem and pinnately lobed leaves bearing stiff creamy spines. Flower heads pink-purple, subtended by involucral spines (bracts).

Distribution: East Afghanistan, Pakistan, Kashmir.

Stony slopes etc., from 3000-4200 m, mainly in northern areas of Pakistan. July-August.

44. Onopordum 2 spp.

Spiny herbs with cobwebby hairiness. Leaves pinnately lobed, spinescent. Flower heads discoid, generally solitary. Involucral bracts many-seriate, the outer ones sometimes recurved. Flower heads pink or purplish-pink. Pappus hairs deciduous.

347. **O. acanthium** L. Pl. 55

Robust spiny plants up to 120 cm tall with winged stems. Lower leaves elliptic, upper ones narrower, margin spinose. Spines yellow. Outer involucral bracts (spines) reflexed.

Distribution: Central Asia, west Europe, Pakistan, Kashmir. Introduced in many places.

Waste places, roadsides from 1000-2000 m mainly N.W.F.P., Swat and Kashmir. March-April.

45. Cnicus 1 sp.

Annual with branched stems. Leaves alternate. Involucral bracts many-seriate. Receptacle with bristles. Flower heads discoid, yellow. Achenes glabrous, many-ribbed. Pappus with 2 rows of bristles.

348. **C. benedictus** L. Pl. 56

Plants 3-30 cm tall with deeply lobed runcinate leaves, the upper most amplexicaul. Involucral bracts leafy, with spinose margin, longer than flowers.

Distribution: Mediterranean, Caucasus, Iran, Afghanistan, Pakistan.

A weed of roadsides, field edges, in dry river regions from plains to 1300 m. March-April.

46. Cirsium c. 10 spp. *Thistle*

Perennial herbs. Leaves with spinose margins, upper surface sometimes hairy. Flowers discoid. Involucral bracts many-seriate. Flowers tubular, white to purplish-pink. Achenes smooth, pappus plumose.

349. **C. arvense** (L.) Scop. Pl. 56
Syn.: *Breea arvensis* (L.) Less

A perennial with leafy stems up to 120 cm tall with cobwebby leaves and weak marginal spines. Flower heads light purple, solitary.

Distribution: Europe, Caucasus, Iran, Afghanistan, Pakistan, north America.

Waste places, ditches. Often gregarious from plains to 1100 m. February-May.

47. Carthamus 3 spp.

Branched herbs with alternate spinescent leaves. Flower heads discoid, in corymbose clusters. Involucral bracts many-seriate. Flowers yellow to orange-red, tubular. Achenes glabrous, 4-angled. Pappus usually of scales.

350. **C. oxycantha** M. Bieb. *Pohli* Pl. 56

A spinose branched herb 30-50 cm tall. Stem whitish. Leaves oblong to oblanceolate, densely hairy on undersurface. Flower heads 15-30 mm broad, yellow. Outer bracts spinescent.

Distribution: Pakistan westwards to Caucasus.

Very common weed of fields, up to 700 m especially Balochistan, N.W.F.P. and Punjab. March-June.

48. Centaurea 5-6 spp.

Herbs with spiny branches, often tomentose. Leaves alternate, radical, pinnatifid or -partite. Flower heads radiate or discoid. Involucral bracts many-seriate, often spinose. Flowers yellow, blue, pink or purplish.

351. **C. iberica** Trev. ex Sprengel Pl. 56

A spiny dichotomously branched annual up to 90 cm tall. Leaves pinnatifid to pinnatipartite. Some of the involucral bracts spinose, straw coloured. Flowers pinkish-red.

Distribution: South east Europe, Pakistan, Kashmir, west Asia.

Waste places, roadsides from 300-2500 m. Common in Balochistan, northern Punjab, N.W.F.P., Gilgit and Salt Range. April-May.

49. Tricholepis 4-5 spp.

Annual or perennial herbs with alternate, gland dotted leaves. Flower heads discoid, solitary, terminal. Involucral bracts spiny. Anthers tailed. Achenes smooth.

352. **T. stewartii** C. B. Clarke ex Hook. f. Pl. 56

Tufted glabrous herbs with ascending branches to 50 cm. Leaves ovate-lanceolate to lanceolate, 30-60 x 10-24 mm. Involucral bracts acicular. Flower heads 20-25 mm broad. Flowers pale white. Achenes 3-4 mm long. Pappus golden yellow.

Distribution: Pakistan, Kashmir.

A cliff plant from 1200-2800 m. Common in Murree hills and Gilgit. July-September.

50. Amberboa 1 sp.

Annual glabrous herbs. Flower heads heterogamous, radiant. Receptacle with smooth bristles. Flowers pinkish. Achenes hairy, laterally compressed.

353. **A. moschata** (L.) DC. Fig. between p. 102-103

Plant up to 50 cm tall, branched. Leaves both basal and cauline, upper ones pinnatilobed or pinnatifid. Flower heads on long peduncles. Ray florets radiant.

Distribution: Turkey, Russia.

Introduced in many places. Appears to be wild in upper Balochistan. March-April.

51. Serratula 1 sp.

Unarmed perennial herbs. Leaves amplexicaul, simple or divided. Flower heads homogamous, discoid, usually in corymbose panicles. Flowers purple-pink or white. Receptacle hairy.

354. **S. pallida** DC. Pl. 57

Glabrous herb up to 60 cm tall. Flower heads 20-35 mm broad, pinkish-purple on long leafless peduncles. Leaves variable from subentire to runcinate-pinnatifid.

Distribution: Kashmir to Uttar Pradesh (India).

Grassy banks or gravelly slopes from 1800-2500 m in Salt Range, Hazara, Murree hills, Khyber, Swat, Kaghan and Kashmir. May-June.

52. Saussurea c. 29-30 spp.

Perennial herbs with radical, unarmed pinnately lobed leaves. Flower heads discoid. Involucral bracts many-seriate. Flowers purple or a lighter shade. Achenes compressed.

1. + Erect herbs, with well developed stems. Leaves both basal and cauline. Flower heads in clusters. 2

   – Prostrate stemless plants with leaves in rosettes. Flower heads solitary. 355. **S. atkinsonii** 356. **S. simpsoniana**

2. + Flower heads surrounded by prominent large yellowish-white bracts. 357. **S. obvallata**

   – Flower heads not as above. 3

3. + Leaf under surface and stems white canescent with cottony hairs. Involucral bracts not dark margined. Flower heads many. 358. **S. heteromalla**

   – Leaf and stems glabrous. Involucral bracts conspicuously black margined. Flower heads few. 359. **S. jacea**

355. **S. atkinsonii** C. B. Clarke Pl. 57

Stemless perennial with a rosette of leaves. Leaves broadly elliptic-oblong, toothed. Flower heads solitary, arising from the centre of the rosette, pale purple.

Distribution: Pakistan to Kashmir.

In alpine areas from 3000-4300 m in northern parts of country. Open places. July-September.

356. **S. simpsoniana** (Field & Gardn.) Lipsch. Pl. 57
Syn.: *S. sacra* Edgew.

A densely woolly plant with stems up to 10 cm tall. Flower heads exposed from the wool, purplish, arranged dome-like.

Distribution: Pakistan, (Karakoram) to Sikkim, Tibet.

Dry stony slopes in Karakoram from 4200-5500 m. July-August.

357. **S. obvallata** (DC.) Sch.-Bip. Pl. 57

Perennial up to 40 cm tall, characterised by inflated inflorescence with large papery yellowish-white bracts, 8-13 cm long. Flower heads purplish, in clusters surrounded by the bracts.

Distribution: Pakistan, Kashmir to south west China.

A high elevation alpine from 3300-4300 m in Gilgit, Baltistan and Babusar pass. Not common in Pakistan. July-September.

358. **S. heteromalla** (D. Don) Hand.-Mazz. Pl. 57
Syn.: *S. candicans* (DC.) Sch.-Bip.

Erect herb up to 120 cm tall; stem branched above. Basal leaves lyrately lobed. Upper ones smaller, toothed. Flower heads pale purplish, many, 20-30 mm broad. Involucral bracts lanceolate. Pappus hairs white.

Distribution: Afghanistan, Pakistan, Kashmir to Bhutan.

Common from plains to 2000 m in northern parts of Pakistan. March-September.

359. **S. jacea** (Klotzsch) C. B. Clarke Pl. 57

A glabrous branched perennial up to 100 cm tall. Leaves ovate to obovate. Flower heads few, 12-18 mm broad. Involucral bracts ovate or lanceolate, close appressed, black margined.

Distribution: Afghanistan, Pakistan, Kashmir.

Open slopes from 2900-4700 m in northern maintainous regions. July-August.

CAMPANULACEAE *Bell Flower Family*

Annual or perennial herbs with usually alternate leaves. Flowers regular. Corolla, campanulate, funnelform. Stamens 5. Ovary 3-5-locular. Styles 3. Fruit capsular.

\+ Capsule beaked, dehiscing by 3 valves. **1. Codonopsis**

– Capsule not beaked, dehiscing by pores. **2. Campanula**

1. Codonopsis 3 spp.

Perennial, twining, decumbent erect plants. Flowers solitary terminal or axillary. Corolla with 5 shallow lobes. Capsule ultimately dry and hard. Stigmas 3.

308. *Inula cuspidata* Margalla hills, S. A. Sultan

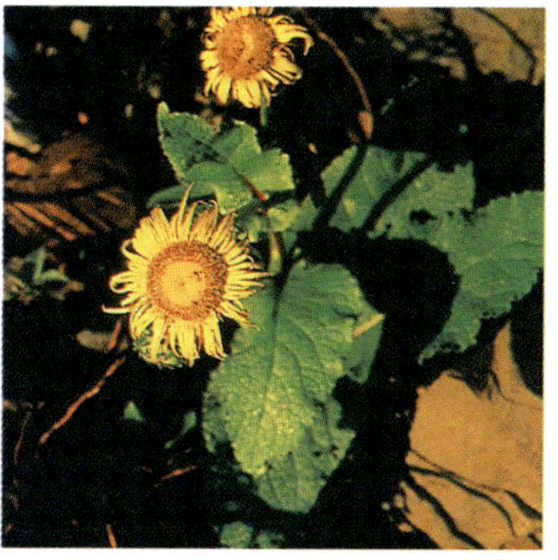

306. *Inula grandiflora* Kaghan, Y. J. Nasir

302. *Epilasia acrolasia* Quetta, Rubina A. Rafiq

303. *Tridax procumbens* Karachi, Y. J. Nasir

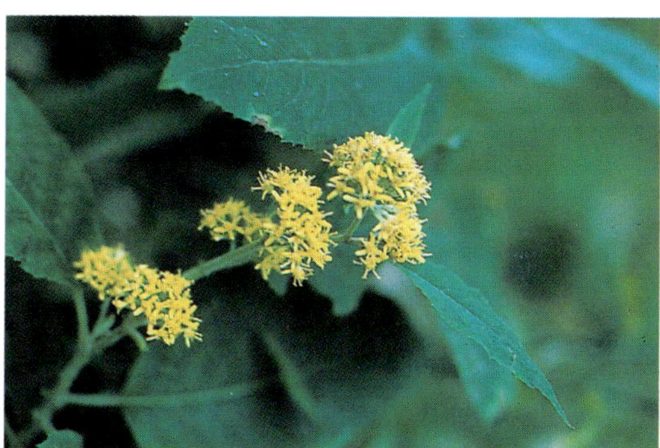

307. *Inula cappa* Murree hills, S. A. Sultan

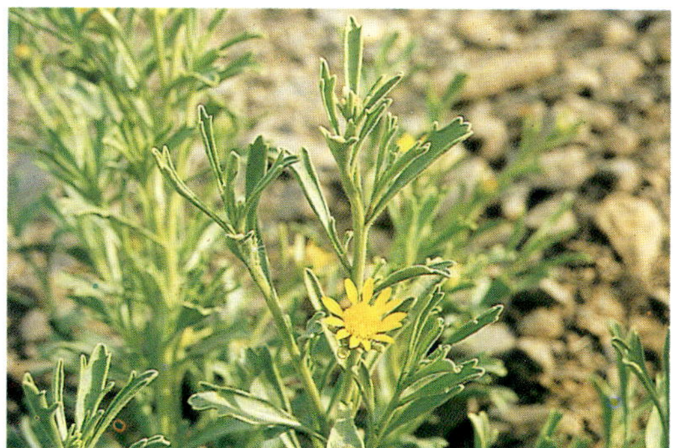

305. *Inula grantioides* Karachi, Khan Muhammad Khan

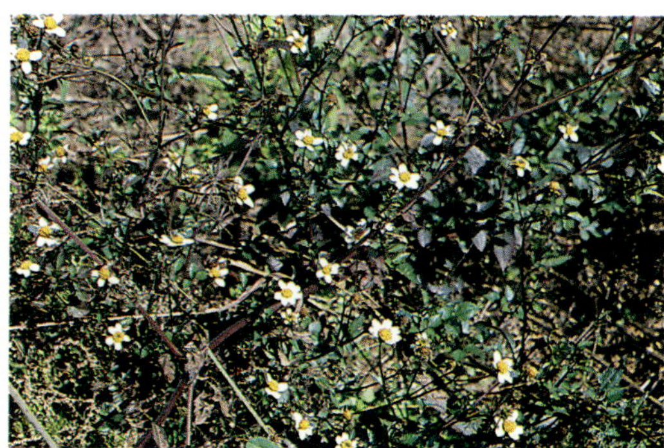

304. *Bidens biternata* Islamabad, Y. J. Nasir

309. *Francoueria undulata* Makran, Y. J. Nasir

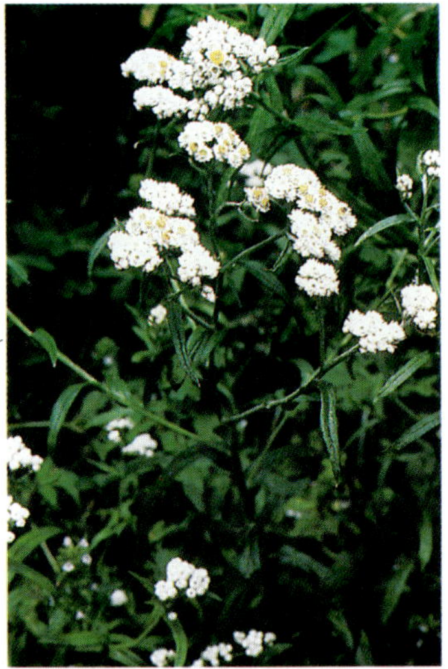

312. *Anaphalis margaritacea* Thandiani, Y. J. Nasir

311. *Anaphalis triplinervis* Upper Swat, Fritz Berger

313. *Pentanema vestitum* Islasmabad, S. A. Sultan

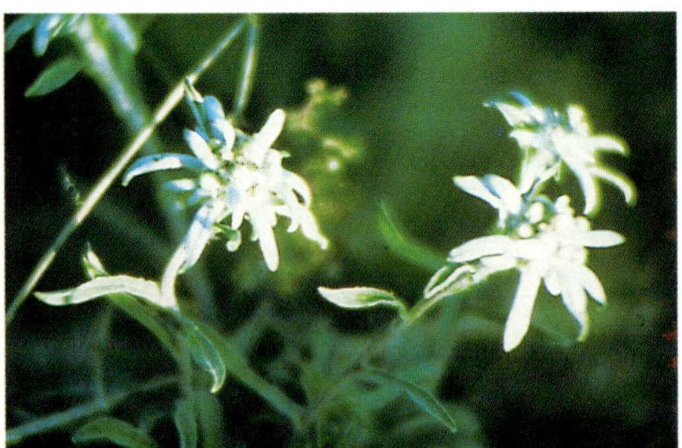

310. *Leontopodium himalayanum* Swat, Fritz Berger

314. *Cyathocline purpurea* Lower Murree hills, Y. J. Nasir

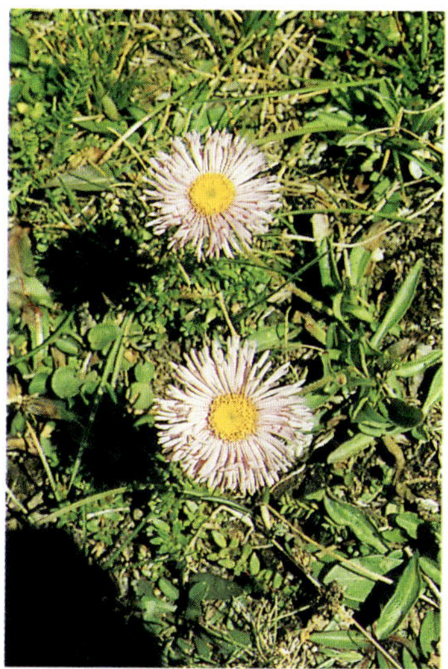

320. *Erigeron multiradiatus* Hunza, Y. J. Nasir

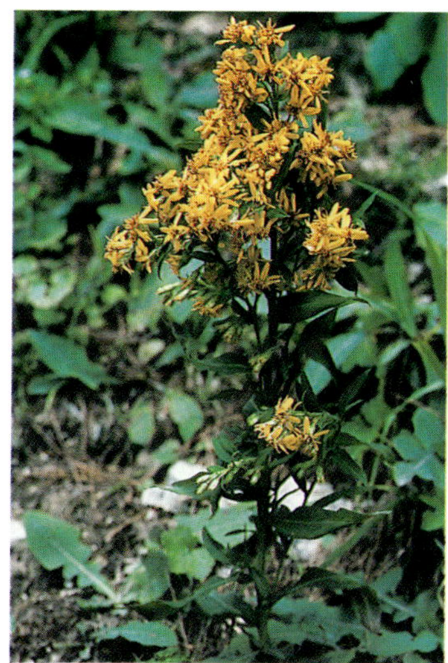

315. *Solidago virga-aurea* Dunga Gali, S. A. Sultan

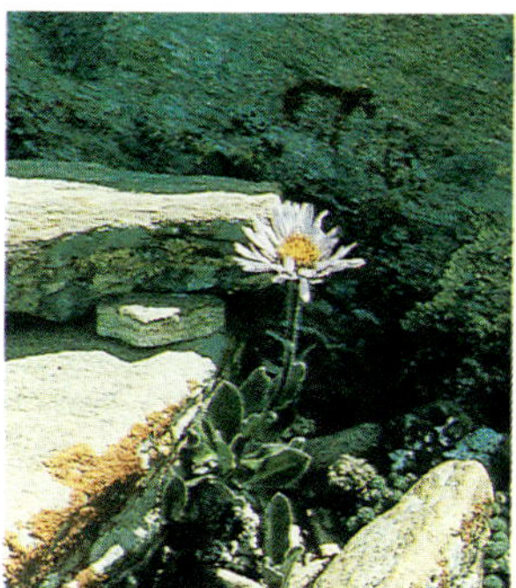

316. *Aster flaccidus* Babusar pass, T. J. Roberts

317. *Aster altaicus* Hunza, Y. J. Nasir

318. *Aster molliusculus* Dunga Gali, S. A. Sultan

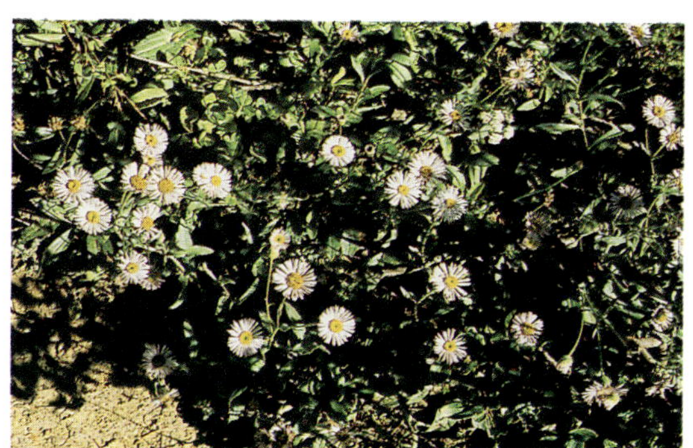

319. *Erigeron bellidioides* Kalam, Y. J. Nasir

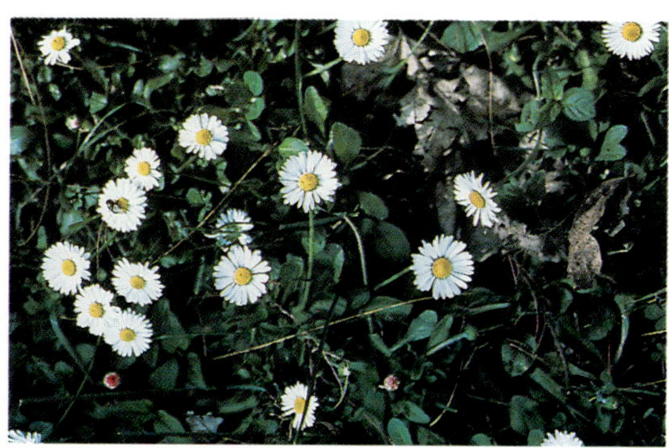

321. *Bellis perennis* Murree hills, Y. J. Nasir

326. *Hertia intermedia* Quetta, Y. J. Nasir

325. *Ligularia thomsonii* Naran, Kaghan, Rubina A. Rafiq

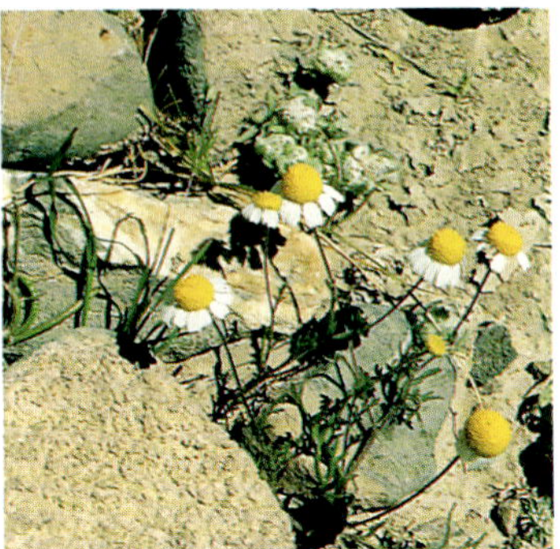

327. *Anthemis odontostephana* Kalat, Rubina A. Rafiq

328. *Anthemis rhodocentra* Chiltan hills, Rubina A. Rafiq

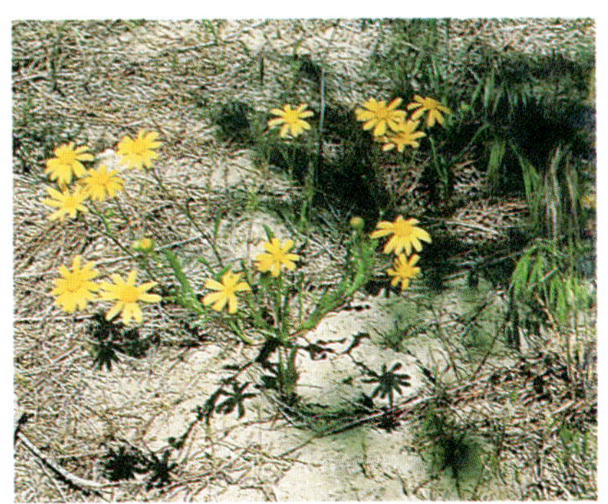

324. *Senecio desfontanei* Chiltan hills, Rubina A. Rafiq

323. *Senecio chrysanthemoides* Changla Gali, S. A. Sultan

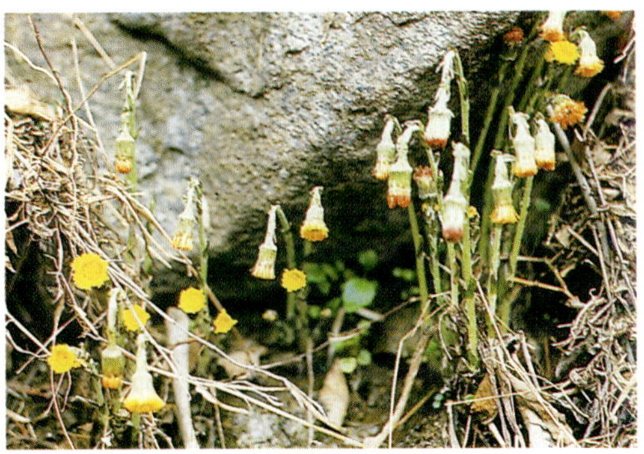

322. *Tussilago farfara* Swat Kohistan, Fritz Berger

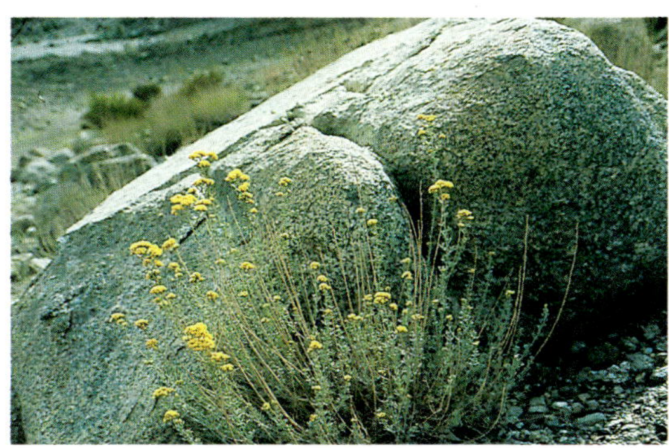

335. *Tanacetum gracilis*
Passu, Hunza, Fritz Berger

334. *Artemisia brevifolia* Gilgit, T. J. Roberts

331. *Leucanthemum vulgare* Dunga Gali, Rubina A. Rafiq

332. *Calendula arvensis* Khanpur, Hazara, S. A. Sultan

333. *Matricaria aurea* Margalla hills, Y. J. Nasir

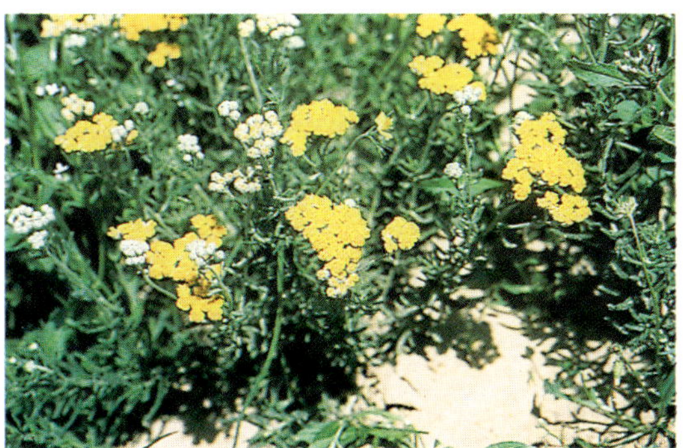

329. *Achillea wilhelmsii* Surgaz, Quetta, Rubina A. Rafiq

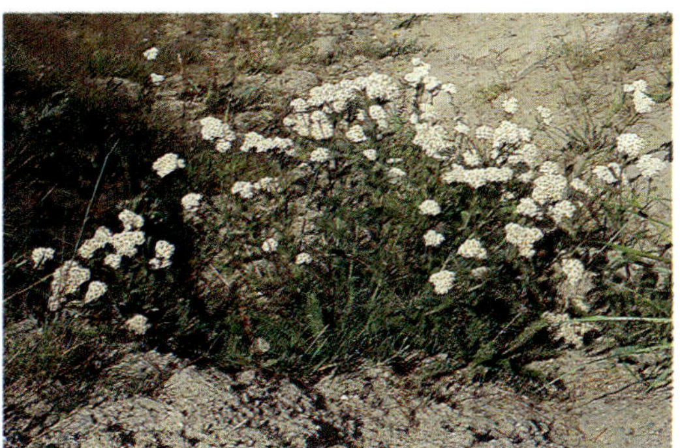

330. *Achillea millefolium* Changla Gali, S. A. Sultan

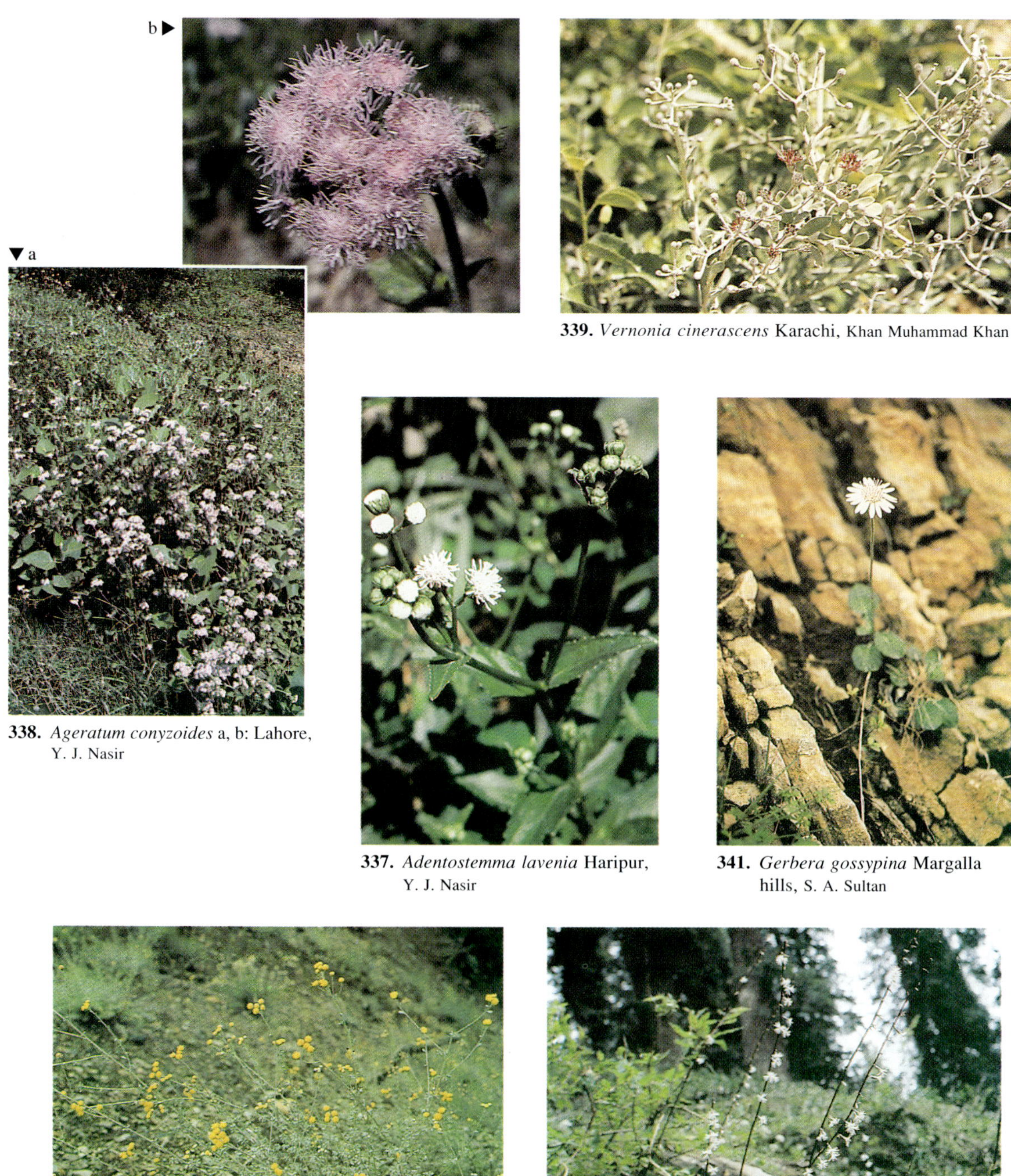

b ▶

▼ a

339. *Vernonia cinerascens* Karachi, Khan Muhammad Khan

338. *Ageratum conyzoides* a, b: Lahore, Y. J. Nasir

337. *Adentostemma lavenia* Haripur, Y. J. Nasir

341. *Gerbera gossypina* Margalla hills, S. A. Sultan

336. *Tanacetum artemisioides* Hunza, Y. J. Nasir

340. *Ainsliaea aptera* Murree hills, T. J. Roberts

345. *Cousinea prolifera* Islamabad, S. A. Sultan

346. *Cousinea thomsonii* Lowari pass, T. J. Roberts

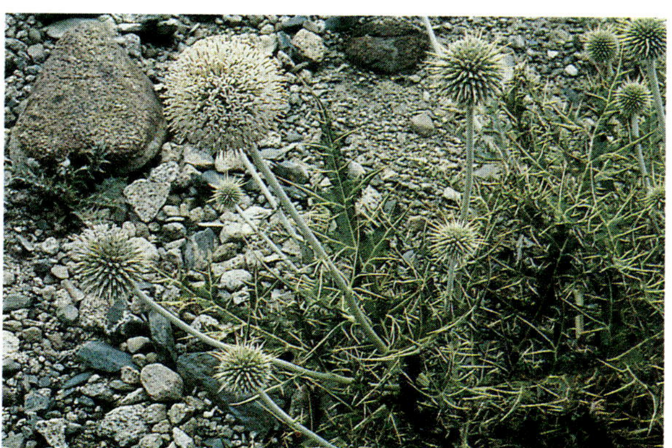

343. *Echinops cornigerus* Batura glacier, Hunza, T. J. Roberts

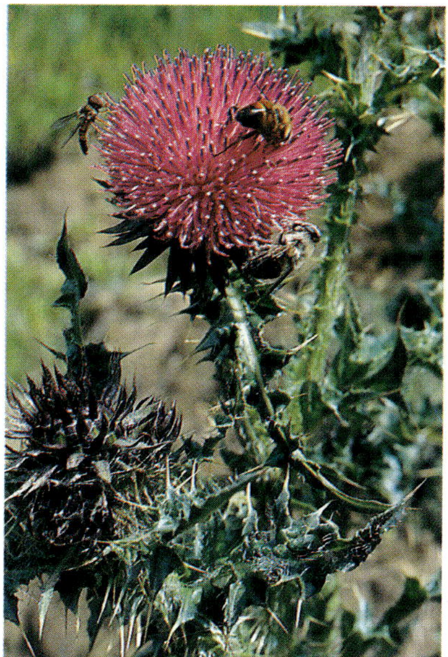

347. *Onopordum acanthium* Upper Swat, Y. J. Nasir

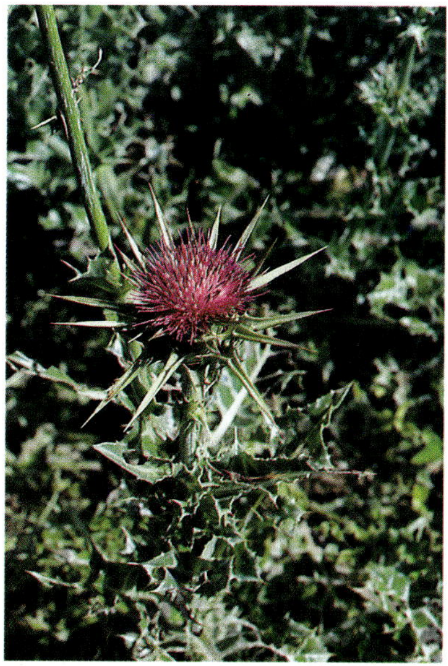

344. *Silybum marianum* Islamabad, S. A. Sultan

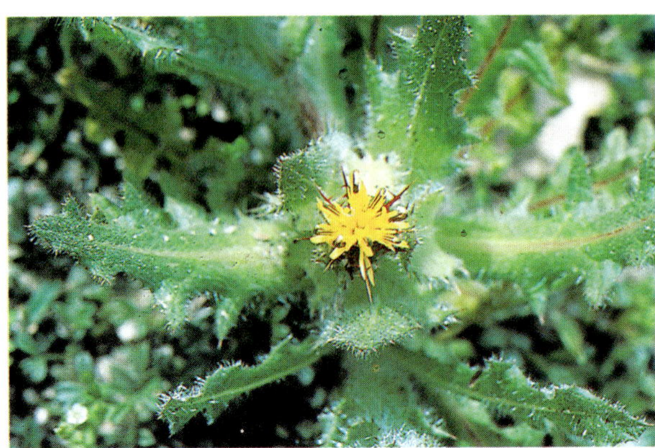

348. *Cnicus benedictus*
Attock hills, S. A. Sultan

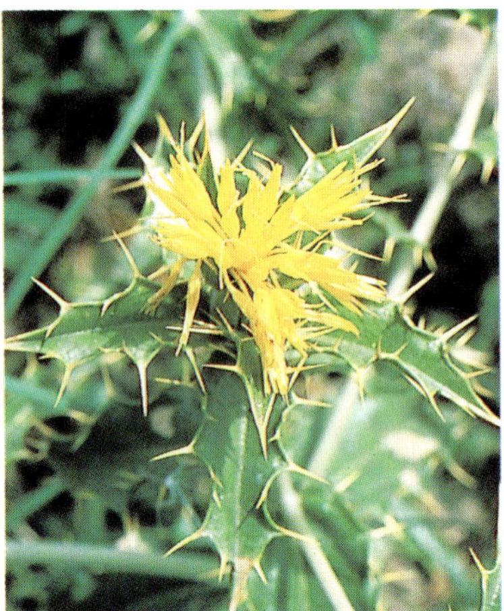

350. *Carthamus oxycantha* Attock hills, S. A. Sultan

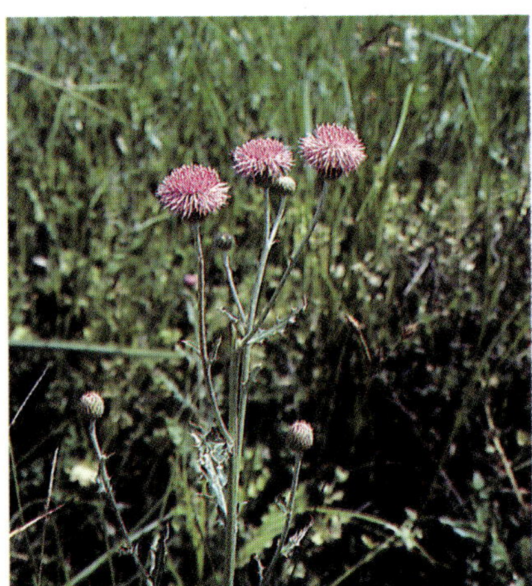

349. *Cirsium arvense* Islamabad, S. A. Sultan

352. *Tricholepis stewartii* Dunga Gali, Y. J. Nasir

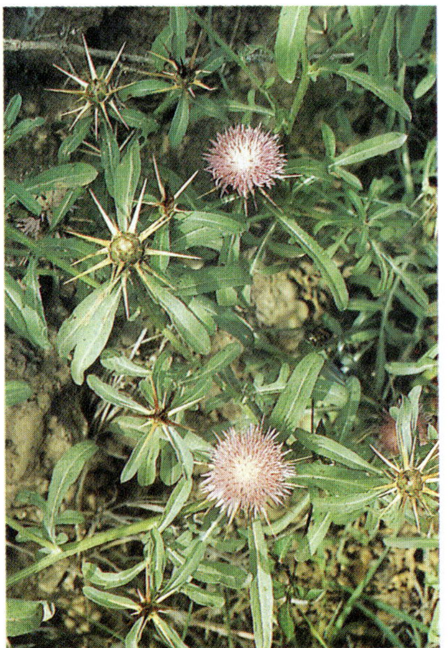

351. *Centaurea iberica* Margalla hills, Y. J. Nasir

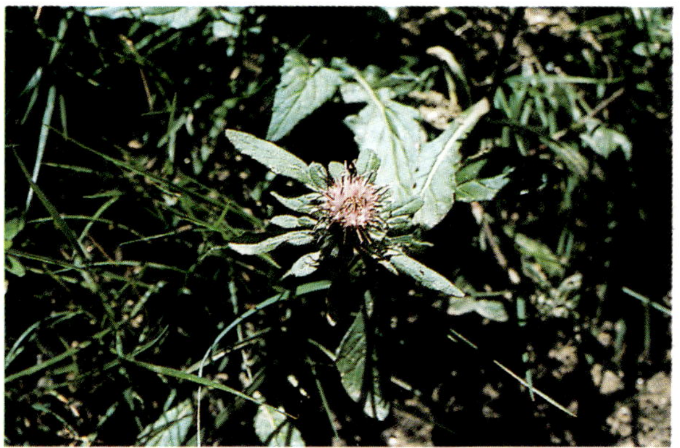

355. *Saussurea atkinsonii* Kaghan, Y. J. Nasir

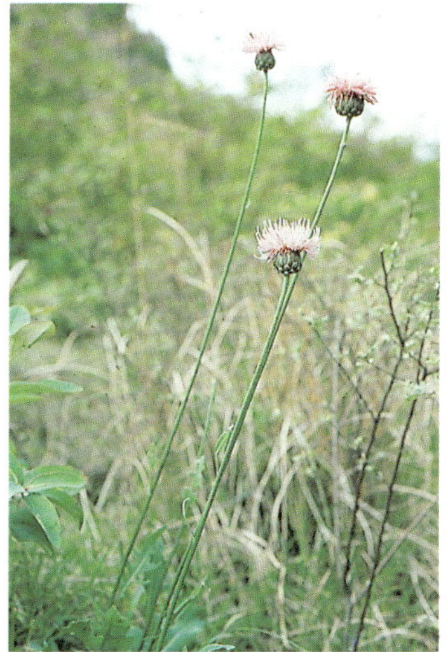

354. *Serratula pallida* Murree hills, Y. J. Nasir

358. *Saussurea heteromalla* Taxila, Y. J. Nasir

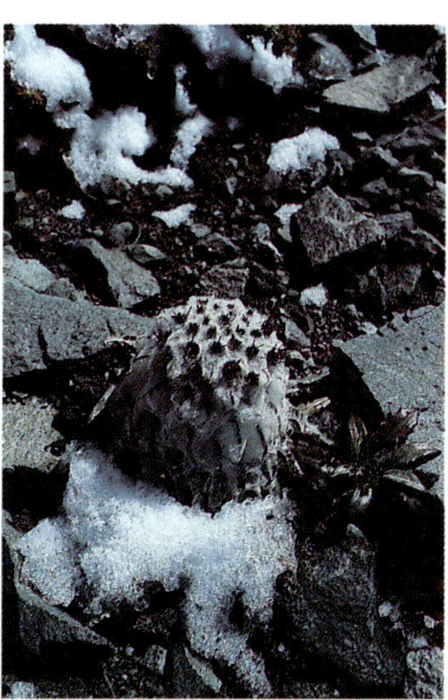

356. *Saussurea simpsoniana* NW Himalaya, Chris Chadwell

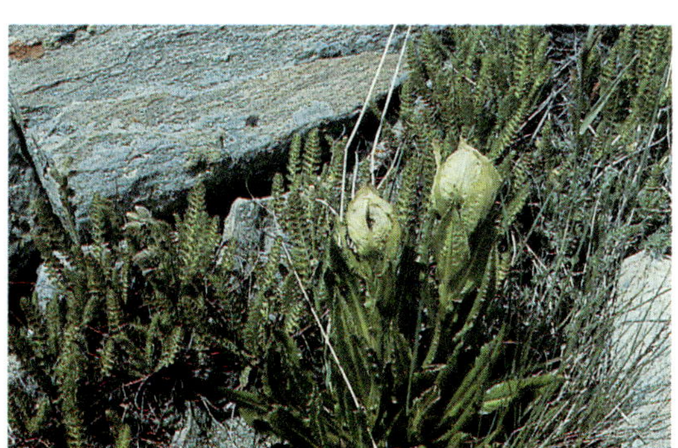

357. *Saussurea obvallata* Babusar, T. J. Roberts

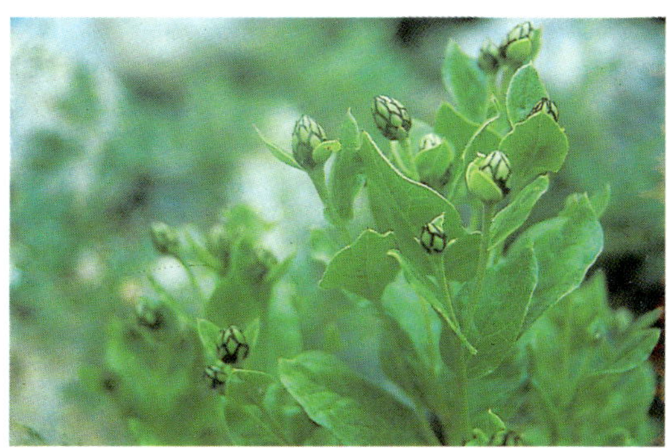

359. *Saussurea jacea* Deosai, Fritz Berger

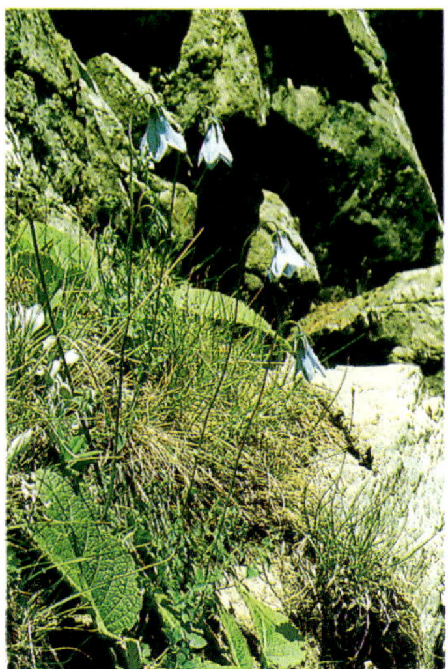

361. *Codonopsis obtusa* Kaghan valley, Y. J. Nasir

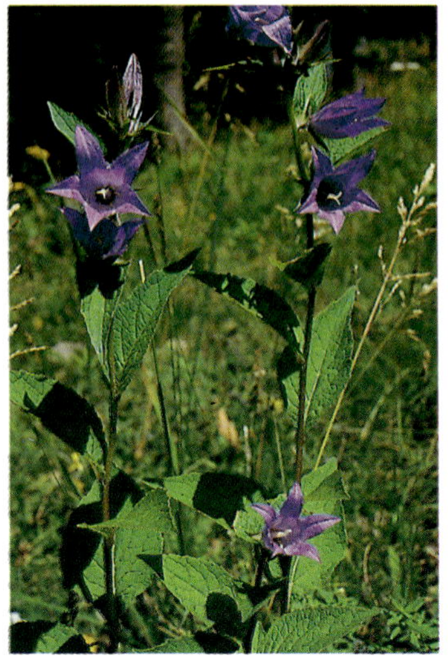

363. *Campanula latifolia* Shogran, Kaghan, Y. J. Nasir

360. *Codonopsis rotundifolia* Kaghan, Rubina A. Rafiq

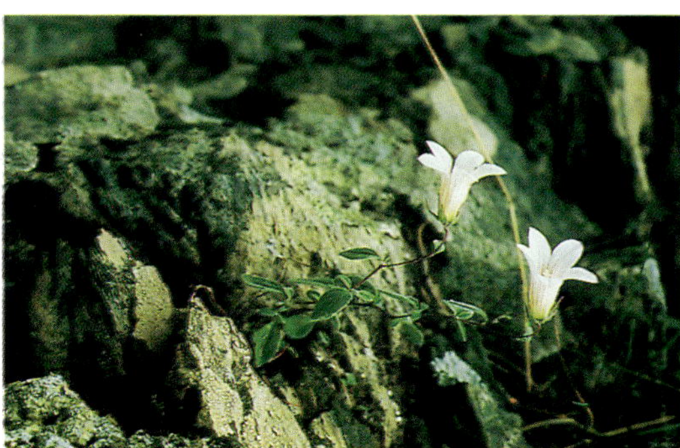

365. *Campanula tenuissima* Kalam, Y. J. Nasir

364. *Campanula pallida* Murree hills, S. A. Sultan

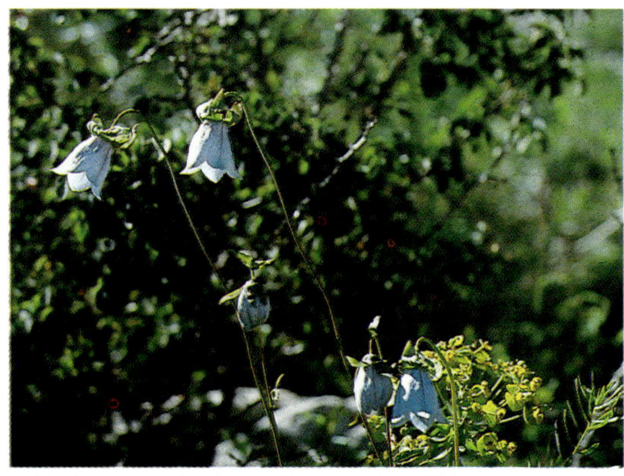

362. *Codonopsis clematidea* Shandur, Chitral, Maj Gen M. Shuaib Quraishi

368. *Acantholimon longiflorum* Chiltan hills, Rubina A. Rafiq

367. *Plumbago zeylanica* Islamabad, S. A. Sultan

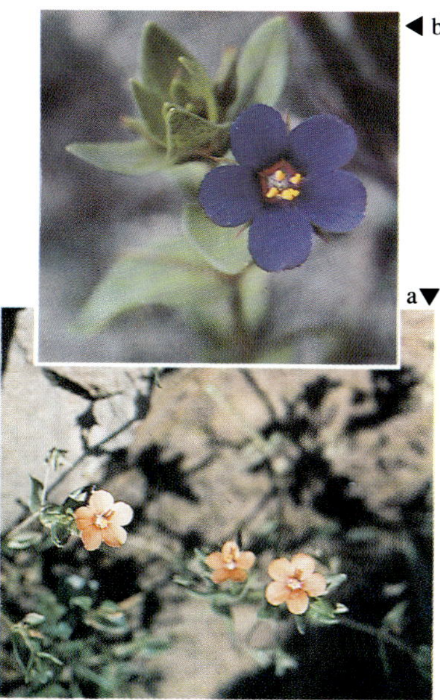

371. *Anagallis arvensis* a, b: Islamabad, S. A. Sultan

366. *Rhododendron arboreum* a, b: Siran valley, Hazara, Y. J. Nasir

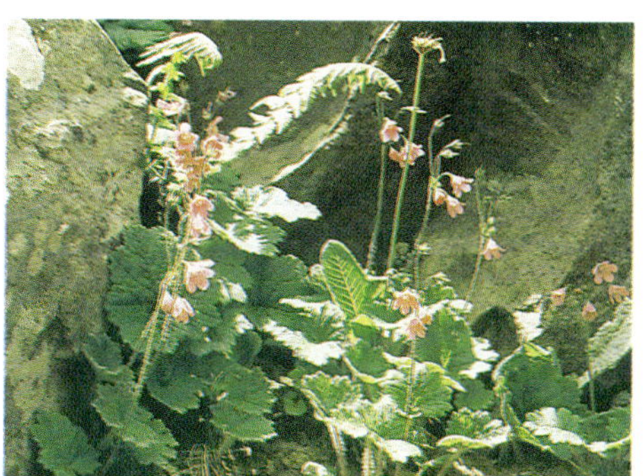

370. *Cortusa brotheri* Upper Kaghan, Y. J. Nasir

369. *Acantholimon lycopodioides* Mohodand, Swat, Fritz Berger

376. *Androsace hazarica* Thandiani, Y. J. Nasir

374. *Androsace foliosa* Changla Gali, Fritz Berger

375. *Androsace himalaica* Astor, Tim Hurrell

377. *Primula denticulata* a, b: Thandiani, S. A. Sultan

372. *Androsace mucronifolia* Deosai, Fritz Berger

373. *Androsace rotundifolia* Mokshpuri, Y. J. Nasir

382. *Olea glandulifera* Margalla hills, Y. J. Nasir

◀b

a▼

379. *Primula rosea* Kaghan, Y. J. Nasir

383. *Olea ferruginea* Margalla hills, Y. J. Nasir

381. *Myrsine africana* a, b: Margalla hills, Y. J. Nasir

378. *Primula macrophylla* Deosai, Fritz Berger

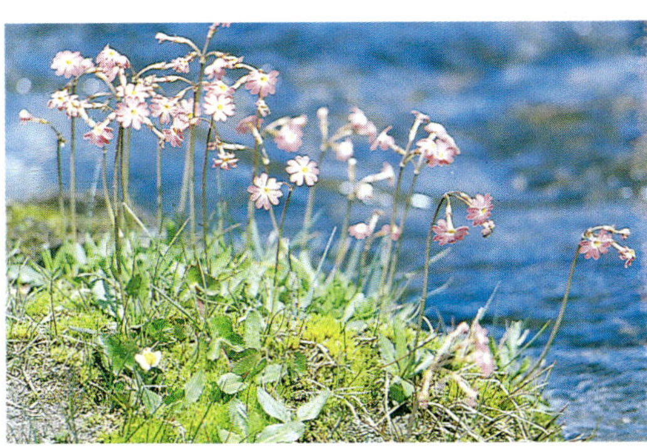

380. *Primula elliptica* Swat, Fritz Berger

384. *Jasminum humile* Abbottabad, Y. J. Nasir

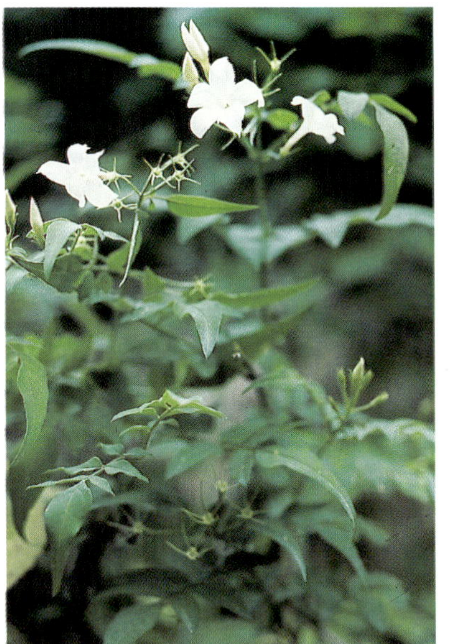

385. *Jasminum officinale* Thandiani, Y. J. Nasir

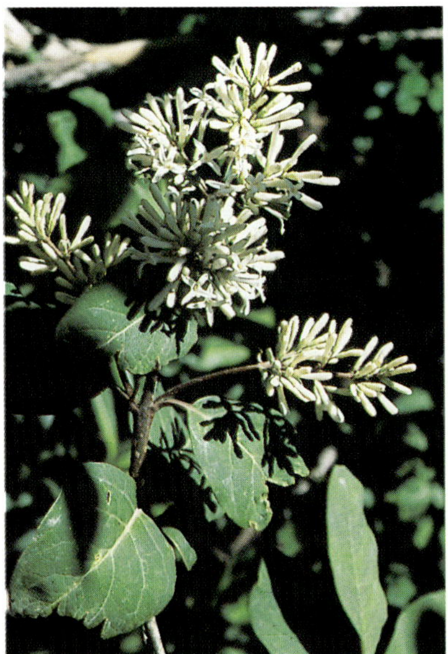

386. *Syringa emodi* Mokshpuri, Y. J. Nasir

389. *Vinca major* Islamabad, Y. J. Nasir

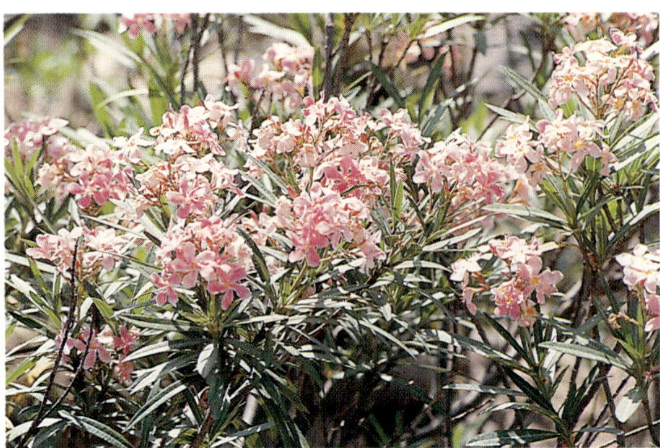

387. *Nerium oleander* Margalla hills, S. A. Sultan

388. *Carissa opaca* Margalla hills, S. A. Sultan

392. *Glossonema varians* Karachi, Y. J. Nasir

394. *Leptadenia pyrotechnica* Turbat, Makran, Y. J. Nasir

393. *Oxystelma esculentum* Thatta, Khan Muhammad Khan

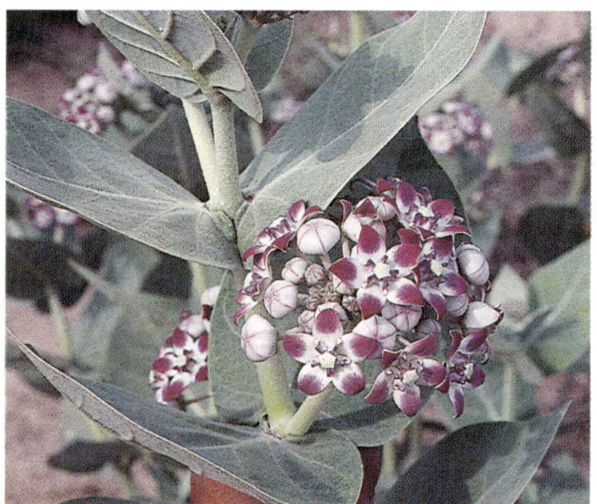

391. *Calotropis procera subsp. hamiltonii* Karachi, Y. J. Nasir

390. *Rhazya stricta* a, b: Thal Desert, Y. J. Nasir

399. *Swertia ciliata* Thandiani, Y. J. Nasir

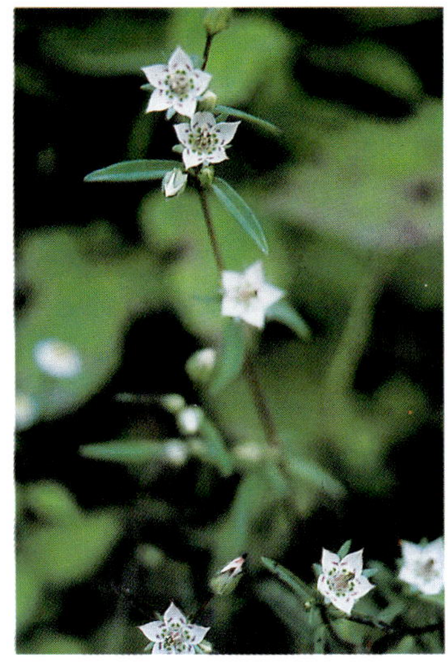

400. *Swertia paniculata* Murree hills, Y. J. Nasir

396. *Buddleja crispa* Quetta, Rubina A. Rafiq

398. *Swertia cordata* Thandiani, Y. J. Nasir

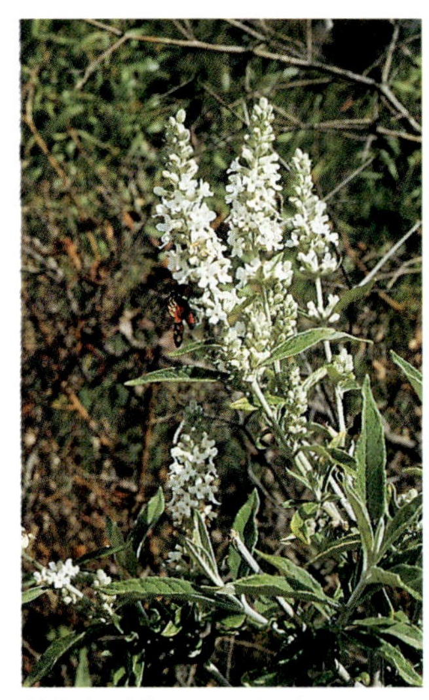

395. *Buddleja asiatica* Lower Murree hills, Y. J. Nasir

397. *Swertia speciosa* Kaghan, Y. J. Nasir

1. + Twiners. Flowers green with purplish veins. Leaves broadly ovate-cordate. 360. **C. rotundifolia**

– Erect to decumbent plants. Flowers pale blue. 2

2. + Calyx lobes lanceolate. Plants generally 1m or less. Corolla lobes not recurved. 361. **C. obtusa**

– Calyx lobes ovate-lanceolate. Plants 1m tall or more. Corolla lobes recurved. 362. **C. clematidea**

360. **C. rotundifolia** Benth. Pl. 58

Glabrous twiners with green bell-shaped nodding flowers reticulated with deep purple veins. Leaves 3-7 x 4-5 cm, toothed. Corolla 2-3 x 1.5-2 cm.

Distribution: Pakistan, Kashmir to central Nepal.

In forests from 1800-3200 m. Common in Chitral, Swat and Hazara district. July-August.

361. **C. obtusa** (Chipp) Nannf. Pl. 58
Syn.: *C. ovate* Bth. var. *obtusa* Chipp

Perennial with pretty pale blue bell shaped flowers. Leaves ovate to lanceolate. Calyx lobes lanceolate, bases not overlapping. Corolla 8-10 mm broad, lobes recurved. Stigma 3-lobed.

Distribution: Turkestan, Afghanistan, Pakistan.

Forest shrubberies from 2000-3100 m in Dir, Chitral and Kaghan. July-August.

362. **C. clematidea** (Schrenk) C. B. Clarke Pl. 58

Like *C. obtusa* but generally taller and with recurved corolla lobes which are less acute, grey-blue.

Distribution: Afghanistan, Pakistan to Kashmir, Himachal Pradesh (India).

Alpine meadows, irrigated places from 2400-3600 m in northern hilly district of the country. July-August.

2. Campanula 14 spp. *Bell Flower*

Like *Codonopsis* but differing in the non beaked capsules which dehisce by irregular pores at the base or sides.

1. + Flowers 4-5 cm long, purple-blue.
Plants robust, 1 m or more tall. 363. **C. latifolia**

– Flowers 0.5-1 cm long, white to mauve.
Plants slender, up to 50 cm tall. 2

2. + Annual. Stems erect, 3-4 mm thick. Corolla
8-10 mm long, mauve or mauvish-white. 364. **C. pallida**

– Perennial. Stems thin up to 1 mm thick.
Corolla c. 5 mm long, white. 365. **C. tenuissima**

363. **C. latifolia** L. Pl. 58

An attractive plant with leafy stems and pretty purple-blue flowers in short racemes. Leaves 8-12 x 5-7 cm, ovate-lanceolate, toothed, upper smaller and sessile.

Distribution: Europe, Caucasus, Turkey, Iran, Pakistan, Kashmir to central Nepal.

Forest banks, slopes from 2200-3000 m in Murree hills, Kaghan and Swat Kohistan. July-August.

364. **C. pallida** Wall. Pl. 58
Syn.: *C. colorata* Wall.

Erect annual up to 50 cm. Leaves lanceolate to narrow ovate, 3-6 x 0.5-2 cm, toothed. Calyx lobes lanceolate.

Distribution: Afghanistan, Pakistan, Kashmir to south west China.

Forests and plains in cultivated or rocky places from 600-3000 m. March-July.

365. **C. tenuissima** Dunn Pl. 58

A slender perennial with well developed rootstock and many wiry stems. Corolla white, c. 5 mm long. Capsule narrowed towards base.

Distribution: Pakistan.

Usually on rocky limestone cliffs from 800-3000 m in Murree hills, Swat and Hazara. July-August.

ERICACEAE

Rhododendron Family

Trees, shrubs or perennial herbs with simple exstipulate leaves. Flowers axillary solitary or in terminal or axillary clusters, racemes or panicles, 4-7-merous. Stamens twice the number of corolla lobes. Fruit a capsule.

Rhododendron 6 spp.

Evergreen or deciduous trees or shrubs with usually leathery leaves. Flowers in terminal clusters, showy, 5-merous. Capsule dehiscing by 5-10 valves.

366. **R. arboreum** Sm. *Ardawal* Pl. 59

Evergreen tree up to 9 m tall. Leaves 10-20 cm long, elliptic-lanceolate, leathery, dark green above, silvery on undersurface. Flowers in large dense terminal clusters, 5-6 cm long, scarlet-red or red.

Distribution: Pakistan, Kashmir to south east Tibet, Sri Lanka.

Rare in Pakistan occurring mainly in Mansehra district (N.W.F.P.) and occasionally in Murree hills. Usually found with oak (*Quercus leucotrichophora*) or pine (*Pinus roxburghii*) at lower elevations (up to 1600 m) and at higher elevation (up to 2500 m) with blue pine (*P. wallichiana*) and *Q. dilatata*. A very attractive tree when in blossom. March-April.

PLUMBAGINACEAE *Plumbago Family*

Generally shrubs with simple whorled leaves, often in basal rosettes. Flowers regular, 5-merous, in panicles or spikes. Calyx persistent, forming a dispersal unit with the fruit. Stamens 5. Fruit 1-seeded.

+ Calyx herbaceous, glandular. Styles united throughout their length. **1. Plumbago**

– Calyx membranous, eglandular. Styles free or united by their bases only. **2. Acantholimon**

1. Plumbago 2 spp.

Calyx tubular, prominently beset with stipitate glands. Corolla limb rotate, 5-lobed, tube exceeding calyx length.

367. **P. zeylanica** L. Pl. 59

A branched shrub with more or less erect weak habit. Leaves pale yellow-green, ovate, 4-7 x 1-3 cm. Flowers pale blue, disposed on branched spikes 10-30 cm long. Corolla tube c. twice length of calyx.

Distribution: Tropical Africa and Asia.

Wild in the sub-Himalayan tract up to 1500 m, also cultivated in the plains. July-September.

2. Acantholimon 8 spp.

Cushion-like perennial shrubs with narrow leaves. Calyx 5-nerved. Petals united at the base. Filament bases dilated and adnate to the petals.

+ Leaves of two types; spring leaves fleshy, summer leaves subulate, persistent. Scape branched, exceeding the foliage. Calyx tubular. 368. **A. longiflorum**

– Leaves flat or linear-angled, persistent. Scape simple, not exceeding foliage. Calyx broadly funnel-shaped. 369. **A. lycopodioides**

368. A. longiflorum Boiss Pl. 59

Early leaves glaucous-green, linear, deciduous. Scape slender. Flowers paniculate. Calyx 5-lobed.

Distribution: Afghanistan, Pakistan (Balochistan.)

Common on dry exposed slopes from 1500-2400 m. Restricted to northern Balochistan. July-September.

369. **A. lycopodioides** (Girard) Boiss. Pl. 59

A more compact plant than the above species, with homomorphic leaves and simple scape not exceeding the foliage leaves. Calyx broadly funnel-shaped.

Distribution: Afghanistan, Pakistan to Kashmir, central Asia.

More widespread than the previous species, forming large cushions in association with *Astragalus* and *Artemisia* especially in juniper tracts. Dry open slopes from 2500-3600 m in northern Balochistan, Chitral, Kaghan, Gilgit, Baltistan and Kashmir. July-September.

PRIMULACEAE *Primrose Family*

Perennial or annual herbs, usually scapigerous. Leaves simple, in basal rosettes or cauline. Flowers regular, 5-merous, sometimes quite showy. Corolla bell-shaped or funnel like or rotate. Stamens 5, epipetalous. Fruit capsular, dehiscing by 5 valves or a lid.

1. + Flowers nodding. Anthers apiculate. **1. Cortusa**

   – Flowers usually not nodding. Anthers not apiculate. 2

2. + Plants without a scape; stems leafy. Capsule dehiscence operculate. **2. Anagallis**

	–	Plants usually with a scape.Leaves radical or in basal rosettes. Capsule dehiscence valvular.	3
3.	+	Plants often cushion-like or of a loose caespitose habit. Corolla tube short, throat constricted. Capsule included in calyx.	**3. Androsace**
	–	Plants rarely cushion-like. Corolla tube long. Capsule usually exserted.	**4. Primula**

1. Cortusa 2 spp.

Scapose perennials with radical, long petioled leaves. Flowers showy, umbellate, nodding. Corolla bell-shaped with a short tube; capsule dehiscing by valves.

370. **C. brotheri** Pax ex Lipsky Pl. 59

Leaves orbicular-reniform, 24-50 x 30-60 mm, margin crenate-dentate. Petiole up to 15 cm long, slender. Scapes 3-10, up to 23 cm long. The pretty flowers are pink-red.

Distribution: Central Asia, Afghanistan, Pakistan, Kashmir to Uttar Pradesh (India).

A shade loving alpine species, growing in damp places and rock crevices near melting snow from 3000-4000 m in Kaghan and Galis. May-July.

2. Anagallis 1 sp.

Annual or perennial diffuse herbs. Corolla rotate, lobes minutely pubescent at the apex. Capsule dehiscing by a lid.

371. **A. arvensis** L. *Dhabbar, Pimpernel* Pl. 59

Prostrate annual with square stems. Leaves opposite, ovate-oblong, gland dotted. Flowers solitary axillary, red or blue, closing in shade. Filaments reddish-pink, dilated, glandular. Capsule globose, 4-5 mm broad.

Distribution: North west Africa, Europe to west Asia, Australia, north America, eastern south America.

The *blue pimpernel* is very common as a weed in abandoned fields, roadside ditches, meadows from plains to 2400 m. The scarlet flowered variety is confined to the hills. Flowering mostly throughout the year.

3. Androsace 24 spp.

Corolla urceolate, with a short tube; throat constricted, limb rotate. Stamens and style included in corolla tube. Capsule ovoid to globose.

1. + Plants cushion-like. Leaves small, closely appressed to the stem. — 372. **A. mucronifolia**
   – Plants of an erect to suberect habit. Leaves well developed. — 2

2. + Petiole c. 2 times lamina length. Lamina reniform-cordate. — 373. **A. rotundifolia**
   – Petiole not exceeding lamina length. Lamina linear-lanceolate to elliptic. — 3

3. + Leave pubescent to villous. Margin not strongly setoseciliate. — 374. **A. foliosa**
   – Leaves glabrous or sparse pubescent. Margin setoseciliate. — 4

4. + Leaves elliptic. Scapes 3-8 flowered. — 375. **A. himalaica**
   – Leaves linear-lanceolate. Scapes 10-22-flowered. — 376. **A. hazarica**

372. **A. mucronifolia** Watt — Pl. 60

Perennial alpine plant, forming cushion-like patches 10-20 cm broad. Leaves in dense rosette, 3-8 x 1.5-3 mm long, glabrous, pale green. Scape c. 12 mm long, 2-6-flowered. Flowers pale pink with a red eye.

Distribution: Afghanistan, Pakistan, Kashmir, west Tibet.

Alpine slopes, screes etc. from 3300-4700 m in Gilgit, Chitral, Swat Kohistan and Baltistan. June-August.

373. **A. rotundifolia** Hardwicke — Pl. 60

Lamina 12.4-20 x 20-42 mm, crenate-dentate. Petiole 30-60 mm long, slender. Scapes 2-20 in number, 6-32-flowered. Flowers white, fading to pink. Bracts quite conspicuous. Capsule glabrous, c. 3 mm broad.

Distribution: Afghanistan, Pakistan, Kashmir to west Nepal.

Variable species as regards the indumentum and degree of incision of calyx lobes and bracts. More tolerant of dryer areas. Common on forest slopes and banks in Kurram, Chitral, Sakesar in Salt Range, Swat and Hazara from 800-3300 m. April-August.

374. **A. foliosa** Decne. ex Duby Pl. 60

A perennial herb with well developed stolons and leafy stems. Leaves 2-11 x 1.1-4 cm, elliptic-ovate. Scapes up to 22 cm long. The pretty flowers are pink. Capsule subglobose.

Distribution: Pakistan, Kashmir.

Forest banks etc. from 2300-3200 m in Chitral, Swat, Kaghan, Hazara and Murree hills. June-August.

375. **A. himalaica** (Knuth) Handel-Mazzetti Pl. 60

Leaves cartilaginous, in basal rosettes, 8-20 x 3-6 mm, elliptic, margin with stiff white hairs. Scapes to 12 cm long, 3-15-flowered. Flowers pink.

Distribution: Pakistan, Kashmir.

Generally drier areas from 1700-3000 m especially, Chitral, Swat Kohistan and Galis. May-June.

376. **A. hazarica** R. R. Stewart ex Y. Nasir Pl. 60

As the above species but a mesophytic plant with larger leaves, scapes and more flowers per scape.

Distribution: Pakistan, Kashmir.

Shady forest banks, slopes from 2400-2900 m. July-August.

4. Primula 23 spp. *Primrose*

Perennial herbs related to *Androsace,* but generally of an erect or more loose habit, flowers which are heterostylous (long style or short style), the corolla tube which is elongated and exceeds calyx length, the non-constricted corolla throat and the capsule which is usually well exserted from the calyx.

1.	+	Flowers many, in dense globose heads.	377. **P. denticulata**
	–	Flowers up to 60 in number, in lax umbels.	2
2.	+	Capsule cylindrical. Leaves 12-29 x 1.2-4.5 cm, entire, white mealy on undersurface.	378. **P. macrophylla**

– Capsule globose to ovoid. Leaves 3-9 x 1-2.5 cm, crenulate to denticulate, undersurface glabrous. Flowers 2-10 in number. 3

3. + Flowers appearing when leaves are very young, pink or rose. Leaves crenulate-denticulate. 379. **P. rosea**

– Flowers appearing with the mature leaves, pink-violet, blue or purple. Leaves irregularly dentate. 380. **P. elliptica**

377. **P. denticulata** Sm. *Dog-tooth Primula* Pl. 60

A pretty herb flowering when the leaves are young. Scape rather stout. Young leaves sheathed at the base with prominent brownish-red bud scales. Flowers in globose heads, mauve to pinkish.

Distribution: Afghanistan, Pakistan, Kashmir to Assam.

Common in forest clearings, slopes from 1300-4300 m throughout Murree hills, Hazara and Azad Kashmir. Flowers appear when the snow melts. April-June.

378. **P. macrophylla** D. Don Pl. 61

A stout leafy perennial. Leaves quite large, white mealy on the undersurface. Scape up to 32 cm long, 8-60-flowered. Flowers purple, pink-purple, violet or lilac. Capsule 12-17 mm long, broadly cylindrical.

Distribution: Afghanistan, Pakistan, Kashmir to Bhutan.

Alpine meadows. Gregarious in moist place from 3300-4500 m at higher elevations in Hunza, Deosai plateau and higher passes of Chitral and Swat. May-August.

379. **P. rosea** Royle Pl. 61

Flowers pink to rose, rarely white, appearing when the leaves are young. Leaves elliptic, obtuse. Scape up to 8 cm, 4-10-flowered. Capsule subglobose.

Distribution: Afghanistan, Pakistan, Kashmir.

Alpine slopes, meadows from 2600-4300 m in moist situations especially Azad Kashmir, Hazara area. May-June.

380. **P. elliptica** Royle Pl. 61

A taller plant than *P. rosea* with larger coarsely toothed leaves and purple-blue flowers. Leaves and flowers appear fully developed.

Distribution: Pakistan, Kashmir to Garhwal (India).

In moist places or near melting snow from 3300-4900 m in Swat Kohistan, Hazara and Karakoram. July-August.

MYRSINACEAE

Myrsine Family

A tropical family of trees or shrubs with simple, alternate leaves. Flowers small, 4-5-merous, disposed in racemes or fascicles. Stamens 4-5, epipetalous. Fruit a drupe or berry.

Myrsine 2 spp.

Plants often dioecious. Leaves coriaceous. Flowers in short axillary fascicles. Corolla rotate. Fruit a drupe.

381. **M. africana** L. Pl. 61

A dioecious evergreen shrub up to 1.5 m tall. Leaves elliptic-lanceolate to obovate, serrulate. Flowers minute, greenish. Drupe globose, 3-4 mm broad, purple when ripe.

Distribution: Africa, Arabia, Afghanistan, Pakistan, Kashmir to Nepal, Tibet, China, Taiwan.

Sub-Himalayan tracts also Salt Range and northern Balochistan. Gregarious from 900-2500 m. March-May.

OLEACEAE

Olive Family

A family of trees or shrubs distributed mainly in tropical and temperate Asia. Leaves simple or compound, opposite. or alternate, exstipulate. Flowers bisexual, usually 4-merous. Fruit a samara, capsule, drupe or berry.

1.	+	Fruit a drupe. Corolla tube up to 3 mm long.	**1. Olea**
	–	Fruit a berry or capsule. Corolla tube 5-20 mm long.	2
2.	+	Fruit a berry. Corolla lobes 5-10.	**2. Jasminum**
	–	Fruit a capsule. Corolla lobes 4.	**3. Syringa**

1. **Olea** 2 spp.

Olive

Evergreen trees with opposite, simple and entire leaves. Flowers in panicles. Corolla lobes 4. Stamens 2, epipetalous.

+ Leaves 7-10 x 3-5 cm, oblanceolate, dark green and shiny above, undersurface usually not scaly. Trees 7-20 m tall. 382. **O. glandulifera**

– Leaves 1.5-5 x 1-2.5 cm, oblong to ovate-oblong or oblanceolate, green above, undersurface with reddish-brown scales. Trees up to 9 m tall. 383. **O. ferruginea**

382. **O. glandulifera** Wall. ex G. Don *Barkao* Pl. 61

A large tree, 7-20 m tall with dark green shiny oblanceolate leaves. Flowers creamish, in terminal or lateral cymes. Drupe 10-14 x 5-7 mm, ovoid.

Distribution: Pakistan, Kashmir to Nepal, India.

Usually in moist ravines from 900-1500 m especially Margalla hills and sub-Himalayan tracts. April-May.

383. **O. ferruginea** Royle *Kao, Wild Olive* Pl. 61

A smaller tree of drier habitats and lower elevations. Leaves smaller. Flowers white. Drupe 6-10 x 5 mm, ovoid, black.

Distribution: Afghanistan, Pakistan, Kashmir to Nepal.

Very common, forming an ecological climax with *Acacia modesta* (phulai) in warmer foothill tracts from 450-2000 m, in N.W.F.P. and northern Punjab. April-May.

2. **Jasminum** 5 spp. *Chambeli, Jasmine*

Erect to scrambling or twining shrubs with simple or imparipinnate leaves. Flowers often fragrant, solitary or in cymes, white or yellow. Fruit a berry.

+ Flowers yellow. Leaves alternate. 384. **J. humile**

– Flowers white. Leaves opposite. 385. **J. officinale**

384. **J. humile** L. *Yellow Jasmine* Pl. 62

Shrub up to 2 m tall. Leaflets 3-7, dark green above, elliptic-ovate or lanceolate. Flowers yellow, in terminal cymes. Corolla tube 10-20 mm long, lobes ovate. Berry 4-5 mm, black.

Distribution: Central Asia, Afghanistan, Pakistan, Kashmir to west China, India.

Temperate forests from 1600-2700 m. Common in Murree hills, often in assocation with *Quercus leucotrichophora.* A form of this, cultivar *revolutum* is grown in gardens. April-June.

385. **J. officinale** L. *White Jasmine* Pl. 62

Twining or scrambling shrub with scandent branches. Leaflets 3-7. Flowers fragrant, white, 1-10 in number. Corolla tube 10-20 mm long. Berries globose or ovoid, 8-10 mm, black.

Distribution: South Europe, central Asia, Afghanistan, Pakistan, Kashmir to Bhutan, Tibet, west China.

Temperate forests from 1600-2700 m. Found at higher elevations and on more open slopes than *J. humile.* A cultivated form of this, var. *grandiflorum* is the national flower of Pakistan. May-July.

3. Syringa 2 spp. *Lilac*

Deciduous shrubs with opposite leaves. Calyx 1-2 mm long. Corolla funnel or salver-shaped. Fruit capsular.

386. **S. emodi** Wall. ex Royle Pl. 62

Scandent shrub up to 4 m tall. Leaves 4-9 x 3-5 cm, elliptic-orbicular, dark green and glabrous above. Flowers fragrant, white in terminal panicles. Capsule 12-15 x 4 mm, cylindrical, dehiscing by 2 valves.

Distribution: Afghanistan, Pakistan, Kashmir to Nepal.

Temperate forests in Swat, Kaghan and Murree hills from 2000-3000 m. Occasionally cultivated. May-July.

APOCYNACEAE *Oleander Family*

A tropical family of trees or shrubs, often twining, with milky juice. Leaves opposite or whorled. Flowers bisexual, regular 4-merous. Fruit a drupe, berry, samara or follicle.

1.	+	Corolla with corona appendages present.	**1. Nerium**
	–	Corolla without corona appendages.	2
2.	+	Plants spiny.	**2. Carissa**
	–	Plants not spiny.	3

3. + Herbs with opposite leaves. Flowers axillary, solitary, blue, purple or white. **3. Vinca**

– Undershrubs with alternate leaves. Flowers in terminal or axillary cymes, white. **4. Rhazya**

1. Nerium 1 sp.

Glabrous shrubs with whorled leaves. Flowers showy, in terminal cymes. Corolla funnel form; lobes spreading, overlapping. Carpels 2, free. Follicles paired.

387. **N. oleander** L. *Kanair, Oleander* Pl. 62
Syn.: *N. indicum* Mill.

Evergreen shrubs with a milky juice. Leaves lanceolate, 80-140 x 10-20 mm, coriaceous, midrib prominent on undersurface. Flowers pink to almost white. Follicles cylindrical, 12-20 cm long.

Distribution: Mediterranean, Iran, Afghanistan, Pakistan, Kashmir to Nepal, India, China, Japan.

Widespread, usually gregarious in stony stream-beds in lower hills up to 1600 m throughout the country. Occasionally cultivated. Leaves very poisonous. April-October.

2. Carissa 1 sp.

Branched spiny shrubs. Leaves opposite. Corolla tube cylindrical. Fruit a berry.

388. **C. opaca** Stapf ex Haines *Garanda* Pl. 62

Shrub up to 3 m or more tall. Spines usually straight, 20-30 cm long. Young shoots with milky juice. The sweet scented flowers in terminal cymes, white, limb of corolla c. 8-10 mm broad. Berry black, 5-8 mm long, elliptic-oblong, purple, with a white juice within.

Distribution: Pakistan, Kashmir, India, Burma, Sri Lanka.

Often the dominant undershrub in Murree foothills and Salt Range up to 1400 m. Fruit edible. May.

3. Vinca 1 sp. *Periwinkle*

Perennial, erect to prostrate herbs. Shoots with a milky juice. Leaves opposite. Flowers solitary, blue, purple or white. Stamens 5, included in corolla. Follicles paired.

389. **V. major** L. *Creeping Periwinkle* Pl. 62

A much branched prostrate suffruticose herb, often rooting at the nodes. Leaves opposite, ovate. Flowers 30-40 mm broad. Corolla lobes obovate.

Distribution: West Mediterranean, Turkey. Naturalized elsewhere.

Sometimes cultivated and has become naturalized in Parachinar, Abbottabad and Murree hills. Found from 600-2300 m. December-March.

4. Rhazya 1 sp.

Glabrous undershrubs with alternate and thickish slender leaves. Flowers small, in terminal or axillary cymes. Corolla tube cylindrical. Stamens included in corolla. Follicles paired.

390. **R. stricta** Decne. *Vena* Pl. 63

An evergreen shrub up to 1 m tall. Leaves narrow, yellow when dry. Flowers 3-4 mm broad, white, fragrant. The conspicuous follicles 40-70 x 4-5 mm, slightly recurved, yellow-brown to orange.

Distribution: Arabia, Afghanistan, Pakistan, India.

Stony plains and slopes. Usually gregarious up to 1800 m. A characteristic shrub of the drier foothill regions, especially Waziristan, Balochistan and Salt Range. December-March.

ASCLEPIADACEAE

Milkweed Family

A family of tropical and sub-tropical twining herbs or shrubs, often with a milky juice. Leaves simple, opposite or radical. Flowers bisexual, regular, 5-merous, disposed in compound cymes. Corolla rotate, bell-shaped or tubular. Carpels 2. Fruit of 2 paired follicles.

1.	+	Plants of a climbing or twining habit.	**3. Oxystelma**
	–	Plants erect.	2
2.	+	Shrubs 1.5-2.5 m tall. Follicles smooth.	3
	–	Perennial herbs up to 40 cm tall. Follicles spinose.	**2. Glossonema**
3.	+	Stems leafy. Leaves large, thick, ovate to broadly ovate-oblong. Follicles large and turgid.	**1. Calotropis**
	–	Stems not leafy. Leaves small and linear, sometimes absent. Follicles narrow, not turgid.	**4. Leptadenia**

1. **Calotropis** 1 sp.

Shrubs sometimes assuming tree-like proportions. Leaves with a waxy bloom, opposite, thick. Flowers in terminal or axillary cymes. Follicles prominent, turgid. Seeds with long hairs.

391. **C. procera** (Willd.) R. Br. *Aak, Sodom's Apple* Pl. 63
subsp. **hamiltonii** (Wight) Ali

Erect shrub, with branches ascending from the base. Plant parts with milky juice. Leaves 4-12 x 2-11 cm, thick with a cottony or waxy bloom. Flowers 2-2.5 cm broad, white, tips purple suffused, in terminal showy clusters. Follicles 6-10 x 3-5 mm, green when unripe, tip recurved. Seeds with fluffy hairs for dispersal.

Distribution: Afghanistan, Pakistan, India.

Very common in plains and hills up to 1600 m especially in desert areas. Medicinal. The juice is an irritant and can cause blindness. Also used as a purgative and for treatment against guinea worms. The copious seed hairs are used for stuffing pillow mattresses. Flowering throughout the year.

2. **Glossonema** 1 sp.

Herbs, branched from the base with opposite leaves. Flowers in lateral cymes. Corolla 5-lobed, tube short. Follicles minutely spinose.

392. **G. varians** (Stocks) Hook. f. *Munga* Pl. 63

Plants densely pubescent. Leaves fleshy, upto 7 cm long, lower ovate-rotund. Flowers 2-8 in number, white or pale yellow. Follicles 2-5 cm long, ellipsoid, pubescent and shortly spinose.

Distribution: Iran, Pakistan, Rajputana (India).

A desert plant in sandy or stony places mainly lower Sindh and southern Balochistan. December-January.

3. **Oxystelma** 1 sp.

Twining herbs with narrow leaves and showy flowers, in solitary cymes at the nodes. Corolla more or less campanulate. Follicles quite thick. Seeds with long hairs.

393. **O. esculentum** (L.f.) R. Brown Pl. 63

Stem branched, slender, with a milky juice. Leaves 3-9 cm long, linear-lanceolate. Flowers pale rose to pinkish-white, drooping. Pedicels 12-22 mm long. Follicles up to 7 cm long, glabrous.

Distribution: Iraq, Pakistan, India, Burma, Indonesia, Sri Lanka, south China.

Generally in dry places from plains to 1000 m in all four provinces. Plant medicinal. Fruit is edible. December.

4. Leptadenia 1 sp. *Khip*

Shrubs with dark green, apparently leafless ascending stems. Flowers small, in lateral cymes. Corolla deeply divided, rotate.

394. **L. pyrotechnica** (Forssk.) Decne. Pl. 63
Syn.: *L. spartium* Wight.

Leaves 2-5 mm long, linear to linear-lanceolate. Flowers yellowish-red, in dense lateral sessile clusters on the branches. Follicles narrow, up to 11 cm long. Seeds with long hairs.

Distribution: Africa, Arabia, Pakistan, India.

A desert species of medicinal importance. Common in Thal and Cholistan deserts. December-February.

BUDDLEJACEAE *Buddleia Family*

Trees or shrubs with simple, opposite leaves. Flowers bisexual, regular 4-5-merous, in terminal or axillary cymes. Carpels 2, united. Fruit a capsule or berry.

Buddleja 2 spp.

Tomentose shrubs with flowers disposed in cymes on branched spikes. Corolla tube cylindrical, lobes 4, spreading. Capsule dehiscing by 2 valves.

+	Flowers white, in continuous spikes.	395. **B. asiatica**
–	Flowers lilac to pinkish-blue, in interrupted spikes.	396. **B. crispa**

395. **B. asiatica** Lour. *Banna* Pl. 64

A bushy shrub up to 2 m tall with white tomentose parts. Leaves lanceolate, 5-9 cm long, white hoary on undersurface. Spikes 10-16 cm long. Flowers many, white, fragrant. Capsule 3-4 mm long, glabrous.

Distribution: Pakistan, Kashmir, Burma, Indo-China, Malaysia, China, Taiwan.

Sub-Himalayan tracts and adjacent plains up to 1600 m. Wood used for making walking sticks. February-April.

396. **B. crispa** Benth. Pl. 64

Shrubs with dense tomentose parts. Leaves oblong or ovate, with a wavy or dentate margin. Flowers mauve or pinkish-blue, in short interrupted spikes.

Distribution: Afghanistan, Pakistan, Kashmir to Bhutan, south Tibet, north west China.

Found from 800-2300 m in Balochistan, N.W.F.P. and Chitral. Wood used as a fuel. April-May.

GENTIANACEAE *Gentian Family*

Annual or perennial glabrous herbs with opposite, entire leaves. Flowers regular, bisexual, in terminal compound cymes. Corolla 3-5-lobed. Stamens as many as and alternating with corolla lobes. Fruit a 2-valved capsule.

1. + Corolla tube shorter than the lobes. **1. Swertia**

– Corolla tube longer than the lobes. 2

2. + Flowers in sessile axillary clusters. Stigma capitate. **2. Enicostemma**

– Flowers in terminal cymes. Stigmas 2-lobed. 3

3. + Styles slender, 2-fid. **3. Centaurium**

– Styles short and stout, not 2-fid. **4. Gentianodes**

1. Swertia 11-12 spp.

Erect herbs with opposite leaves. Flowers in showy paniculate or corymbose cymes, 5-merous. Stamens as many as the corolla lobes. Style very short, stigma 2-lobed.

1. + Plants robust, 60-120 cm tall. Leaves elliptic-lanceolate to elliptic-oblong, large, 3-12 x 3-7 cm, amplexicaul, basal. 397. **S. speciosa**

– Plants slender, up to 50 cm tall. Leaves smaller, 1-3.5 x 0.5-2 cm. 2

2. + Leaves cordate. Petals with purplish specks all over. 398. **S. cordata**

	–	Leaves linear-lanceolate to lanceolate. Petals with 2 purple blotches or ring at base.	3
3.	+	Petals reflexed, with a purplish ring.	399. **S. ciliata**
	–	Petals not reflexed, with 2 blotches at base of each petal.	400. **S. paniculata**

397. **S. speciosa** D. Don — Pl. 64
Syn.: *S. perfoliata* Royle ex G. Don

A robust perennial with large basal leaves and stem leaves that are clasping. Creamy yellow flowers c. 20-25 mm broad, disposed in spike-like clusters.

Distribution: Pakistan, Kashmir to Bhutan.

In moist open places and river banks etc. in Kaghan valley from 2500-4000 m. July-August.

398. **S. cordata** Wall. — Pl. 64

Leaves broadly ovate to cordate, sessile. Flowers about 8 mm broad, petals white, with purple specks with a greenish-yellow spot at the base, broadly ovate.

Distribute: Pakistan, Kashmir to Bhutan, Khasia, Burma, South Tibet.

Widespread in forests from 2000-3100 m from Kurram, Chitral, Gilgit, Baltistan and Murree hills. July-August.

399. **S. ciliata** (D. Don ex G. Don) B. L. Burtt — Pl. 64
Syn.: *S. purpurascens* Wall.

Leaves linear-lanceolate. Petals white, reflexed, with a purplish ring at the base.

Distribution: Afghanistan, Pakistan, Kashmir to Sikkim.

In forests from 1800-3000 m. August-September.

400. **S. paniculata** Wall. ex C. B. Clarke — Pl. 64

Like *S. ciliata,* but petals not reflexed and with 2 purple blotches at base of each petal. Leaves linear.

Distribution: Pakistan, Kashmir to Bhutan.

In forests from 1800-3000 m in Swat, Kaghan, Hazara and Murree hills. July-August.

2. Enicostemma 1 sp.

Perennial leafy herbs with opposite narrow leaves. Corolla funnel-form, 5-lobed. Stamens 5. Fruit a capsule dehiscing by 2-valves.

401. **E. verticellatum** (L.) Engler Pl. 65
Syn.: *E. littorale* Blume

Procumbent branched herb. Leaves 2-6 x 3-7 mm, linear-lanceolate. Flowers white, 5-7 mm long, in axillary clusters. Capsule c. 4 mm long, ellipsoid.

Distribution: Tropical America, Africa, Pakistan, India, Sri Lanka, Malaysia.

Seashores, rocky or sandy places in southern Balochistan, lower Sindh, Salt Range, and south Punjab. February-March.

3. Centaurium 1 sp.

Erect annual or biennial herbs with opposite leaves. Calyx 5-partite. Flowers 4-5-merous, cymose. Style bifid, stigmas 2. Capsule oblong.

402. **C. centaurioides** (Roxb.) R. Rao & Hemadri Pl. 65
Syn.: *Erythraea roxburghii* D. Don

Erect slender herb up to 25 cm tall with basal rosette of leaves. Leaves oblong or obovate, 15-24 x 4-6 mm. Flowers pink, in dichotomous cymes, lobes spreading, elliptic. Capsule c. 8 mm long, oblong.

Distribution: Pakistan, India, Nepal.

Damp places, cultivated areas from plains to 1600 m especially Rawalpindi area. March-April.

4. Gentianodes 20 spp.

A genus closely related to and often included in *Gentiana,* but differing in the anthers which are free from each other, basal nerved leaves, corolla provided with subsidiary small lobes that alternate with the corolla lobes.

1. + Perennial or biennial plants with rhizomes. 2

 – Annual lacking rhizomes. 3

2. + Radical leaves lanceolate. Flowers deep blue with white specks, subsidiary corolla lobes obtuse. 403. **G. kurroo**

– Radical leaves oblanceolate. Flowers bluish-purple, without white specks. Subsidiary corolla lobes acute. 405. **G. olivieri**

3. + Flowers pale blue, 6-8 mm long. Leaves with a silvery shine. 404. **G. argentea**

– Flowers deep blue, 10-12 mm long. Leaves not so. 406. **G. marginata**

403. **G. kurroo** (Royle) Omer, Ali & Qaiser *Nilkanth* Pl. 65
Syn.: *Gentiana kurroo* Royle

The largest flowered of our gentians, flowering after the monsoons. Flowers showy, 1-3 per stalk. Corolla blue, speckled with white. Radical leaves linear, upper ones shorter. Capsule c. 20 mm long.

Distribution: Pakistan, Kashmir to Uttar Pradesh (India).

The autumn gentian is locally common on grassy slopes. 1700-2500 m especially Murree hills and Abbottabad district. September-November.

404. **G. argentea** (Royle ex D. Don) Omer, Ali & Qaiser Pl. 65
Syn.: *Gentiana argentea* (Royle ex D. Don) Griseb

Plants compact, low growing, 3-4 cm tall. Leaves narrow, recurved, with a silvery shine. Flowers pale blue, 6-8 mm long. Capsule ovoid.

Distribution: Pakistan, Kashmir to Nepal.

Very common especially in the Murree and Hazara hills, forming carpets on the ground and open slopes from 1500-3500 m. April-May.

405. **G. olivieri** (Griseb.) Omer, Ali & Qaiser Pl. 65
Syn.: *Gentiana olivieri* Griseb.

Perennial herb with basal rosette of oblanceolate leaves 12-16 cm long. Flowering stalk 1-3, up to 30 cm long, bearing 1-4 flowers. Corolla purple-blue, throat white streaked; subsidiary lobes acute.

Distribution: North Iraq, Iran, Afghanistan, Pakistan, central Asia.

The only gentian found in Balochistan with graceful flowers born on long stalks. Dry slopes, screes. 1800-3200 m. April-May.

406. **G. marginata** (D. Don) Omer, Ali & Qaiser Pl. 65
Syn.: *Gentiana marginata* D. Don

A compact annual. Flowers sessile, dark blue or purplish-blue, in clusters. Young leaves somewhat reflexed. Corolla tube without scales; lobes 5, broadly ovate, obtuse; subsidiary lobes 5, as long as the lobes, apex obtusely 2-3-lobed.

Distribution: Afghanistan, Pakistan to Uttar Pradesh (India).

Sub-alpine zone especially Hazara area from 2400-3600 m. June-July.

POLEMONIACEAE *Phlox Family*

Annual or perennial herbs. Leaves alternate, entire or pinnate. Flowers bisexual, regular, 5-merous, in terminal cymes. Carpels 3. Fruit a capsule.

Polemonium 1 sp.

Perennial herbs with alternate leaves. Flowers in terminal corymbose cymes. Stamens epipetalous.

407. **P. coeruleum** L. *Jacobs Ladder* Pl. 66

Erect leafy plants up to 40 cm tall. Leaves pinnate. Leaflets many, 10-32 x 4-8 mm. Flowers showy, blue to pale blue, funnel shaped, 15-20 mm broad. Capsule 4-5 mm broad, ellipsoid, dehiscing by 3 valves.

Distribution: Europe, north America, Pakistan to Nepal.

A striking plant with decorative leaves and tall spires of blue flowers. Forest slopes and shrubberies. Not very common, 2400-3200 m especially in Swat and Kaghan valley. June-July.

BORAGINACEAE *Forget-Me-Not Family*

Trees, shrubs or herbs, often variously hairy. Leaves alternate simple. Flowers in coiled cymes, clusters or sometimes solitary, regular, bisexual, 5-merous. Corolla 5-lobed, the throat often provided with hairs. Style arising from the ovary base or terminal on it. Fruit a drupe or of 2 or 4 nutlets.

1. + Trees or shrubs. Fruit a drupe. **1. Ehretia**
   – Herbs. Fruit of 2 or 4 nutlets. 2
2. + Style arising from the top of ovary. **2. Heliotropium**

	–	Style arising from ovary base.	3
3.	+	Anthers twisted, apex elongated.	**3. Trichodesma**
	–	Anthers not twisted, apex obtuse or slightly elongated.	4
4.	+	Corolla throat without any scales, but sometimes hairy to villous.	**4. Onosma** **5. Arnebia**
	–	Corolla throat with scales.	5
5.	+	Nutlets prickly.	6
	–	Nutlets tubercled, hairy or glabrous but without prickles.	8
6.	+	Nutlets globose or napiform. Fruiting calyx not reflexed.	7
	–	Nutlets pyramidal or oblong-truncate. Fruiting calyx reflexed.	**6. Hackelia**
7.	+	Stamens exserted.	**7. Solenanthus** **8. Lindelofia**
	–	Stamens included.	**9. Cynoglossum**
8	+	Corolla with throat scales fimbriate or densely pubescent.	**10. Nonea**
	–	Corolla with throat scales not as above.	9
9.	+	Perennial. Flowers long tubular.	**11. Pseudomertensia**
	–	Annual or perennial. Flowers shortly tubular.	**12. Myosotis** **13. Gastrocotyle**

1. Ehretia 3 spp.

Flowers in terminal or axillary cymes, 5-merous. Corolla tube short, lobes 5, spreading. Anthers exserted. Style terminal. Drupe 2 or 4-seeded.

408. **E. obtusifolia** Hochst ex DC. Pl. 66

Shrubs or small trees, up to 3 m. Leaves obovate to oblong-ovate, 25-70 x 20-40 mm, obtuse. Corolla white, 8-10 mm broad, lobes obtuse. Drupe subglobose, c. 5 mm broad.

Distribution: Africa, Afghanistan, Pakistan, India.

Dry foothill zone, coastal hills of Sindh and Balochistan, Salt Range, sub-Himalayan tract and N.W.F.P. up to 800 m. March-May.

2. Heliotropium 23 spp.

Annual or perennial hairy herbs, sometimes with a woody base. Flowers small, white, in terminal or axillary cymes. Corolla lobes erect or incurved, sometimes with dentitions or smaller lobes between. Fruit of 2-4 nutlets.

1.	+	Plants glabrous.	409. **H. curassavicum**
	–	Plants hairy.	2
2.	+	Stigmas 2, linear-recurved, exserted.	410. **H. ophioglossum**
	–	Stigma solitary, capitate or obscurely 2-fid, included.	3
3.	+	Flowers bracteate.	411. **H. strigosum**
	–	Flowers ebracteate.	4
4.	+	Corolla lobes acuminate.	412. **H. subulatum**
	–	Corolla lobes obtuse to ovate.	5
5.	+	Leaves elliptic-ovate, 15-60 x 10-40 mm; margin entire.	413. **H. europaeum**
	–	Leaves ovate-lanceolate, 20-55 x 5-16 mm; margin wavy.	414. **H. crispum**

409. **H. curassavicum** L. Pl. 66

Branched perennial with glabrous, oblanceolate leaves. Stems and branches hollow. Flowers in simple or bifurcate cymes, white; lobes c. 1 mm long, obtuse. Nutlets c. 2 mm long, glabrous.

Distribution: America, Europe, Pakistan, India, Australia.

Salt loving. Sea level to 300 m especially in lower Sindh, Makran coast, also in saline areas of Punjab. March-April.

410. **H. ophioglossum** Stocks ex Boiss. Pl. 66

Leaves lanceolate to elliptic-lanceolate, 10-30 x 6-12 mm. Flowers white. Corolla tube less than 3 mm. Stigma 2, linear-recurved, exserted.

Distribution: Africa, Afghanistan, Pakistan, India.

Stony or sandy places in lower Sindh and southern Balochistan. April-July.

411. **H. strigosum** Willd. Pl. 66

Stems procumbent to suberect, up to 30 cm tall. Leaves linear-lanceolate, 5-30 x 3-4 mm. Inflorescence up to 10 cm long, sometimes bifurcate. Flowers bracteate. Stigmas 2, linear-recurved, exserted. Nutlets 4, dark brown.

Distribution: Afghanistan, Pakistan, Nepal, Burma, Malaya.

Plains to 1200 m, including northern Balochistan, Waziristan, Salt Range and Attock district. July-September.

412. **H. subulatum** (DC.) Vatke Fig. between p. 102-103

A perennial up to 50 cm tall distinguished from the other species in the corolla lobes which are acuminate. Leaves lanceolate, 18-50 x 6-12 mm. Flowers yellowish-white, corolla lobes acuminate. Fruit globose-depressed. Nutlets brown.

Distribution: Africa, Pakistan, India.

Dry sandy places in the plains throughout the country. November-February.

413. **H. europaeum** L. *Hathi Sunda* Pl. 66
var. **lasiocarpum** (Fisch. & Mey.) Kazmi

Annual, 50 cm or more tall, branched. Leaves elliptic-ovate to ovate. Inflorescence terminal, up to 15 cm tall. Corolla lobes more or less orbicular to oblong. Nutlets 2 mm long, ovoid.

Distribution: Trans-Jordan, Syria, Iraq, Iran, Russia, Afghanistan, Pakistan, India.

Common from sea level to 2400 m in Gilgit. April-May.

414. **H. crispum** Desf. Fig. between p. 102-103

Distinguished from other species of *Heliotropium* by the wavy leaf margin. Corolla white, 3-3.5 mm long. Nutlets c. 2 mm long, with few stiff hairs.

Distribution: Senegal, Arabia, Iraq, Trans-Jordan, Afghanistan, Pakistan, India.

Variable in size of plant and pubescence. Waste or sandy places. January-April.

3. Trichodesma 4 spp.

Perennial herbs with hairy parts. Flowers showy, solitary or in terminal short racemes. Corolla with throat scales absent. Anthers twisted together and cone-like, tip acuminate.

415. **T. indica** (L.) R. Br. Pl. 67

An annual semiprostrate hairy herb. Leaves oblanceolate to lanceolate, 30-80 x 5-20 mm. Flowers mauve to pinkish-blue. Anthers hairy. Nutlets 4-5 mm long, smooth.

Distribution: Afghanistan, Pakistan, India, Philippines, Mauritius.

Common almost throughout the country in plains and lower hills up to 1400 m. Flowering sporadic, but usually March-August.

4. Onosma 8 spp.

Perennial or biennial herbs with hairy parts and stems often woody. Flowers nodding, yellow, white or blue. Corolla gradually broadening upwards, throat glabrous or hairy.

416. **O. hispida** Wall. ex G. Don Pl. 67

Perennial, branched hairy herb up to 70 cm tall. Basal leaves 8-35 x 0.5-1 cm, linear to oblanceolate. Flowers from orange-yellow to almost white, tubular, gradually broadening upwards, 20-30 mm long. Nutlets shiny.

Distribution: Afghanistan, Pakistan, Kashmir.

Generally edge of fields, cultivated areas from 1000-3400 m in Balochistan and N.W.F.P., Kaghan and Swat. May-June.

5. Arnebia 11 spp.

Flowers long styled or short styled, white, yellow or bluish. Corolla lobes spreading, throat glabrous. Style arising from the ovary base. Stigma 2-fid or simple. Nutlets 4.

\+ Plants densely hairy. Leaves lanceolate.
Corolla 10-15 mm long; lobes without spots. 417. **A. hispidissima**

– Plants not densely hairy. Leave broader. Corolla 20-15 mm long; lobes with 5 dark brown spots. 418. **A. griffithii**

417. **A. hispidissima** (Lehm.) A. DC. Pl. 67

Prostrate to suberect, branched herb with villous parts. Flowers yellow; lobes 5, spreading. Style simple, slender. Stigmas 2. Nutlets less than 2 mm, ovoid.

Distribution: Africa, Arabia, Pakistan.

Open gravelly or sandy places from plains to 1500 m in all four provinces. December-April.

418. **A. griffithii** Boiss. *Prophet Flower* Pl. 67

An erect herb up to 35 cm tall, with larger leaves and yellow flowers. The corolla lobes have a dark brown spot at the base.

Distribution: Afghanistan, Pakistan.

Widespread from Karachi to Gilgit. Locally abundant in Kala Chita hills. The pretty spotted flowers are fragrant. Plains up to 1500 m. March-May.

6. Hackelia 2 spp.

Perennials with broad alternate leaves. Calyx lobes reflexed in fruit. Corolla with throat scales. Stamens included in corolla tube.

\+ Leaves cordate-ovate. Flowers pale white, with bluish spot at base of each corolla lobe. 419. **H. macrophylla**

– Leaves elliptic-ovate to ovate. Flowers sky-blue. 420. **H. uncinata**

419. **H. macrophylla** (Brand) I. M. Johnston Pl. 67

A robust herb up to 90 cm tall with cordate leaves and whitish flowers. Leaves 9-20 x 4-14 cm.

Distribution: Pakistan, Kashmir.

Forest slopes. Found from 1850-2900 m in Chitral, Swat, Kaghan and Murree hills. June-July.

420. **H. uncinata** (Royle ex Benth.) Fischer Pl. 68

Similar but more slender, with narrower leaves and pale blue flowers. Nutlets ovoid, winged; wings with appendages 1-2 mm long.

Distribution: Pakistan, Kashmir to Nepal, Bhutan, north Assam, west China.

Forest shrubberies. 2400-3500 m. June-July.

7. Solenanthus 2 spp.

Flowers in dense terminal or axillary racemes. Corolla throat with scales. Stamens exserted from corolla. Nutlets more or less ovoid.

421. **S. circinnatus** Ledeb. Pl. 68

Perennial up to 1 m tall. Basal leaves truncate to cordate, upper ones somewhat amplexicaul. Flowers in loose panicles, pale pink to red, c. 5 mm long. Nutlets 6 mm long, ovoid.

Distribution: Iraq, Iran, Caucasus, Afghanistan, Pakistan, Ladak, central Asia.

Forests or shrubberies. Not very common. Found from 2500-3500 m in Chitral, Swat Kohistan, Kaghan and Astor area. April-May.

8. Lindelofia 3 spp.

Flowers with a long corolla tube, dark blue or purple-blue. Corolla lobes spreading; throat scales present. Style long and slender.

422. **L. longiflora** (Benth.) Baill. Pl. 68

Stem leafy. Upper cauline leaves lanceolate. Calyx lobes more or less obtuse. Corolla lobes ovate.

Distribution: Pakistan, Kashmir to Nepal.

A handsome erect stemmed plant not uncommon in mountainous areas from 2500-4000 m in Chitral, Swat, Gilgit, Deosai and Kaghan. July-August.

9. Cynoglossum c. 3-4 spp.

Flowers in elongated cymes. Corolla funnel-form or rotate. Throat with scales. Stamens included. Nutlets 4.

423. **C. glochidiatum** Wall. ex Benth. Pl. 68

Herb up to 70 cm tall. Stem and branches with hairs 2-3 mm long, some with a broad circular base. Basal leaves obovate, upper generally longer. Corolla blue to bluish-white. Style c. 2.5 mm long, slender.

Distribution: Afghanistan, Kashmir to Nepal, Bhutan.

Common in hilly places from 1500-3300 m in Balochistan, N.W.F.P. and Chitral east to Baltistan. May-August.

10. Nonea 4-5 spp.

Flowers regular, in racemes. Calyx enlarging in fruit. Corolla throat with scales. Stamens included. Style slender, stigma 2-lobed. Nutlets 4.

424. **N. edgeworthii** A. DC. Pl. 68

Branched annual up to 40 cm, branches arising from the base, pubescent hairy. Leaves lanceolate to oblanceolate, 30-90 x 5-15 mm. Flowers creamy white. Corolla tube c. 7 mm long. Nutlets 5 mm long, brown-black.

Distribution: Pakistan, India.

Common weed. Plains and lower hills up to 1800 m from Balochistan, N.W.F.P. and Chitral. March-April.

11. Pseudomertensia c. 11 spp.

Perennial with alternate leaves. Flowers in simple terminal racemes, blue or blue-purple. Corolla tube as long as or much longer than calyx. Stamens exserted or not.

1. + Anthers exserted from corolla tube. Filaments 2-5 mm long. — 425. **P. parvifolia**

   – Anthers more or less included in corolla tube. Filaments less than 2 mm long. — 2

2. + Corolla exceeding calyx length; lobes 2-3 mm long; throat scales present. — 426. **P. trollii**

   – Corolla about calyx size; lobes 1-1.5 mm long; throat scales absent. — 427. **P. moltkioides**

425. **P. parvifolia** (Decne.) Riedl Pl. 68

A rhizomatous herb spreading by suckers. Shoots leafy. Basal leaves 30-50 x 4-8 mm, lanceolate to oblanceolate. Flowers blue. Anthers exserted from corolla. Stigma capitate.

Distribution: Pakistan, Kashmir.

Forest clearings, slopes. Gregarious from 1500-2700 m in upper Kurram, Chitral and Galis. April-May.

426. **P. trollii** (Melchior) Stewart & Kazmi Pl. 69

Like *P. parviflora,* but with larger corollas which exceed calyx length. Corolla lobes oblong. Throat scales well developed.

Distribution: Pakistan, Kashmir.

Alpine and sub-alpine areas of Gilgit, Kaghan and Kashmir from 2400-3700 m. May-June.

427. **P. moltkioides** (Royle ex Benth.) Kazmi Pl. 69

Rhizomatous herb. Shoots leafy. Flowering stalks up to 13 cm tall. Basal leaves petiolate, 20-90 x 4-10 mm, lanceolate to oblanceolate or oblong. Flowers blue or darker shades often deep purple; corolla tube 10-12 mm long. Nutlets 2-3 mm long.

Distribution: Pakistan, Kashmir.

Variable species. Alpine meadows especially Hazara district and Swat from 2400-4500 m. July-August.

12. Myosotis 7 spp.

Flowers blue or white. Calyx barely enlarged in fruit. Corolla 5-lobed; tube smaller than calyx; lobes spreading. Throat scales present. Stamens included. Nutlets smooth, shiny.

428. **M. sylvatica** Ehrh. ex Hoffm. *Forget-Me-Not* Pl. 69

Perennial up to 30 cm tall; stem simple or with few branches. Corolla limb 5-7 mm broad. Basal leaves 30-50 x 15-30 mm; elliptic-oblong to oblanceolate. Nutlets blackish-brown.

Distribution: North Africa, Turkey, Europe, Russia, Pakistan, Kashmir eastward to Bhutan.

A variable species. Forests from 2300-2700 m in Swat, Kaghan and Kashmir. June-August.

13. Gastrocotyle 1 sp.

Annual hispid herbs with solitary axillary flowers. Corolla shortly tubular, with scales in the throat. Nutlets usually 4.

429. **G. hispida** (Forssk.) Bunge Pl. 69

Stem and branches hispid. Basal leaves up to 10 more or less 1.2 cm, oblanceolate, margin wavy. Upper stem leaves smaller. Corolla purplish-blue to blue, 2.5-3 mm long; tube very short. Nutlets 3-4 mm broad, brown, rugulose-tuberculate.

Distribution: North Africa, Arabia, Syria, Iraq, Iran, Afghanistan, Pakistan, India.

Fairly common in dry areas of country from plains to 2300 m. March-May.

CUSCUTACEAE *Dodder Family*

Rootless and leafless herbaceous parasites. Flowers small in globose clusters. The family includes a single cosmopolitan genus closely related to *Convolvulaceae.*

Cuscuta 14 spp. *Dodder*

Climbing leafless total parasites. Stem threadlike, spirally twining sometimes interlaced. It has no green tissues of its own, and draws all the food material from host plant.

430. **C. reflexa** Roxb. *Nila tar, Ghas bel* Pl. 69

Stem branched, thick, succulent. Flowers ivory white, fragrant, fruit 4-seeded capsule, brown.

Distribution: Afghanistan, Pakistan, south west China to south east Asia.

Often very common on *Zizyphus* spp. forming dense masses of yellow threads covering whole tree. Widespread throughout the country. Troublesome weed. August-September.

CONVOLVULACEAE *Morning Glory Family*

A cosmopolitan family of herbs or shrubs, often with a twining or creeping habit. Leaves alternate, entire or not. Flowers bisexual, regular, in terminal or axillary cymose clusters. Corolla 5-lobed, bell or funnel-shaped. Stamens 5, epipetalous. Fruit a dry capsule or berry.

1.	+	Stigmas very slender, ellipsoid to oblong.	2
	–	Stigmas fairly stout, more or less capitate.	3
2.	+	Styles 2, stigmas filiform.	**1. Evolvulus**
	–	Style solitary.	**2. Convolvulus**
3.	+	Sepals enlarging in fruit. Flowers 5-6 mm long.	**4. Porana**

– Sepals not enlarging in fruit.
Flowers more than 10 mm long. **3. Ipomoea**

1. Evolvulus 1 sp.

Annual or perennial herbs, often creeping. Flowers axillary, solitary or several. Corolla rotate, funnel-form or bell shaped. Styles 2. Fruit a globose to ovoid capsule.

431. **E. alsinoides** (L.) L. Pl. 69

A slender prostrate herb with delicate branches and sky blue flowers. Leaves 8-12 mm long, lanceolate, elliptic-oblong. Peduncles equalling the leaves, slender.

Distribution: Central and south America, southern USA, Africa, Iran, Pakistan, India, Nepal, Sri Lanka, Malaysia.

Widespread in the plains but preferring hilly tracts up to 1500 m. Medicinal, used as a tonic for asthma and as a vermifuge. February-October.

2. Convolvulus c. 20 spp.

A cosmopolitan genus of herbs or shrubs, often with twining, prostrate or spiny habit. Flowers solitary axillary or in axillary cymes, bell shaped. Stigmas 2, very slender. Fruit capsuler.

1. + Stem and branches spiny. 433. **C. spinosus**

– Stem and branches not spiny. 2

2. + Leaves basally congested. 432. **C. kotschyanus**

– Leaves not basally congested 3

3. + Leaf base hastate. 435. **C. arvensis**

– Leaf base not hastate. 4

4. + Plants erect. 434. **C. virgatus**

– Plants prostrate. 5

5. + Leaves sessile. Flowers 1-3 in sessile clusters. 436. **C. prostratus**

– Leaves sub-petiolate. Flowers
4-10 in clusters with long stalks. 437. **C. glomeratus**

432. **C. kotschyanus** Boiss. Pl. 69

Small, perennial herb with a woody base and basally congested leaves. Flowers pale pink in compact clusters.

Distribution: Iraq, Iran, Pakistan.

Waste ground or sandy hills, on rocks and in crevices. Uncommon and confined to Iran and neighbouring western belt in Balochistan. April-June.

433. **C. spinosus** Burm. f. Pl. 69

Much branched spinescent shrub. Leaves linear-oblong. Flowers 1-3 in number, white.

Distribution: Iran, Afghanistan, and adjoining area of Pakistan.

This floriferous shrub is common in sandy and rocky places from 850-1800 m in Balochistan and Waziristan. April.

434. **C. virgatus** Boiss. Pl. 70

A branched woody herb up to 40 cm tall, with spreading glabrous branches. Older branches yellowish-brown, stiff. Leaves few, linear-lanceolate, 5-15 mm long. Flowers white to pink, solitary or 2-3. Peduncles 12-20 mm long. Sepals very hairy.

Distribution: South Iran, Muscat, Oman, Pakistan.

A very xerophytic *Convolvulus* found in open sandy places, gravelly beds from plains to 1000 m in southern Balochistan. January-February.

435. **C. arvensis** L. *Hiranpadi, Bindweed* Pl. 70

Prostrate or climbing twining herbs. Stems slender. Leaves 20-40 x 5-22 mm, broadly ovate to oblong to linear-oblong, sagittate to hastate. Petals 9-20 mm long, slender. Flowers pale pink to pink or white, often with darker stripes.

Distribution: Temperate and tropical regions of the world.

Very common and widespread as a weed from 600-2600 m from Sindh up to Baltistan. Variable species. Flowering mostly throughout the year.

436. **C. prostratus** Forssk. Pl. 70
Syn.: *C. pluricaulis* Choisy

A prostrate perennial with spreading, wavy, branches. Leaves 12-20 x 4-6 mm, oblanceolate to linear-oblong. Flowers 1-3, white, in axillary, sessile clusters.

Distribution: Egypt to Pakistan, India.

A rather variable species widespread in the country. Common in open, waste or sandy places from plains to 1700 m. March-April.

437. **C. glomeratus** Choisy Pl. 70

Prostrate branched herb with subpetiolate, ovate-lanceolate leaves. Flowers 4-10 in number, white, in axillary heads. Peduncles up to 45 mm long.

Distribution: North Africa, Arabia, Iran, Afghanistan, Pakistan, India.

In open sandy places in plains up to 700 m especially Balochistan and Salt Range. December-August.

3. Ipomoea c. 10 spp.

Herbs or shrubs, often of a prostrate or twining habit. Flowers in axillary clusters or paniculate, regular, usually funnel- or bell-shaped. Capsule dehiscing by 4 valves.

1.	+	Plants aquatic or of marshy places.	438. **I. aquatica**
	–	Plants terrestrial.	2
2.	+	Erect shrubs up to 2.5 m tall. Leaves ovate to ovate-cordate or ovate-lanceolate, 8-20 cm long.	439. **I. carnea**
	–	Climbing or twining herbs.	3
3.	+	Corolla 7-9 mm long.	440. **I. eriocarpa**
	–	Corolla 50-70 mm long.	4
4.	+	Leaves palmately 5-7-lobed.	441. **I. carica**
	–	Leaves ovate-cordate.	442. **I. purpurea**

438. **I. aquatica** Forssk. Pl. 70

Stem hollow, rooting at the nodes. Leaves hastate, 3-10 cm long. Flowers purplish tinged, 40-50 mm long, solitary or 2-3.

Distribution: Old World tropics.

Mostly in marshy places in the plains. A pernicious weed in rice growing tracts. Used as a pot-herb. Flowering mostly throughout the year.

439. **I. carnea** Jacq. Pl. 70
subsp. **fistulosa** (Mart ex Choisy) D. Austin

A leafy shrub up to 2.5 m high with hollow stems and entire large leaves. Flowers in terminal clusters, pink or pinkish-purple, up to 70 cm long.

Distribution: Tropical America. Cultivated or naturalized elsewhere.

A quick growing shrub often planted around farmsteads as a hedgerow and naturalized in many places from plains to 900 m. July-December.

440. **I. eriocarpa** R. Br. Pl. 71

A slender pubescent twiner with ovate-cordate leaves and small pink flowers in axillary heads.

Distribution: Tropical Africa, Pakistan, India, Malaysia to north Australia.

Common. Plains and lower hills to 1600 m. August-October.

441. **I. carica** (L.) Sweet *Railway Creeper* Pl. 71

Perennial climber with tuberous roots. Leaves palmately 5-7-lobed; lobes ovate-lanceolate to elliptic, 3-9 cm long. Flowers pink with dark purple centres, 45-60 mm long, solitary or 2-3.

Distribution: Native to America. Cultivated and naturalized elsewhere.

Commonly cultivated in the plains to cover fences etc. but also naturalized. July-December.

442. **I. purpurea** (L.) Roth *Morning Glory* Pl. 71

Annual twiners. Leaves ovate-cordate, 4-11 cm long. The showy flowers can vary from deep purple to red or pink, often striped with similar shades, 30-35 mm long.

Distribution: Native to America. Cultivated and naturalized elsewhere.

Plains and lower hills especially sub-Himalayan tracts up to 1600 m. July-September.

4. Porana 1 sp.

Twiners with a woody base. Leaves ovate-cordate. Flowers in showy paniculate clusters, regular. Style simple; stigma globose. Fruit a capsule.

443. **P. paniculata** Roxb. *Safed Bel, Bridal Wreath* Pl. 71

An extensive climber or creeper covered with pretty white clusters of flowers in the autumn. Flowers c. 5 mm long. Leaves ovate-cordate, acuminate.

Distribution: Pakistan, north India to Upper Burma.

The sub-Himalayan tract especially the Margalla hills up to 900 m. An extensive straggler. October-March.

SOLANACEAE

Potato Family

Herbs, shrubs or trees with simple, alternate leaves. Flowers regular, bisexual, 5-merous. Corolla rotate, bell-or trumpet-shaped. Stamens 5, epipetalous. Fruit a capsule or berry.

A family of mildly varying to highly poisonous plants and having many species of economic importance.

1.	+	Anthers adnate to one another, cone-like.	**1. Solanum** **2. Withania**
	–	Anthers not so.	2
2.	+	Fruit capsular.	**3. Hyoscyamus** **4. Datura**
	–	Fruit a berry.	3
3.	+	Plants spiny.	**5. Lycium**
	–	Plants not spiny.	4
4.	+	Flowers in clusters. Fruiting calyx not 3-angled or enclosing the berry.	**6. Atropa**
	–	Flowers solitary. Fruiting calyx 3-angled completely enclosing the berry.	**7. Physalis**

1. Solanum 8 spp.

Herbs, shrubs or small trees, often prickly. Flowers solitary. Corolla rotate, 5-lobed. Stamens exserted. Fruit a berry.

1.	+	Plants prickly. Berries yellow or yellow-red.	2
	–	Plants not prickly. Berries scarlet red.	444. **S. pseudocapsicum**

2.	+	Plants prostrate, leaves dark green.	445. **S. surratense**
	–	Plants erect or scrambling. Leaves greyish-green.	3
3.	+	Plants scrambling. Leaves subcordate. Berries yellow.	446. **S. cordatum**
	–	Plants erect. Leaves ovate, margin repand. Berries yellow.	447. **S. incanum**

444. **S. pseudo-capsicum** L. *Jerusalem Cherry* Pl. 71

Glabrous shrub up to 120 cm tall. Leaves lanceolate or elliptic lanceolate, 35-75 x 8-18 mm. Flowers white, in clusters of 1-3, 10-12 mm broad. Berry 8-12 mm broad, shiny, scarlet-red.

Distribution: Native of south America. Cultivated or naturalized elsewhere.

Cultivated and naturalized as an escape, mainly Hazara and Murree foot hills from 700-2200 m. April-July.

445. **S. surratense** Burm. f. *Kundiari, Mokri* Pl. 71

A prostrate branched herb with prickly dark green leaves. Leaves 20-60 x 20-40 mm, elliptic-oblong, margin repand to wavy. Flowers purple-mauve, 2-4 in number. Anthers yellow. Berries 15-20 mm broad, globose, yellow.

Distribution: North Africa, south and southeast Asia, Australia, Polynesia.

Common in plains from Makran, lower Sindh upto northern Punjab and lower hills up to 1300 m. Plant parts medicinal. Flowering throughout the year.

446. **S. cordatum** Forssk. Pl. 72
Syn.: *S. gracilipes* Decne.

A prickly scrambler with large subcordate leaves. Flowers 1-3 in number, purplish, in axillary or lateral clusters, 5-8 mm long. Berries yellow-red, 5-7 mm broad.

Distribution: North Africa, Arabia, Pakistan, India.

Dry rocky places up to 1000 m. Juice from the fruit and leaves is medicinal. March-April.

447. **S. incanum** L. Pl. 72

Branched shrub up to 110 cm or more tall. Stem and branches prickly and with a yellowish-white tomentum. Leaves large, greyish-green, tomentose. Flowers purplish-blue. Berries 20-30 mm broad, yellow.

Distribution: Iran, north Africa, Arabia, Pakistan, India.

Common in plains and lower hills up to 1400 m in all provinces. Plants medicinal. Flowering mostly throughout the year.

2. **Withania** 2 spp.

Unarmed shrubs with alternate entire leaves. Flowers in axillary clusters. Corolla slightly exceeding the calyx size. Fruit a berry.

448. **W. somnifera** (L.) Dunal *Aksan* Pl. 72

A perennial suffruticose herb up to 90 cm tall. Leaves and shoots tomentose. Leaves 3-12 x 2-8 cm, elliptic-ovate or broadly ovate. Flowers subsessile, small, greenish-yellow, in axillary clusters of 2-5. Berries globose, 6-8 mm, red.

Distribution: Canary Isls., Africa, Mediterranean, Africa, Iraq, south Iran, Syria, Turkey, Palestine, Arabia, Pakistan, India.

Common in waste places etc. from plains to 2300 m, almost throughout the country. Plant parts alkaloid and with a sedative effect. Fruit diuretic. Roots used in debility and rheumatism. Flowering mostly throughout the year.

3. **Hyoscyamus** 4 spp. *Henbane*

Perennial or biennial herbs. Leaves variously lobed. Flowers often with dark centres and reticulate patterns. Calyx enlarging in fruit. Corolla 5-lobed. Capsule dehiscing by a lid.

1. + Calyx lobes obtuse, shortly tipped. Flowers white, purple streaked. 449. **H. insanus**

– Calyx lobes toothed to dentate. Flowers pale yellow with dark throat. 450. **H. squarrosus**

– Margin of leaves subentire. Flowers pale yellow, with brown-purple reticulate veins. 451. **H. niger**

449. **H. insanus** Stocks Pl. 72
Syn.: *H. muticus* auct. non L.

Perennial bushy plant. Leaves ovate to rhombic-ovate, sinuate, 4-9 x 3-7 cm. Flowers showy, secund, white, mauvish or pale yellowish, purple streaked. Calyx lobes obtuse.

Distribution: South Iran, south east Afghanistan, Pakistan.

Rocky slopes etc., from 700-2000m, especially southern Balochistan. February-April.

450. **H. squarrosus** Griffith Pl. 72

Branched biennial up to 30 cm tall. Leaf margin prominently toothed. Calyx lobes triangular, long pointed. Flowers 20-25 mm long, yellow, throat purplish-brown. Stamens included.

Distribution: Iran, Turcomania, Afghanistan, Pakistan.

Rocky places from 1500-2000 m especially Balochistan. March-May.

451. **H. niger** L. Pl. 72

Leaves ovate-oblong, semi-amplexicaul. Calyx lobes broadly triangular. Flowers pale yellow, with brown-purple reticulations. Corolla lobes obtuse.

Distribution: North America, north Africa, Europe, Caucasus, Siberia, south west and central Asia, Iraq, Iran, Afghanistan, Pakistan, Kashmir to south west and central China, Japan.

Widespread in mountainous regions of country. River banks, open places in forests etc from 1800-2800 m in Balochistan, Chitral, Gilgit, Baltistan, Kaghan. June-July.

4. Datura 3 spp.

Annual herbs with entire to wavy or toothed leaves. Flowers axillary solitary, trumpet-like, white or purplish. Stamens 5. Fruit a capsule, dehiscing by valves.

+ Leaves with a wavy, dentate margin.
  Capsule erect, dehiscing by 4 valves. 452. **D. stramonium**

– Leaves with a wavy or entire margin.
  Capsule nodding. Dehiscing irregularly. 453. **D. innoxia**

452. **D. stramonium** L. *Thorn Apple* Pl. 73

Branched pubescent herb up to 130 cm tall. Leaves 6-15 x 4-10 cm, ovate or broadly so, sinuate-dentate. Flowers large, trumpet-shaped, 7-11 cm long, white. Capsule 3-4 cm long, spinose.

Distribution: Native to tropical America. Naturalized elsewhere.

Highly poisonous weed in waste places, roadsides from 900-2300 m. Plant contains the alkaloid hyoscyamine. June-July.

453. **D. innoxia** Miller Pl. 73

Similar but leaves entire-wavy and capsules with longer spines (5-10 mm) and dehiscing irregularly.

Distribution: Native to tropical America. Widely naturalized.

Common along roadsides, open places from plains to 1600 m. Plant parts including seeds are poisonous. May-October.

5. Lycium 7 spp.

Armed shrubs with shoots modified into spines. Flowers white to light purple, in axillary clusters of 1-6. Corolla tubular; lobes obtuse or acutish. Filaments glabrous or villous. Berries red or black.

+	Shoots glabrous. Corolla lobes obtuse.	454. **L. edgeworthii**
–	Shoots pubescent. Corolla lobes acutish.	455. **L. shawii**

454. **L. edgeworthii** Dunal Pl. 73
Syn.: *L. europeum* auct. non L.

A spiny much branched glabrous shrub. Leaves elliptic-oblong, 5-20 x 3-7 mm. Flowers white; corolla lobes c. 3 mm long. Berries red, 3-4 mm broad.

Distribution: South Iran, Pakistan, India.

Common and gregarious in open dry hot places in the plains. October-January.

455. **L. shawii** Roemer & Schultes Pl. 73

Similar to the above species, but shoots pubescent. Corolla tube narrower and with acutish lobes. Found in Sindh, Balochistan and Punjab. November-January.

6. Atropa 1 sp. *Nightshade, Belladona*

Perennial glabrous herbs with showy flowers. Calyx enlarging in fruit. Corolla bell shaped, with overlapping lobes. Fruit a berry.

456. **A. acuminata** Royle ex Miers. *Deadly Nightshade* Pl. 73

Branched herb up to 2 m tall. Leaves 6-15 x 4-8 cm, elliptic-lanceolate, petiolate. Flowers pale yellow, 20-24 mm long. Stamens included. Berries 10 mm broad, globose, black.

Distribution: East Iran, east Afghanistan, Pakistan, Kashmir, Mongolia.

Forest openings from 1800-3100 m in Kurram, Swat, Dir, Hazara and Murree hills. Plant parts contain the alkaloids atropine, hyoscyamine and belladonnine. Berries very poisonous. June-August.

7. **Physalis** 2 spp. *Cape Gooseberry*

Annual or perennial herbs. Flowers solitary, white or yellow. Calyx enlarged, bladder-like and angled in fruit. Fruit a berry, enclosed in the calyx.

457. **P. divaricata** D. Don Pl. 73

A suberect diffuse annual. Leaves 3-8 x 2-5 cm, ovate, wavy to subentire. Flowers yellow, c. 5 mm long. Berries 10 mm broad, orange, globose.

Distribution: Afghanistan, Pakistan, Kashmir to Nepal.

Common field weed, waste places from 600-1000 m in Sindh, Punjab and N.W.F.P. up to Chitral. August-October.

SCROPHULARIACEAE *Snap Dragon Family*

Generally herbs with opposite or alternate leaves. Flowers irregular, 4-5-merous. Corolla often 2-lipped. Stamens 4, in two groups of 2 long and 2 short, epipetalous. Fruit a capsule.

1.	+	Stamens 5	**4. Verbascum**
	–	Stamens 2 or 4	2
2.	+	Stamens 2	3
	–	Stamens 4	5
3.	+	Corolla tube 1-2 times as long as broad.	**1. Veronica**
	–	Corolla tube 3-4 times as long as broad.	4
4.	+	Calyx 5-lobed, lobes linear.	**2. Wulfenia**
	–	Calyx 2-lobed, lobes not linear.	**3. Lagotis**
5.	+	Corolla tubular, tube saccate or spurred.	**5. Misopates** **6. Kickxia** **7. Scweinfurthia**
	–	Corolla tubular, tube neither saccate nor spurred.	6
6.	+	Plants of moist places.	**8. Mazus** **9. Bacopa** **10. Mimulus**

– Plants of drier habitats. 7

7. + Corolla 5-lobed. Calyx 5-toothed. **11. Leptorhabdos**
12. Lindenbergia

– Corolla 2-lipped. Calyx 4-lobed. 8

8. + Flowers 4-12 mm long, white with streaks of yellow. **13. Euphrasia**

– Flowers 14-30 mm long, red, pink or yellow or shades of them. **14. Pedicularis**
15. Scrophularia

1. Veronica 21-23 spp. *Speedwell*

Perennial or annuals with opposite, entire leaves. Flowers solitary or many, in axillary or terminal racemes or spikes. Calyx lobes unequal. Corolla rotate; the limb longer than tube with 4 unequal lobes. Stamens 2.

1. + Plants of marshy places. 458. **V. anagallis-aquatica**

– Plants of drier situations. 2

2. + Annual prostrate herb. Flowers solitary terminal. Bracts leafy. 459. **V. polita**

– Erect herbs. Flowers in racemes or spikes. Bracts not leafy. 3

3. + Flowers blue or purple-blue, in axillary and terminal racemes. 460. **V. laxa**

– Flowers pale lavender, with a pink ring at the base, in terminal racemes. 461. **V. lanosa**

458. **V. anagallis-aquatica** L. *Water Speedwell* Pl. 74

Perennial branched herb up to 100 cm tall. Leaves 20-60 x 8-20 mm, ovate to elliptic-oblong, uppermost ones sessile. Flowers pale lilac to almost white, 4-5 mm broad, in lateral and terminal racemes.

Distribution: Africa, temperate Europe, west and central Asia, Pakistan, Kashmir to Bhutan, China, Korea.

Common in marshy places from plains to 3000 m. March-May.

459. **V. polita** Fries Pl. 74
Syn.: *V. didyma auct.* non Tenore

A prostrate branched annual herb with opposite, ovate, toothed leaves. Bracts leafy. Flowers pale violet, 5-7 mm broad.

Distribution: South west Asia, temperate Europe, north Africa.

A common weed from plains to 2500 m. January-May.

460. **V. laxa** Benth. Pl. 74
Syn.: *V. melissaefolia auct.* non Poiret

Perennial with suberect stems up to 60 cm tall, leafy. Leaves opposite, ovate or broadly so, coarsely toothed. Flowers blue or purple-blue, 6-8 mm broad, in axillary and terminal racemes.

Distribution: Pakistan, Kashmir to central Nepal, Tibet, China, Japan.

In Himalayan moist temperate forests from 1900-3200 m. May-July.

461. **V. lanosa** Royle ex Benth. Pl. 74

Stem simple, up to 60 cm tall. Leaves elliptic-lanceolate, sessile. Flowers showy,10-12 mm broad, in terminal racemes, pale lavender with a pink ring at the base.

Distribution: Pakistan, Kashmir to Himachal Pradesh.

Alpine and sub-alpine. Rocky places from 2700-3800 m from Safed Koh, Chitral east to Baltistan. July-August.

2. Wulfenia 1 sp.

Perennial herbs, usually with leaves in rosettes. Flowers in spike-like racemes. Calyx 5-partite into linear lobes. Corolla 2-lipped. Stamens 2.

462. **W. amherstiana** Benth. Pl. 74

Flowering stalk 12-18 cm long with flowers dispersed on one side. Leaves spathulate to elliptic-oblong, 5-10 cm long, in rosettes. Corolla purple-blue, slightly curved, 2-lipped; stamens 4.

Distribution: Afghanistan, Pakistan, Kashmir to Nepal.

Steep rocks, bank sides, shady places in forested areas from 1600-3100 m especially Murree and Hazara hills. July-August.

3. **Lagotis** 4-5 spp.

Perennial glabrous herbs. Flowers spicate. Calyx 2-lobed. Corolla 2-lipped. Stamens 2, subexserted.

463. **L. cashmeriana** (Royle) Rupr. Pl. 74

Flowers purple-blue, on short stout spikes. Leaves 30-65 mm long, elliptic-oblong, crenate, semi-amplexicaul.

Distribution: Pakistan, Kashmir to Himachal Pradesh (India).

Alpine meadows. 3100-4000 m especially Kaghan, Hunza valley and Deosai plains. June-August.

4. **Verbascum** 4 spp.

Annual or perennial tomentose herbs with alternate leaves. Flowers yellow, regular. Corolla rotate. Stamens usually 5. Capsule globose to cylindrical. Stem simple. Basal leaves up to 25 cm long.

464. **V. thapsus** L. *Gidhar Tambaku, Mullein* Pl. 75

Perennial up to 45 cm tall. Stem simple. Basal leaves up to 25 cm long. Leaves and stems covered with dense tomentum. Flowers yellow, 20-25 mm broad, in stout spikes up to 35 cm. Filaments dense hairy.

Distribution: Temperate Eurasia, Afghanistan, Pakistan, Kashmir to south west China.

From plains to 3600 m. Common in Chitral. Slopes, rocky places. June-August.

5. **Misopates** 1 sp.

Annual. Calyx lobes very unequal. Fruit capsular. Seeds compressed, one face smooth and keeled.

465. **M. orontium** (L.) Rafin. Pl. 75
Syn.: *Antirrhinum orontium* L.

Erect herb 10-50 cm tall. Leaves linear, 20-40 x 2-4 mm. Bracts linear, longer than the flowers. Flowers pinkish-mauve to pale lilac, pouched at the base, 10 mm long. Capsule c. 8 mm long, pouched at the base.

Distribution: West and central Europe, Mediterranean region, south Russia, south west Asia.

Weed of grain fields from plains to 1100 m, especially N.W.F.P. and northern Punjab. March-April.

6. Kickxia 3 spp.

Annual or perennial prostrate herbs. Leaves shortly petiolate, at least some hastate. Flowers axillary solitary. Corolla spurred. Capsule dehiscing by means of a lid.

+	Plants glabrous, slender. Leaves long stalked. Spur of corolla straight.	466. **K. ramosissima**
–	Plants hairy, quite stout. Leaves short stalked. Spur of corolla curved.	467. **K. incana**

466. **K. ramosissima** (Wall.) Janchen Pl. 75

Leaves ovate-cordate, 10-20 mm long, upper narrower, hastate. Flowers lemon-yellow. Pedicels slender.

Distribution: Afghanistan, Pakistan, Kashmir to Bhutan.

Rock crevices from Sindh Kohistan to Chitral upto 1500 m. March-April.

467. **K. incana** (Wall.) Pennell Pl. 75

A stouter plant with hairy parts, shorter stalked leaves. Flowers orange-yellowish. Spur of corolla curved.

Distribution: Afghanistan, Pakistan, Kashmir to Nepal, Tibet.

Rocky places, slopes from 400-1500 m almost through out the country. March-April.

7. Schweinfurthia 2 spp.

Glabrous annual or perennial herbs with alternate entire leaves. Flowers small axillary. Corolla tube pouched at the base. Stamens 4, in two groups. Fruit capsular, dehiscing irregularly.

468. **S. papilionacea** (Burm. f.) Boiss. Pl. 75

Perennial branched herbs up to 35 cm tall. Leaves ovate to subspathulate, 10-15 mm broad. Flowers creamy orange coloured, stalk deflexed in fruit. Capsule 6-8 mm broad, subglobose.

Distribution: Afghanistan, Pakistan, India.

Sandy or stony desert areas up to 3500 m especially Sindh, Balochistan and Waziristan. Dried plant parts used in local medicines. February-April.

8. Mazus 3 spp.

Prostrate to suberect perennial herbs. Leaves usually in basal rosettes, cauline ones opposite. Flowers small, in terminal racemes. Corolla 2-lipped; tube short, limb 5-lobed. Stamens 4. Capsule globose.

469. **M. pumilus** (Burm. f.) Van Steenis Pl. 75
Syn.: *M. rugosus* Lour.

Stem tufted. Radical leaves 20-50 mm long, narrowed into a petiole, obspathulate. Flowers white, with bluish streaks and throat speckled with yellow. Calyx half cleft into 5 lobes.

Distribution: Tropical Asia, Kashmir to Bhutan.

Wet situations especially rice growing areas. Plains and lower hills to 2000 m, mainly sub-Himalayan tracts. March-May.

9. Bacopa 1 sp.

Low growing, usually succulent herbs with opposite leaves. Flowers small, axillary solitary. Corolla 2-lipped, 5-lobed. Stames 4. Capsule dehiscing by 4 valves.

470. **B. monnieri** (L.) Wettstein Pl. 76
Syn.: *Bramia monnieri* (L.) Pennell

A creeping branched herb rooting at the nodes. Leaves fleshy, 6-20 mm long, oblong or spathulate. Flowers pale lavender to lavender-pink, c. 10 mm long. Capsule ovoid, 6-8 mm long.

Distribution: Tropical and subtropical regions of Old World.

Marshy places from plains to 1200 m throughout the country. March-April.

10. Mimulus 1 sp.

Erect glabrous herbs with opposite leaves. Calyx 5-lobed. Corolla 2-lipped, tube long. Stamens 4, in two groups.

471. **M. strictus** Benth. Pl. 76

Erect herb up to 30 cm tall. Leaves narrow oblong, 25-40 mm long. Stems clasping. Flowers white, throat spotted yellow, on long axillary slender stalks. Corolla tube pink cylindric; limb 5-lobed, white with yellow spotted throat. Stamens included in tube.

Distribution: Tropical Africa, Pakistan, Kashmir to Nepal, India.

Wet places in plains and lower hills mainly in sub-Himalayan tracts up to 1200 m. March-April.

11. Leptorhabdos 1 sp.

Erect annual herbs with opposite narrowly dissected leaves. Flowers racemose. Corolla tube short, limb 5-lobed. Stamens 4, included. Capsule subglobose.

472. **L. parvifolia** (Benth.) Benth. Fig. p. 267

A tall slender plant. Stems up to 100 cm, branched. Leaves divided into linear segments. Flowers pale pink, 4-5 mm long. Capsule oblong, more or less included in the calyx.

Distribution: Caucasus, central Asia, Afghanistan, Pakistan, Kashmir to Bhutan, Tibet, west China.

Forest slopes and clearings from 1200-3000 m. Widespread from juniper tracts in Balochistan, Chitral, Dir, Gilgit, Baltistan, Hazara and Murree hills. August-September.

12. Lindenbergia 3 spp.

Annual or perennial herbs, often woody below. Leaves opposite, toothed. Flowers yellow. Calyx 5-fid. Corolla 2-lipped. Fruit a capsule.

473. **L. macrostachya** (Benth.) Benth. Pl. 76

Erect branched perennial 90 cm or more tall. Leaves elliptic-ovate, serrate. Flowers in rigid spikes.

Distribution: Pakistan, India, China, Indonesia.

Foothills and adjacent plains up to 1800 m. Often weedy. March-April.

13. Euphrasia c.10-12 spp. *Eye Bright*

Annual herbs with opposite leaves. Corolla 2-lipped, white or lilac. Calyx 4-lobed. Stamens 4, in two groups.

474. **E. himalayica** Wetts. Pl. 76

A branched suberect annual. Leaves ovate, 5-12 mm, toothed. Flowers white and purplish with a yellow throat, 6-8 mm long. Corolla limb 5-lobed, lower lip 3-lobed.

Distribution: Afghanistan, Pakistan, Kashmir to Bhutan.

Open or moist places from 2200-3600 m particularly, Kaghan. June-July.

14. Pedicularis c. 22-23 spp. *Lousewort*

Erect perennial herbs with divided leaves. Flowers in terminal spike-like racemes, showy. Calyx 5-lobed. Corolla 2-lipped, the lower 3-lobed, the upper hooded and beaked. Stamens 4, in unequal pairs. Capsule longer than calyx.

1.	+	Flowers yellow.	475. **P. bicornuta**
	–	Flowers red or pink or pink and white.	2
2.	+	Flowers pink with a white centre.	3
	–	Flowers pink-red.	477. **P. pyramidata**
3.	+	Corolla tube more or less equalling the calyx.	476. **P. pectinata**
	–	Corolla tube much exceeding the calyx.	478. **P. punctata**

475. **P. bicornuta** Kl. Pl. 76

An erect perennial up to 60 cm tall. Leaves pinnately divided, up to 30 cm long; pinnae oblong or toothed. Flowers yellow, 15-20 mm broad, in terminal racemes. Corolla with the upper lip coiled.

Distribution: Pakistan, Kashmir to Uttar Pradesh (India).

A showy plant of higher alpine slopes and meadows, often growing gregariously from 2700-4000 m in Chitral, Gilgit, Hazara area. July-August.

476. **P. pectinata** Wall. ex Benth. Pl. 76
subsp. **palans** Prain

A week branched perennial with pretty red flowers with a white throat. Leaves pinnately divided into linear toothed segments.

Distribution: Pakistan, India to west Nepal.

Locally common in forests from Chitral eastwards to Swat, Kaghan, Baltistan, Hazara and Murree hills. A cliff plant, from 1600-2700 m. July-August.

477. **P. pyramidata** Royle Pl. 77

Gregarious perennial up to 90 cm tall. Stem usually simple. Leaves whorled, pinnately divided into lanceolate toothed segments. Flowers pink, in showy racemes. Bracts broadly elliptic.

Distribution: Pakistan, Kashmir to Himachal Pradesh (India).

Alpine and forest meadows throughout northern parts of Pakistan from 2400-4500 m. July-August.

478. **P. punctata** Decne. Pl. 77

Low growing perennial with rose coloured flowers with a white throat. Corolla tube longer than the calyx, tubular; limb with lower lip 3-lobed, the middle lobe notched and shorter than the other two.

Distribution: Pakistan, Kashmir.

Alpine slopes. Common in damp places from 2600-4200 m throughout northern regions of country. July-August.

15. Scrophularia 15-16 spp.

Annual or perennial herbs with usually opposite leaves. Flowers greenish-yellow, red or yellow, in panicles or cymes. Sepals 5. Corolla 2-lipped; upper lip 4-lobed, erect, lower 1-lobed, spreading. Stamens 4. Capsule ovoid.

479. **S. scabiosifolia** Benth. Pl. 77

Branched perennial herbs. Leaves glabrous, irregularly dentate or lobed, mostly basal. Flowers small, crimson and white. Capsule globose c.3 mm long.

Distribution: Afghanistan, Pakistan.

A plant of dry areas from 400-1800 m in Balochistan, N.W.F.P., Salt Range, Hazara and northern areas.

OROBANCHACEAE *Broomrape Family*

Fleshy root parasites with scale-like leaves and bracts, lacking chlorophyll. Flowers attractive, in spikes or racemes. Flowers bisexual, irregular. Corolla 2-lipped or subequally so. Stamens 4, epipetalous. Fruit capsular.

+	Corolla 2-lipped.	**1. Orobanche**
–	Corolla subequally 5-lobed.	**2. Cistanche**

1. Orobanche c. 17-19 spp. *Broomrape*

Flowers in terminal bracteate spikes. Bracteoles present or absent. Corolla bell shaped or funnel form to tubular; lower lip of corolla 3-lobed. Stamens in two groups, included in the corolla tube.

1.	+	Stems branched. Bracteoles present.	480. **O. aegyptiaca**
	–	Stems simple. Bracteoles absent.	2
2.	+	Flowers bluish-violet.	481. **O. cernua**
	–	Flowers reddish-brown to yellowish-brown.	482. **O. alba**

480. **O. aegyptiaca** Pers. *Sabzgul* Pl. 77

Glandular-pubescent annual with stem branched from below, 8-30 cm tall. Corolla limb purplish-blue; tube creamish to the outside, 20-35 mm long, slightly curved.

Distribution: North Africa, Arabia to Pakistan.

Parasitic on various cultivated crops as mustard, potato, tomato and tobacco. From plains to 2400 m. Common in Quetta and Kalat valleys. February-September.

481. **O. cernua** Loefl. Pl. 77

Flowers bluish-violet. Corolla tube 15-20 mm long, curved, lobes somewhat irregular, acutish. Calyx 2-partite.

Distribution: South Europe, north Africa, Arabia, Iraq, Iran, Afghanistan, Pakistan, Kashmir to Nepal, west Tibet.

Parasitic on wormwood (*Artemisia* spp.), tobacco, petunia etc. Quite common from 900-3300 m, extending north to Baltistan.

482. **O. alba** Stephen ex Willd. Pl. 77

Plants 15-20 cm tall with brown scale-like triangular leaves. Flowers reddish-brown to yellowish-brown. Corolla tube curved, glandular, 15-20 mm long.

Distribution: Europe, central and north Asia, Afghanistan, Pakistan, Kashmir to Nepal, Tibet.

Stony slopes. Parasitic on various members of the *Labiatae* (Mint family) as *Thymus* and *Origanum,* from 2400-3800 m, especially Swat and Hazara area. July-August.

402. *Centaurium centaurioides*
Islamabad, Y. J. Nasir

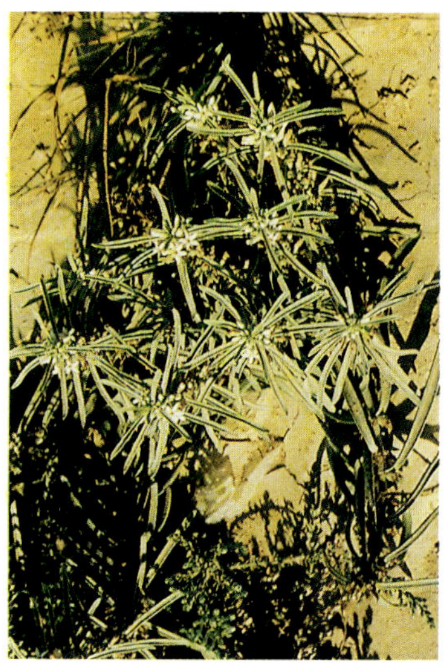

401. *Enicostemma verticellatum*
Karachi, Y. J. Nasir

406. *Gentianodes marginata* Upper Kaghan,
Y. J. Nasir

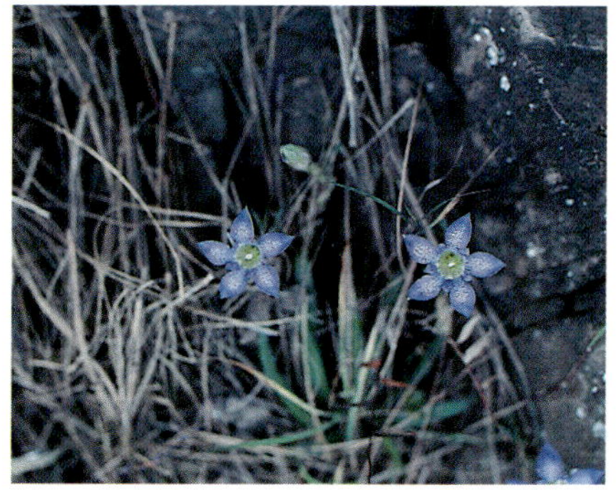

403. *Gentianodes kurroo* Ghora Gali,
S. A. Sultan

405. *Gentianodes olivieri* Chiltan hills, Rubina A. Rafiq

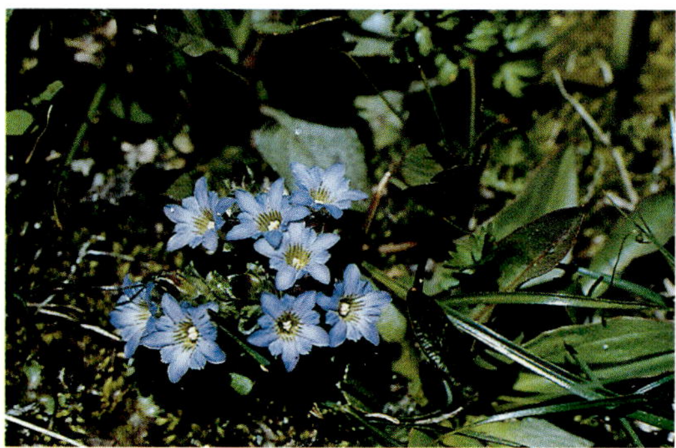

404. *Gentianodes argentea* Kaghan, Mark Mallalieu

411. *Heliotropium strigosum* Karachi, Y. J. Nasir

407. *Polemonium coeruleum* Kaghan, Y. J. Nasir

413. *Heliotropium europaeum var. lasiocarpum* Thal desert, Y. J. Nasir

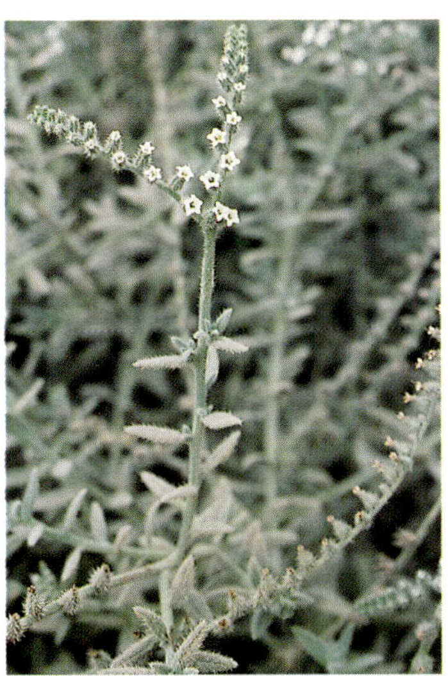

410. *Heliotropium ophioglossum* Karachi, Y. J. Nasir

408. *Ehretia obtusifolia* Margalla hills, S. A. Sultan

409. *Heliotropium curassavicum* Karachi, Y. J. Nasir

418. *Arnebia griffithii* a, b: Salt Range
S. A. Sultan

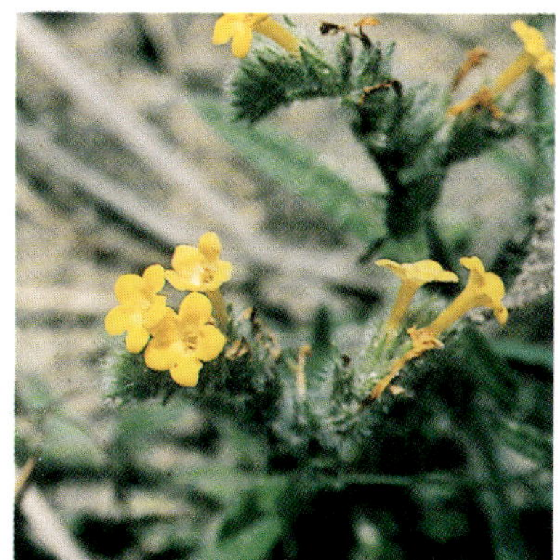

417. *Arnebia hispidissima* Attock,
S. A. Sultan

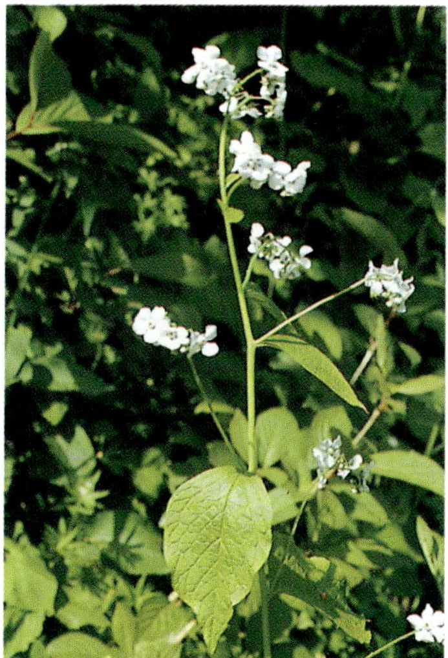

419. *Hackellia macrophylla* Kaghan,
S. A. Sultan

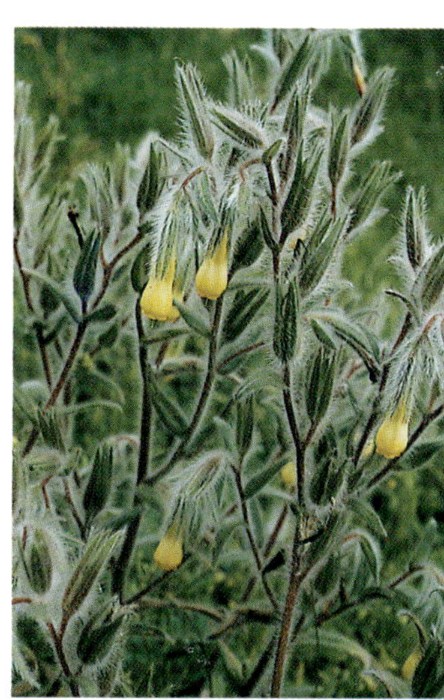

416. *Onosma hispida* Swat,
S. A. Sultan

415. *Trichodesma indica* Margalla hills, S. A. Sultan

422. *Lindelofia longiflora* Kaghan, Y. J. Nasir

421. *Solenanthus circinnatus* Upper Swat, Fritz Berger

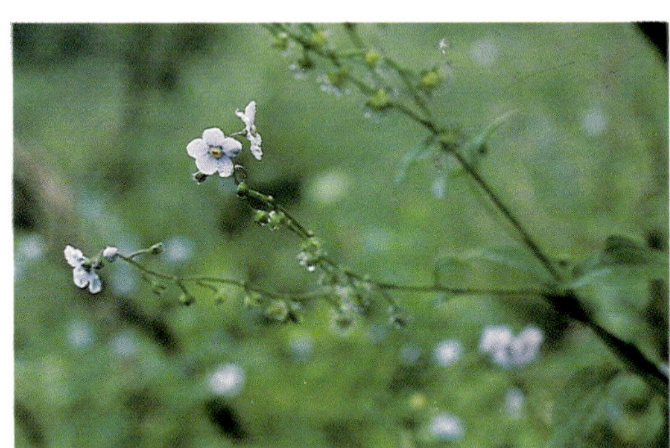

423. *Cynoglossum glochidiatum* Kaghan, Y. J. Nasir

424. *Nonea edgeworthii* Margalla hills, Y. J. Nasir

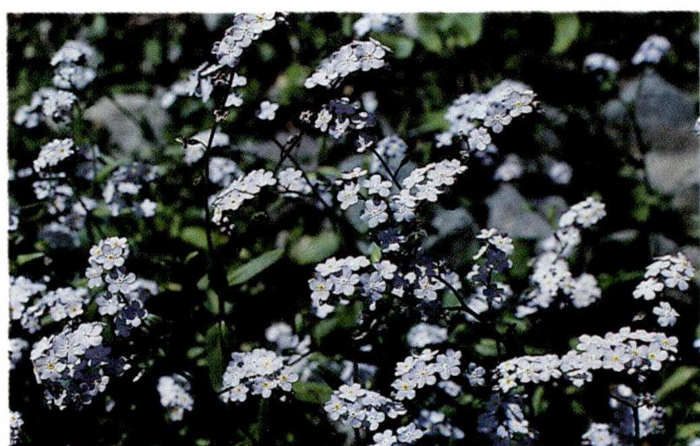

420. *Hackelia uncinata* Ushu, Swat, Fritz Berger

425. *Pseudomertensia parvifolia* Swat, Fritz Berger

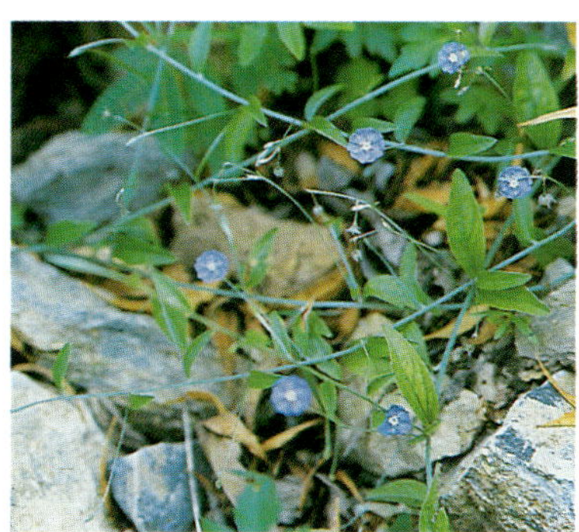

432. *Convolvulus kotschyanus* Zarchi, Kalat, Rubina A. Rafiq

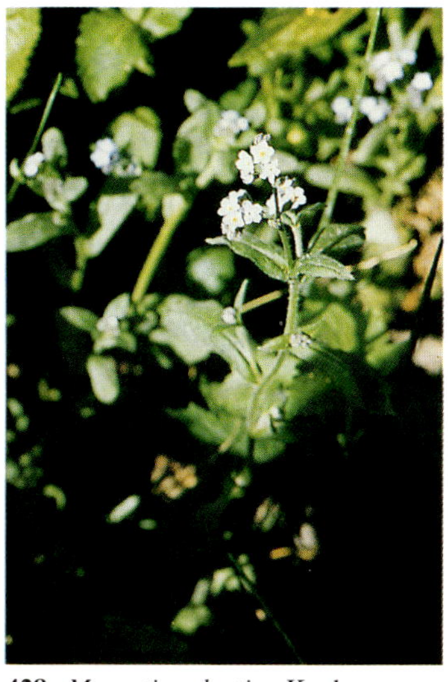

428. *Myosotis sylvatica* Kaghan, Y. J. Nasir

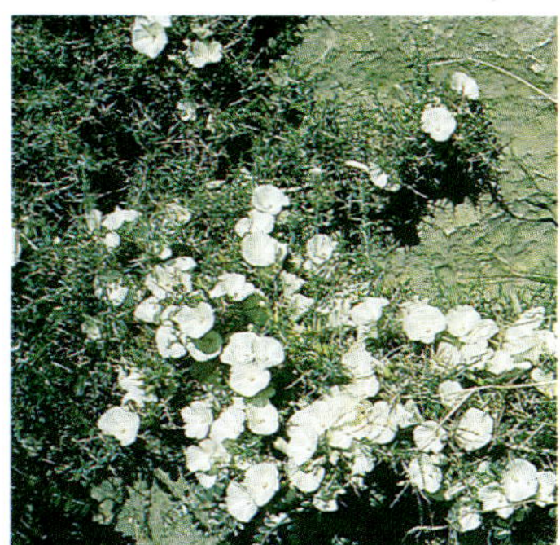

433. *Convolvulus spinosus* Chiltan hills, Rubina A. Rafiq

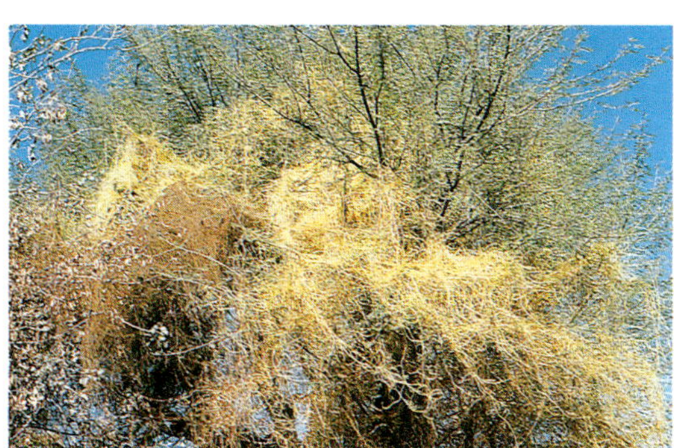

430. *Cuscuta reflexa* Sindh, T. J. Roberts

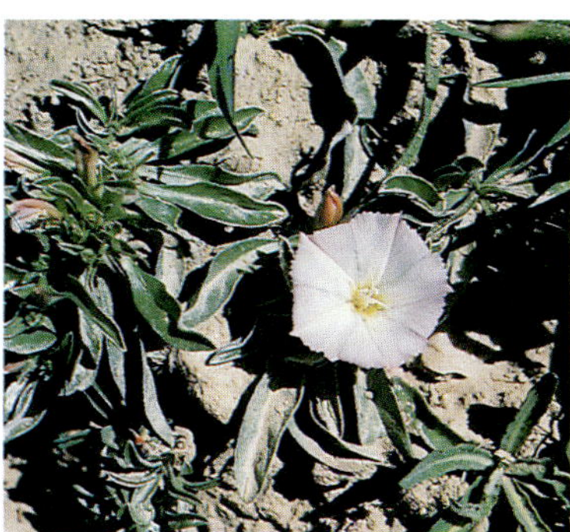

431. *Evolvulus alsinoides* Margalla hills, S. A. Sultan

427. *Pseudomertensia moltkioides* Matiltan, Swat, T. J. Roberts

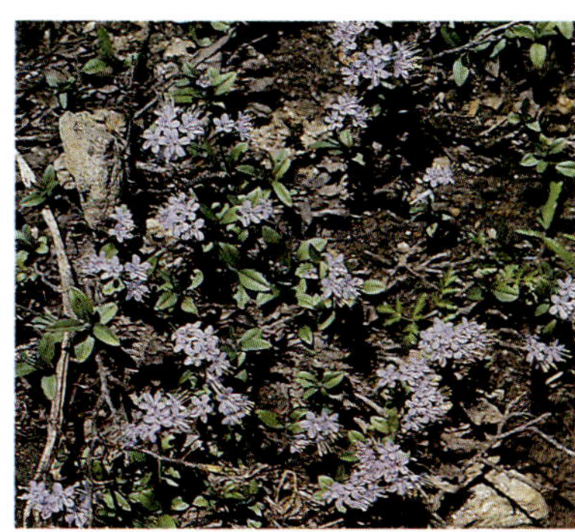

426. *Pseudomertensia trollii* Kaghan, Y. J. Nasir

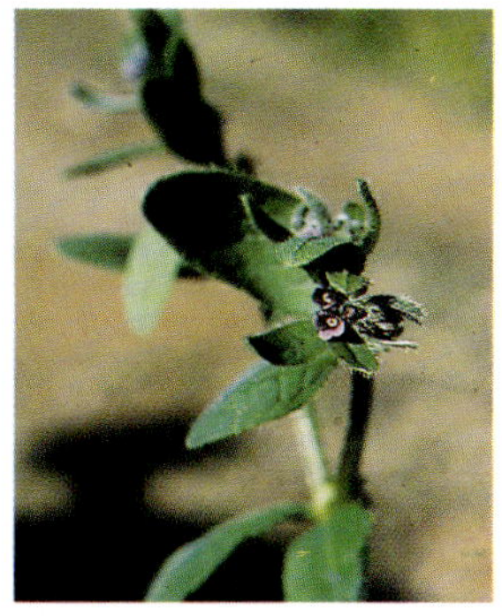

429. *Gastrocotyl hispida* Margalla hills, Y. J. Nasir

434. *Convolvulus virgatus* Makran, Y. J. Nasir

435. *Convolvulus arvensis* Quetta, Rubina A. Rafiq

439. *Ipomoea carnea subsp. fistulosa* Margalla hills, Y. J. Nasir

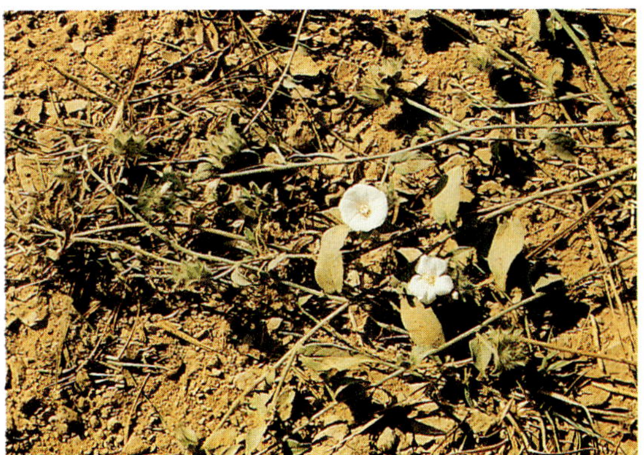

437. *Convolvulus glomeratus* Karachi, Y. J. Nasir

436. *Convolvulus prostratus* Islamabad, Rubina A. Rafiq

438. *Ipomoea aquatica* Thatta, T. J. Roberts

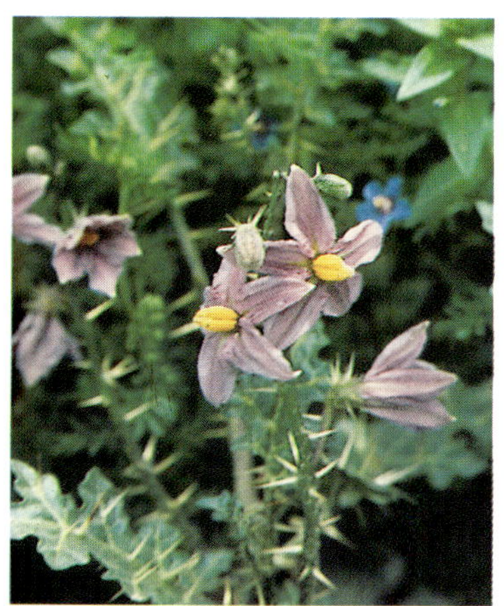

445. *Solanum surratense* Islamabad, S. A. Sultan

444. *Solanum pseudo-capsicum* Siran valley, Hazara, Y. J. Nasir

b▶

▼a

440. *Ipomoea eriocarpa* Karachi, Khan Muhammad Khan

442. *Ipomoea purpurea* a, b: Islamabad, S. A. Sultan

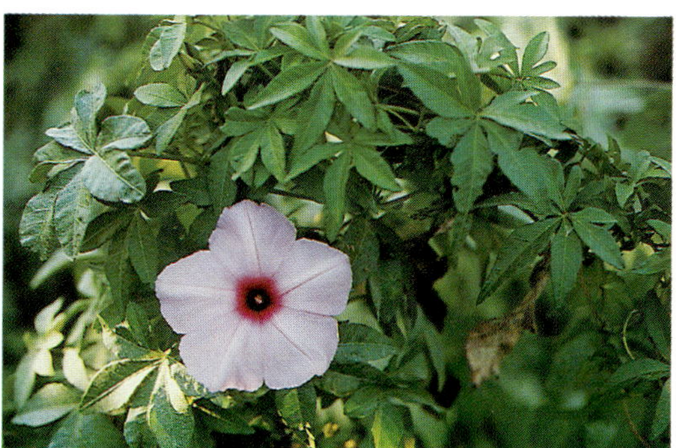

443. *Porana paniculata* Margalla hills, Rubina A. Rafiq

441. *Ipomoea carica* Wah, Y. J. Nasir

450. *Hyoscyamus squarrosus* Urak, Quetta, Rubina A. Rafiq

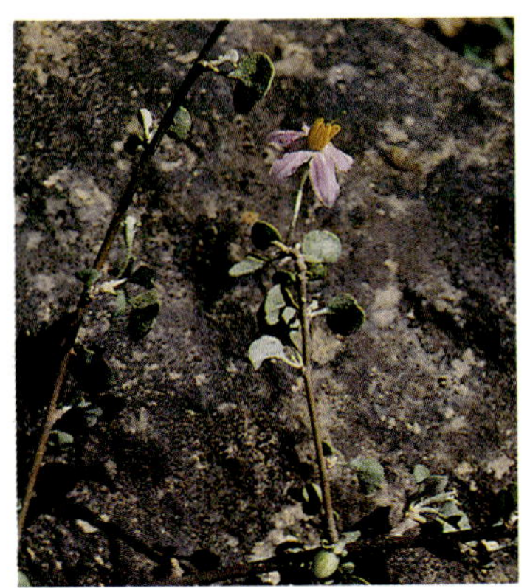

446. *Solanum cordatum* Salt Range, Y. J. Nasir

448. *Withania somnifera* Karachi, Y. J. Nasir

451. *Hyoscyamus niger* Quetta, Rubina A. Rafiq

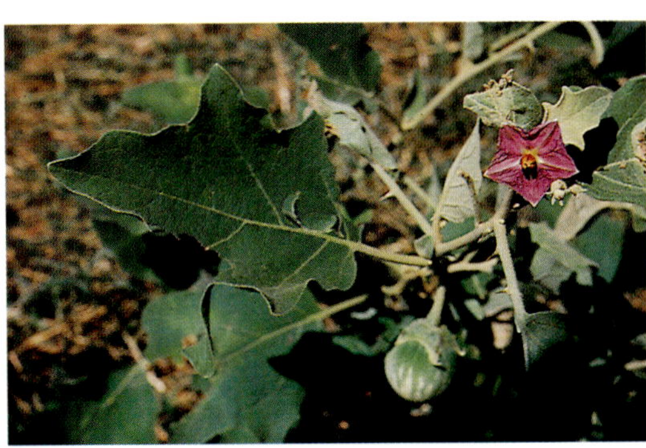

447. *Solanum incanum* Margalla hills, Y. J. Nasir

449. *Hyoscyamus insanus* Malakand pass, S. A. Sultan

454. *Lycium edgeworthii* Makran, Y. J. Nasir

452. *Datura stramonium* Astor, Rubina A. Rafiq

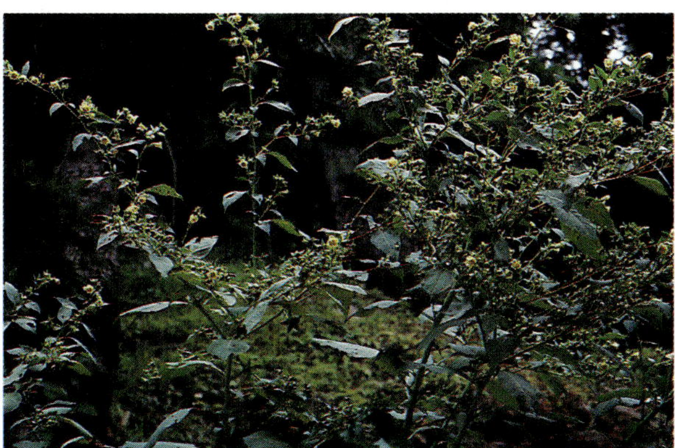

456. *Atropa acuminata* Thanḍiani, Y. J. Nasir

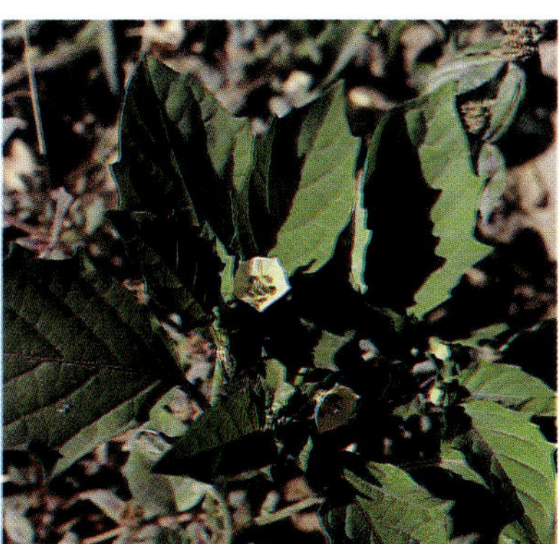

457. *Physalis divaricata* Karachi, Y. J. Nasir

455. *Lycium shawii* Makran, Y. J. Nasir

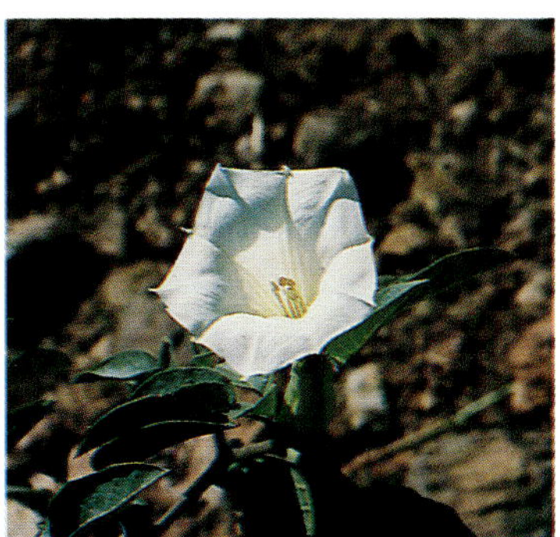

453. *Datura innoxia* Islamabad, Rubina A. Rafiq

460. *Veronica laxa* Changla Gali, S. A. Sultan

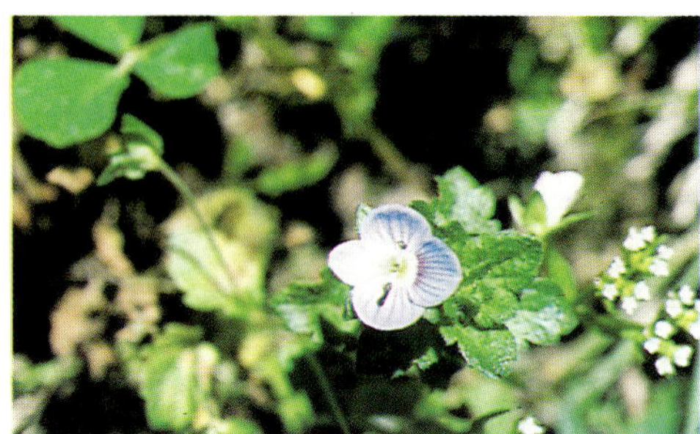

459. *Veronica polita* Margalla hills, S. A. Sultan

463. *Lagotis cashmeriana* Upper Kaghan, Y. J. Nasir

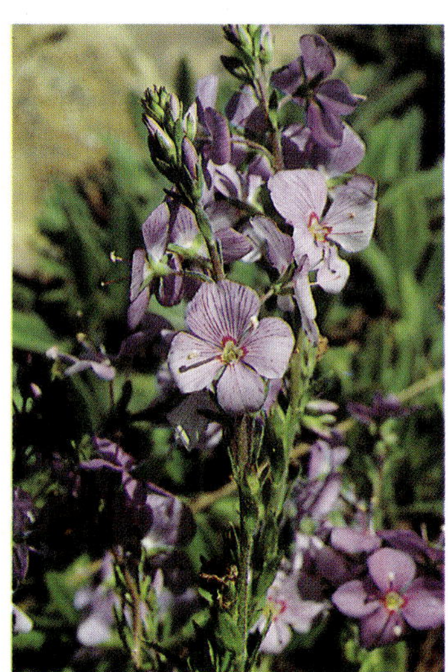

461. *Veronica lanosa* Kalam, Y. J. Nasir

458. *Veronica anagallis-aquatica* Lower Swat, Y. J. Nasir

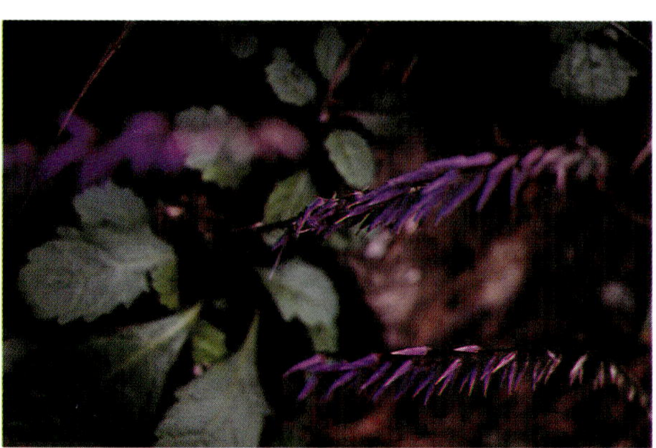

462. *Wulfenia amherstiana* Kalapani, Hazara, S. A. Sultan

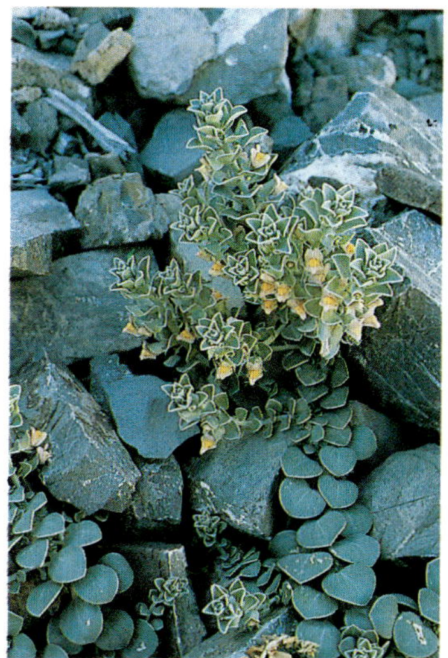

468. *Schweinfurthia papillionacea* Makran, Y. J. Nasir

466. *Kickxia ramosissima* Margalla hills, S. A. Sultan

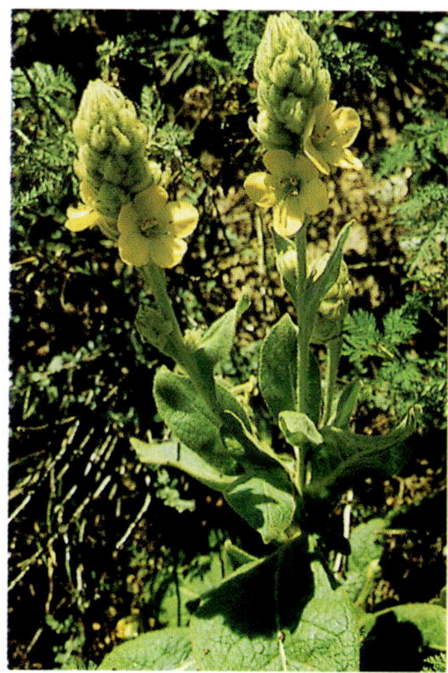

464. *Verbascum thapsus* Kalam, Swat, Y. J. Nasir

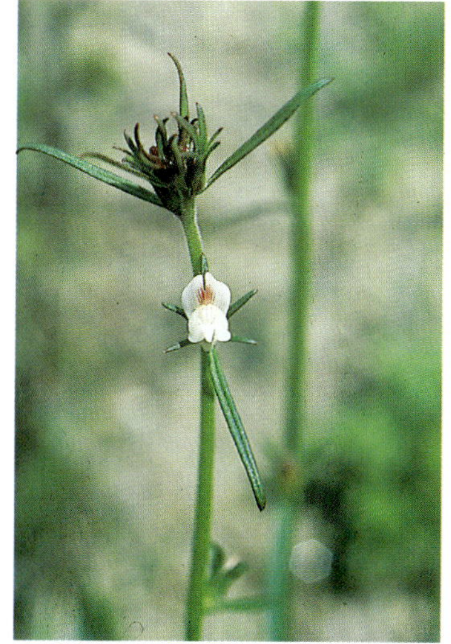

465. *Misopates orontium* Islamabad, Y. J. Nasir

467. *Kicksia incana* Margalla hills, S. A. Sultan

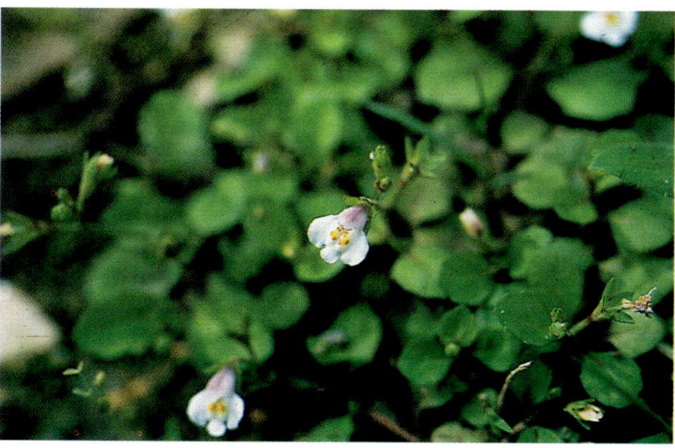

469. *Mazus pumilus* Islamabad, Rubina A. Rafiq

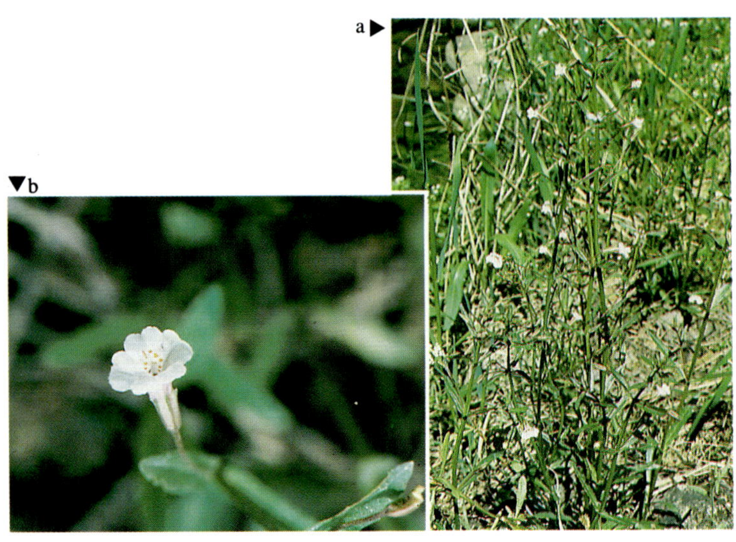

471. *Mimulus strictus*
a, b: Murree hills, Y. J. Nasir

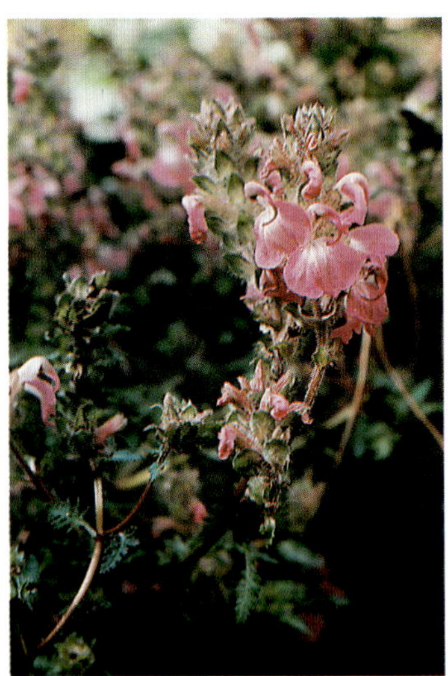

476. *Pedicularis pectinata subsp. palans* Murree hills, S. A. Sultan

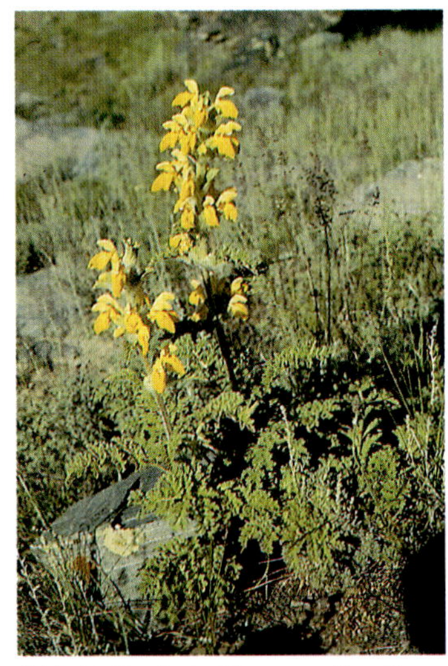

475. *Pedicularis bicornuta* Shandur Plateau, T. J. Roberts

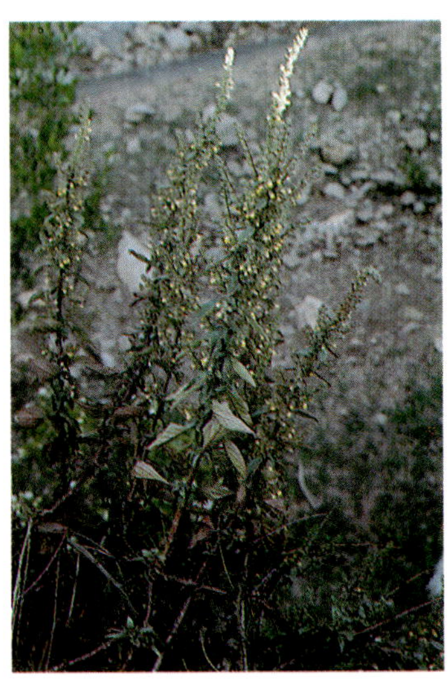

473. *Lindenbergia macrostachya* Margalla hills, Y. J. Nasir

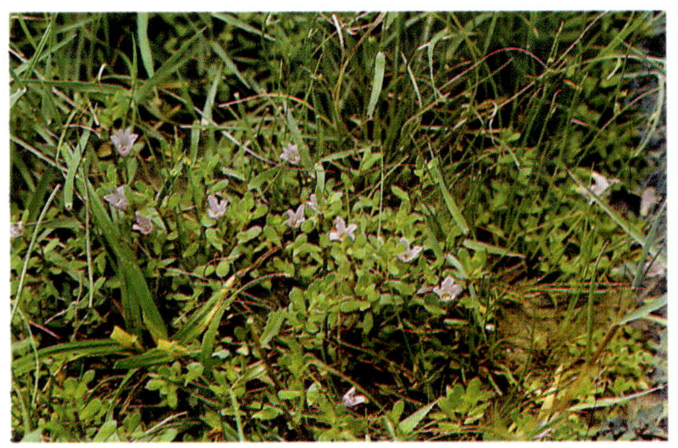

470. *Bacopa monnieri* Islamabad, S. A. Sultan

474. *Euphrasia himalaica* Lalazar, Kaghan, Y. J. Nasir

483. *Cistanche tubulosa* Thal desert, Y. J. Nasir

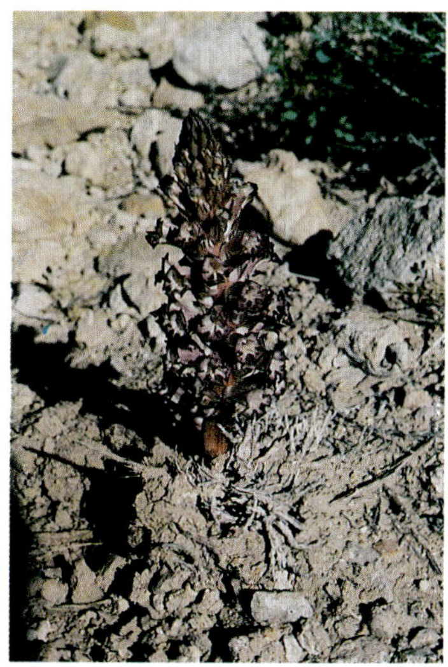

481. *Orobanche cernua* Zarchi, Kalat, Rubina A. Rafiq

479. *Scrophularia scabiosiefolia* Murree hills, Y. J. Nasir

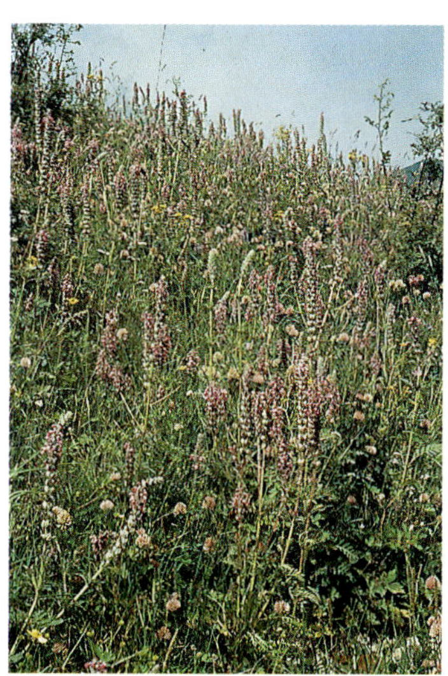

477. *Pedicularis pyramidata* Kaghan, Y. J. Nasir

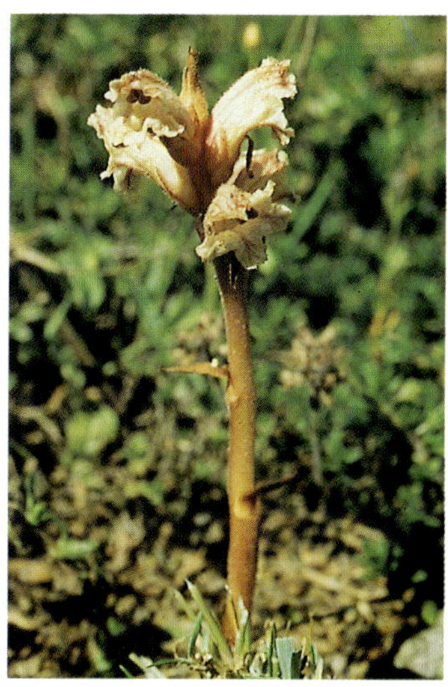

482. *Orobanche alba* Upper Swat, Y. J. Nasir

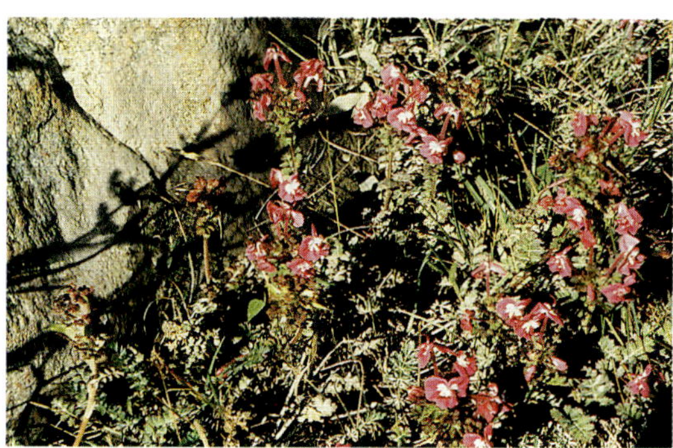

478. *Pedicularis punctata* Kaghan, Y. J. Nasir

480. *Orobanche aegyptiaca* Kalat, Rubina A. Rafiq

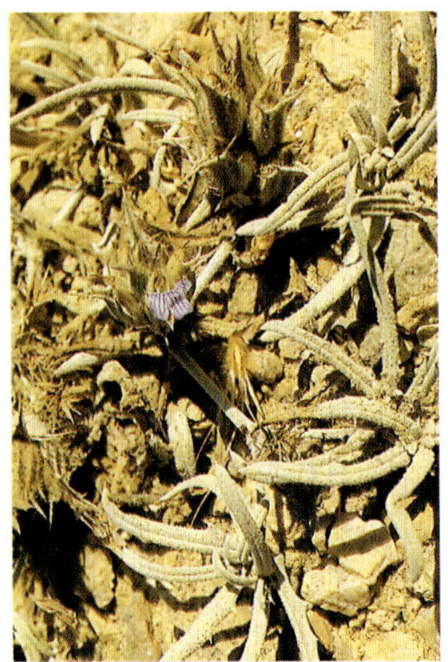

489. *Blepharis sindica* Sindh, Y. J. Nasir

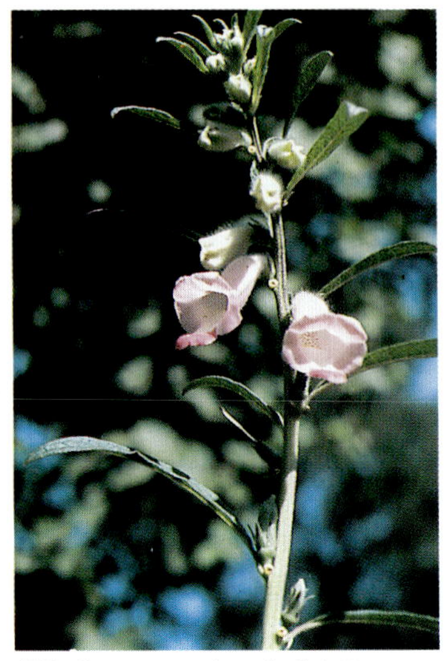

487. *Sesamum orientale* Salt Range, S. A. Sultan

a ▶

▼b

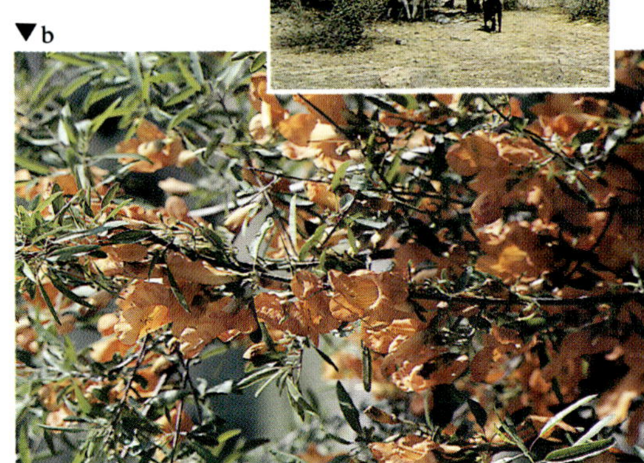

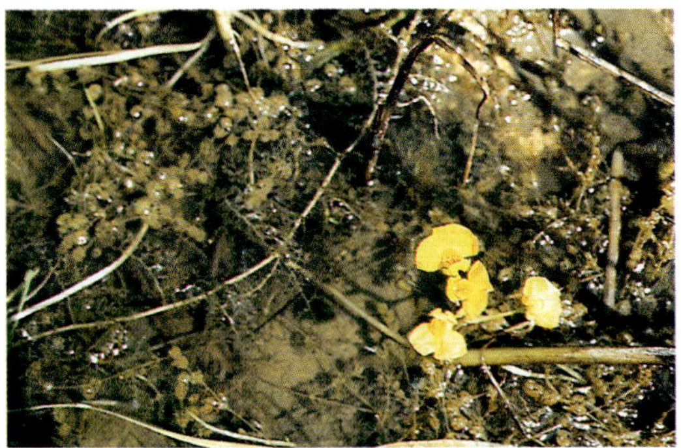

488. *Martynia annua* Islamabad, S. A. Sultan

486. *Tecomella undulata* a. Sindh, T. J. Roberts
b. Attock, S. A. Sultan

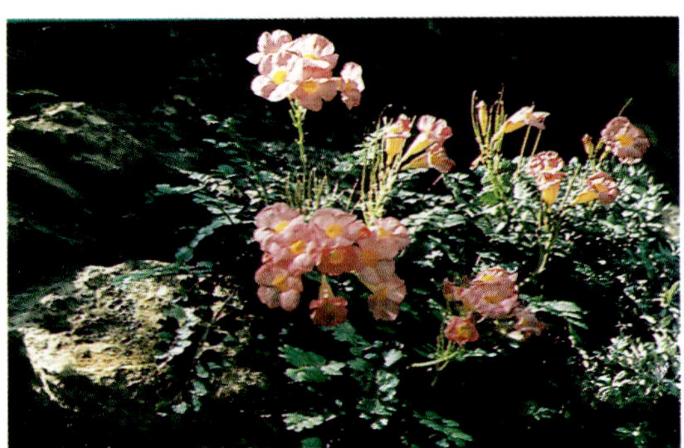

484. *Utricularia aurea* Chashma Barrage, Y. J. Nasir

485. *Incarvillea emodi* Swat, T. J. Roberts

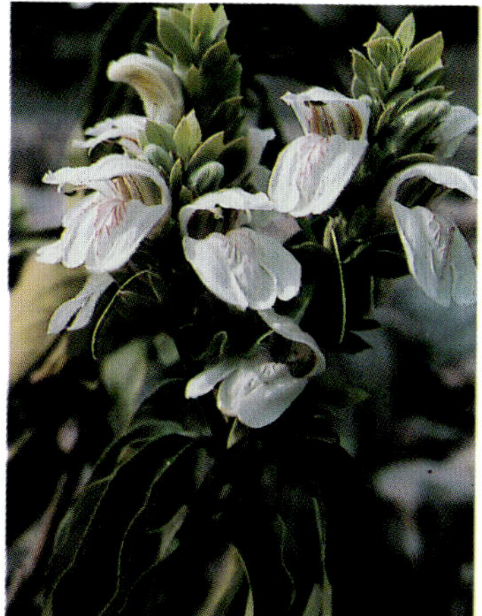

493. *Justicia adhatoda* Islamabad, Y. J. Nasir

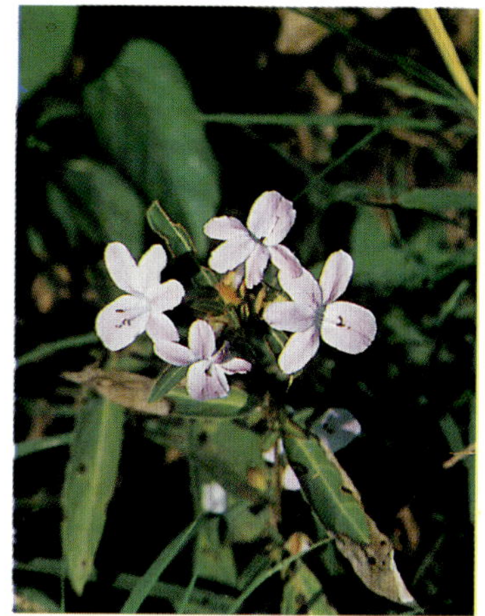

494. *Barleria cristata* Margalla hills, S. A. Sultan

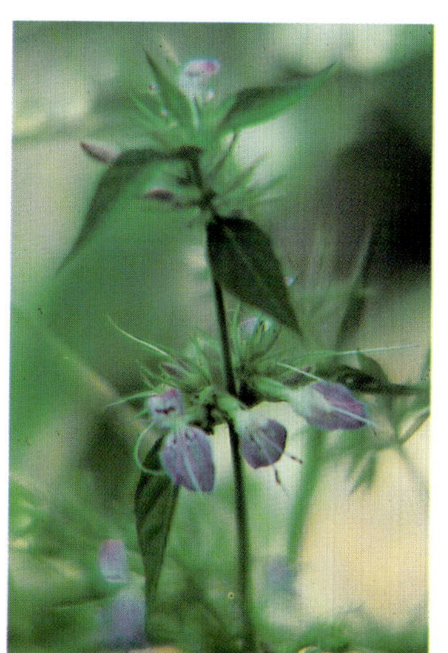

491. *Dicliptera bupleuroides* Islamabad, S. A. Sultan

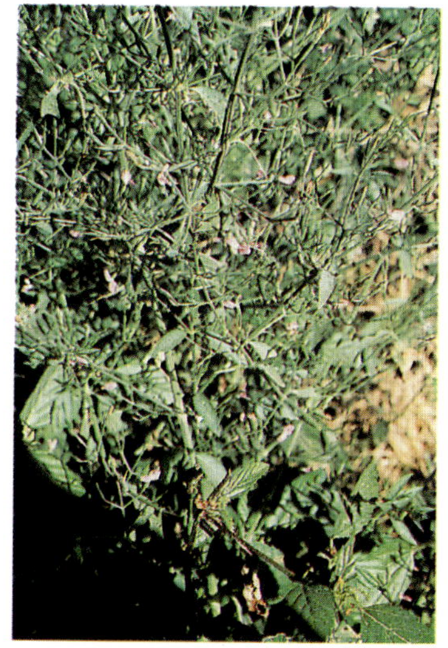

492. *Peristrophe paniculata* Islamabad, Y. J. Nasir

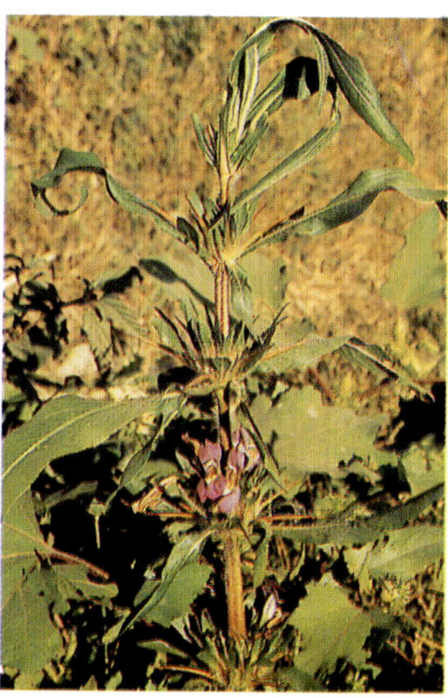

490. *Hygrophila auriculata* Salt Range, Rubina A. Rafiq

496. *Ruellia patula* Karachi, Y. J. Nasir

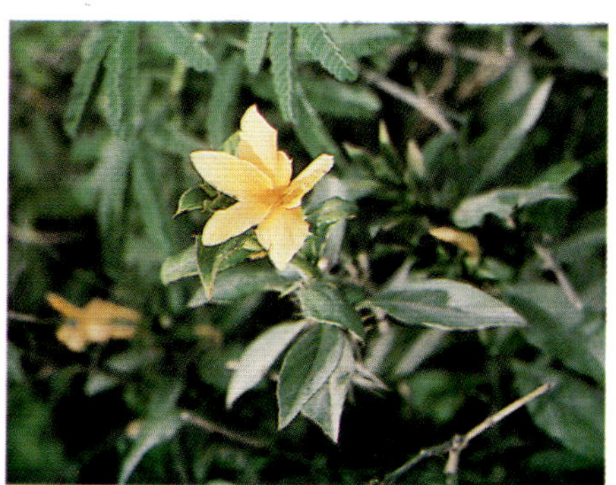

495. *Barleria prionotis* Karachi, Y. J. Nasir

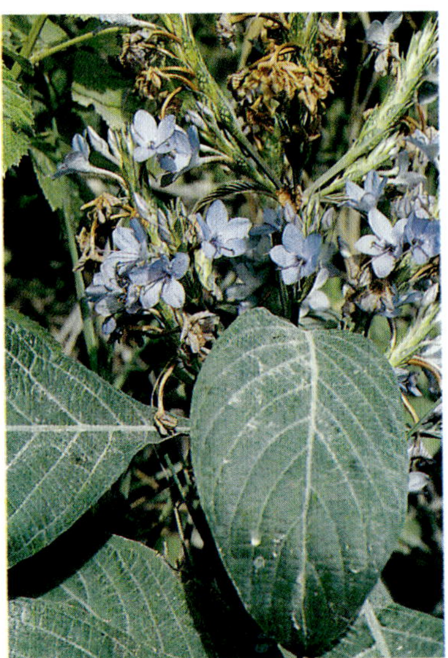

498. *Eranthemum pulchellum* Wah, Y. J. Nasir

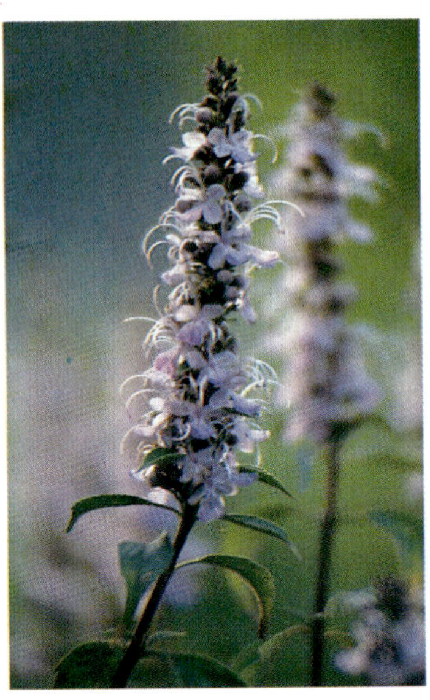

502. *Caryopteris odorata* Swat, Fritz Berger

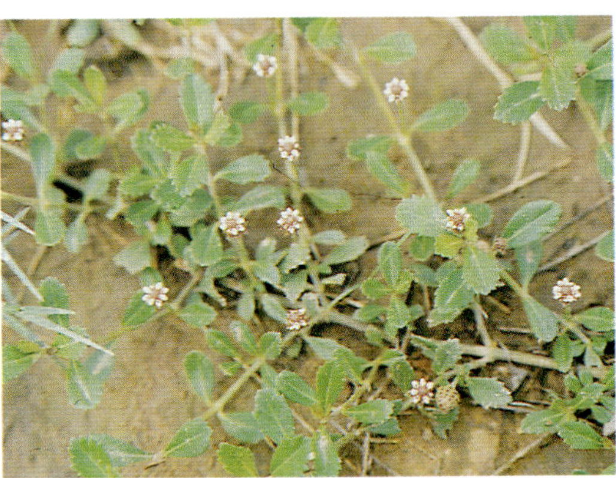

500. *Phyla nodiflora* Islamabad, S. A. Sultan

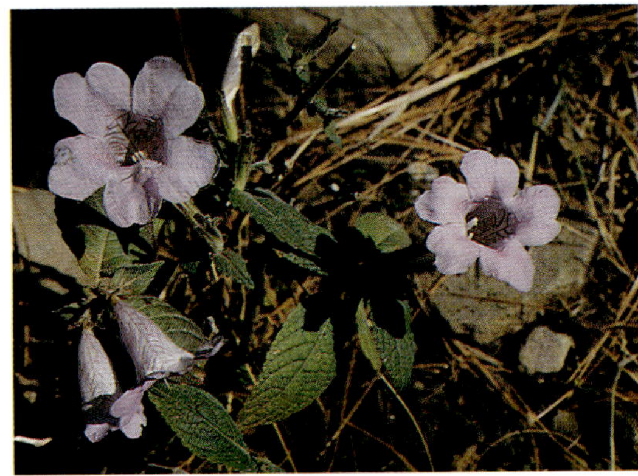

499. *Strobilanthes glutinosus* Margalla hills, Rubina A. Rafiq

501. *Verbena tenuisecta* Islamabad, S. A. Sultan

497. *Pteracanthus urticifolius* Thandiani, Y. J. Nasir

2. **Cistanche** 1 sp.

Resembling *Orobanche,* but flowers sub-regular.

483. **C. tubulosa** (Schrenk) Hook. f. *Broomrape, Kasi* Pl. 77

A spectacular stout fleshy herb up to 70 cm tall. Stem simple, thick. Leaves scale-like, 15-20 x 8-12 mm, triangular to broadly lanceolate. Corolla tube slightly curved, 30-50 mm long, lemon, limb with light purplish lobes, throat yellow.

Distribution: North Africa, west and central Asia, Pakistan, India.

Common on sand dunes. Parasitic on various desert shrubs as *Calotropis procera, Salvadora persica, Calligonum polygonoides* etc. in Sindh and Balochistan. Medicinal, used to stop diarrhoea in Balochistan. February-April.

LENTIBULARIACEAE *Bladderwort Family*

Aquatic or marshy insectivorous herbs. Leaves much dissected and with bladders to entrap insects. Flowers 2-lipped, in racemes or spikes. Corolla with lower lip 2-lobed. Stamens 2. Fruit capsular.

Utricularia 3 spp. *Bladderwort*

Aquatic rootless herbs with much dissected leaves (in our species). Calyx spurred, the spur longer or shorter than wide. Capsule globose.

484. **U. aurea** Lour. Pl. 78

Submerged branched aquatic herb with thread-like dissected leaves. Flowers yellow, in racemes, exserted above water.

Distribution: Tropics of the Old World.

Marshy places in the plains and lower hills to 1000 m, mainly in northern Punjab. March-April.

BIGNONIACEAE *Bignonia Family*

A tropical or subtropical family of trees, shrubs or herbs with opposite or alternate compound or simple leaves. Flowers showy, irregular, often in clusters. Corolla 2-lipped, the upper lip 2-lobed. Stamens 4-5. Fruit capsular (in our genera).

\+ Perennial or annual herbs, often with a woody base. Leaves alternate. **1. Incarvillea**

– Small or medium sized trees. Leaves opposite. **2. Tecomella**

1. Incarvillea 1 sp.

Leaves pinnate. Corolla lobes imbricate. Stamens 4. Capsule linear.

485. **I. emodi** (Royle ex Lindl.) Chatterjee Pl. 78

Perennial herb with a short leafy rootstock. Leaves imparipinnate. Leaflets 7-11, opposite, 8-30 x 5-20 mm, ovate-oblong, toothed or crenate. Flowers rose-pink or mauve, showy, 4-13, in terminal short racemes. Corolla tube 30-35 mm long, limb with 5 spreading obtuse lobes; throat yellow. Capsule 10-20 cm long.

Distribution: Afghanistan, Pakistan, Kashmir to west Nepal, India.

Sub-Himalayan tracts and lower hills from 600-2700 m. A rock plant worth cultivating for its beautiful flowers. April-May.

2. Tecomella 1 sp.

Small or medium sized trees with opposite, simple leaves. Flowers large and showy.

486. **T. undulata** (Roxb.) Seeman *Lahura* Pl. 78

Tree up to 7 m tall. Leaves 30-90 x 8-20 mm, elliptic-oblong or narrowly so. Flowers orange-red or sometimes yellow, 2-7, in short racemes. Corolla campanulate; lobes obtuse. Capsule 15-30 mm, linear-oblong, slightly curved.

Distribution: Arabia, south Iran, Afghanistan, Pakistan, India.

In hot drier hilly tracts up to 1000 m in Salt Range and Balochistan. The bark and seeds are medicinal. Wood used as fuel and for furniture and carving. January-May.

PEDALIACEAE *Sesame Family*

A tropical and subtropical family of annual or perennial herbs with simple, opposite leaves. Flowers irregular, in solitary or few axillary clusters. Corolla 2-lipped, campanulate or broadly tubular. Stamens 4. Fruit a capsule or indehiscent.

Sesamum 1 sp.

Leaves entire or lobed. Flowers shortly pedicellate. Stamens included. Capsule 4-angled.

487. **S. orientale** L. *Til, Sesame* Pl. 78
Syn.: *S. indicum* L.

An erect annual up to 70 cm tall. Leaves elliptic-oblong or narrowly oblong, 3-15 cm long, subentire to lobed. Flowers pretty, tubular, white with rose or purple tinge. Capsule erect, 15-20 mm long.

Distribution: Tropics of the Old world. Widely cultivated or naturalized.

Cultivated for an oil extract from the seeds which is used in soap, confectionery and hair oil. Naturalized in the sub-Himalayan and adjacent plains to 700 m. July-October.

MARTYNIACEAE *Unicorn or Devil's Claw Family*

A family closely related to *Pedaliaceae* but differing in the unilocular ovary with parietal placentation.

Martynia 1 sp.

Annual herbs with subcordate leaves. Flowers large, showy, in terminal or axillary racemes. Corolla 2-lipped. Capsule splitting by 2 valves.

488. **M. annua** L. *Choa, Hath jori, Tiger's Claw* Pl. 78

Branched herb up to 1.5 m tall. Leaves opposite, 5-15 x 4-9 cm, petiolate. Flowers pinkish-white, with red, blue or yellow specks within. Corolla tube c. 25 mm long, glandular, 5-lobed; lobes pale purplish, obtuse. Capsule ovoid, 3 cm long, with 2 curved spiny and woody processes.

Distribution: A native of central America. Naturalized elsewhere.

Waste places and open areas from plains to 1000 m mainly in foothill zone. April-May.

ACANTHACEAE *Barleria Family*

A tropical family of armed or unarmed shrubs or herbs, with simple, entire or lobed leaves. Flowers irregular, variously disposed in spikes, racemes or panicles. Calyx persistent, 4-5-lobed. Corolla 2-lipped, tube cylindrical; limb of 5 subequal lobes. Stamens 2 or 4. Fruit a capsule dehiscing by 2 valves.

1. + Corolla 1-lipped, the upper lip absent or obscure. **1. Blepharis**
   – Corolla 2-lipped, or subequally 5-lobed. 2
2. + Corolla distinctly 2-lipped. 3

	–	Corolla subequally 5-lobed.	4
3.	+	Staminodes present. Fertile stamens 2 or 4.	**2. Hygrophila**
	–	Staminodes absent. Fertile stamens 2.	**3. Dicliptera** **4. Peristrophe** **5. Justicia**
4.	+	Calyx 4-lobed.	**6. Barleria**
	–	Calyx 5-lobed.	5
5.	+	Bracts if present, small.	**7. Ruellia**
	–	Bracts generally leafy.	**8. Pteracanthus** **9. Eranthemum** **10. Strobilanthes**

1. **Blepharis** 3 spp.

Undershrubs or perennial herbs with simple, entire, spinescent leaves. Flowers solitary or in axillary clusters or spikes. Bracts spiny, persistent in fruit. Calyx unequally 4-lobed. Corolla limb 1-lipped, the upper being absent or obscure. Stamens 4, exserted. Capsule laterally compressed.

489. **B. sindica** Stocks ex T. Anders. *Bhangari* Pl. 78

A small undershrub. Leaves linear-oblong, 20-50 x 3-4 mm, pubescent above. Flowers in stout cone-like spikes 30-70 mm long, purplish-mauve, c. 15 mm long. Bracts 15-20 x 8-10 mm, tip spinose, margin with smaller spines. Corolla limb with lower lip 3-lobed, the median larger than the other two. Capsule 6-8 mm long, ellipsoid.

Distribution: Sindh (Pakistan) and adjacent Rajasthan (India).

Rocky or sandy places in Sindh and Salt Range. July-November.

2. **Hygrophila** 1 sp.

Armed perennial herbs of marshy places. Leaves opposite, entire. Flowers in axillary clusters, terminal heads or spikes. Calyx 4-5-partite into narrow lobes. Corolla limb 2-lipped, upper one bilobed. Stamens 4 or only 2-fertile, exserted from corolla tube.

490. **H. auriculata** (Schum.) Heine *Talmakhana* Pl. 79

A suffruticose much branched perennial herb up to 100 cm tall. Stems with long yellow spikes at the nodes. Leaves oblanceolate to lanceolate, 50-140 x 10-30 mm. Flowers pink, in axillary clusters. Stamens 4.

Distribution: Tropical Africa, Pakistan, India, Burma, Indo-China, Malaya.

A marsh plant. Rare from 600-800 m in Salt Range. September-October.

3. **Dicliptera** 2 spp.

Perennial herb or undershrubs with angled stems. Leaves entire, petiolate. Bracts larger than calyx. Calyx 5-lobed; lobes narrow. Corolla slender; limb 2-lipped, the lower larger than the upper, 3-lobed. Stamens 2, exserted.

491. **D. bupleuroides** Nees Pl. 79
Syn.: *D. roxburghiana* Nees *var. bupleuroides* (Nees) Clarke

A suberect herb up to 80 cm tall. Leaves elliptic-ovate to ovate-lanceolate, long petioled. Flowers pinkish-mauve, 10-12 mm long, in terminal and axillary clusters. Bracts 5-8 mm long. Corolla tube 6-8 mm long. Capsule 5-6 mm long, club-like.

Distribution: Afghanistan, Pakistan, Kashmir to west Nepal, India, Thailand, Indo-China, west China.

Very common, plains to 1800 m in hilly tracts from Balochistan, Salt Range, N.W.F.P. up to Swat.

4. **Peristrophe** 1 sp.

Like *Dicliptera,* but bracts free at the base and capsule which is not elastically dehiscent.

492. **P. paniculata** (Forssk.) Brummit Pl. 79
Syn.: *P. bicalyculata* (Retz.) Nees

A diffuse suberect herb up to 1 m tall with angled stems. Leaves petiolate, elliptic-ovate. Bracts 2, 5-12 mm long, narrow. Flowers pink. Corolla tube 4-6 mm long, cylindrical. Corolla limb with lower lip slightly longer than the upper one, 3-lobed. Stamens 2. Capsule 10-12 mm long, ellipsoid.

Distribution: Tropical Africa, Pakistan, India, Burma, Malaya, Indo-China.

Waste places, road sides from plains to 600 m. July-October.

5. **Justicia** 7 spp.

Perennial branched herbs or undershrubs with evil smelling herbage when crushed. Leaves ovate to elliptic-lanceolate. Flowers white, pink or purplish, in panicles or spikes. Bracts often leafy. Corolla tube straight or curved; limb 2-lipped, lower lip 3-lobed. Stamens 2.

493. **J. adhatoda** L. *Bhaikar* Pl. 79

Shrub up to 90 cm tall. Leaves 10-20 x 3-8 cm, elliptic-lanceolate, petiolate, yellow-green. Flowers white, 3 cm long, in terminal and axillary spikes 5-8 cm long. Bracts elliptic-oblong, 10-20 mm long. Capsule 20-25 mm long, club-like.

Distribution: Pakistan, Kashmir to west Nepal, India, Indo-China, Malaya.

Very common in the foothill zone and Salt Range up to 1200 m. Foliage obnoxious to browsing animals. Birds are particularly attracted to nectar of this flower. Waste places. Flowering mostly throughout the year.

6. Barleria 4 spp.

Armed or unarmed undershrubs or herbs. Flowers showy, in axillary or terminal spikes. Bracts often spiny. Calyx 4-partite. Corolla tube slender; limb subequally 5-lobed or barely 2-lipped. Stamens 4, exserted.

+ Flowers purplish-blue, pink or white. Plants unarmed. 494. **B. cristata**

– Flowers orange-yellow. Plants armed. 495. **B. prionotis**

494. **B. cristata** L. *Tadrelu* Pl. 79

Unarmed shrublet up to 100 cm tall. Bract spiny on margin. Flowers 40-50 mm long, limb with unequal lobes. Capsule 12-18 mm long, ellipsoid.

Distribution: Pakistan, Kashmir to west Nepal, India, Burma, Indo-China, south China, Philippines.

Very common in the foothills and Salt Range up to 1800 m. The purple flowered form is usual. Often cultivated in gardens. November-February.

495. **B. prionotis** L. Pl. 79

A branched undershrub with whitish 10-20 mm long axillary spines. Flowers showy, 25-35 mm long, orange-yellow. Corolla nearly as long as the tube. Stamens 2.

Distribution: Tropical Africa, Asia.

Both cultivated and naturalized. Grown as a hedge. Flowering mostly throughout the year.

7. **Ruellia** 3 spp.

Flowers generally showy, purple-violet to yellowish-white, solitary axillary or in terminal panicles or cymes. Bracts absent or small and narrow. Calyx 5-partite into narrow lobes. Corolla tube straight or curved; limb subequally 5-lobed, lobes spreading. Stamens 4. Leaves entire or margin spinescent.

496. **R. patula** Jacq. *Kakori Buti* Pl. 79

Hoary pubescent shrub. Leaves petiolate, 15-50 x 10-40 mm long. Flowers whitish mauve, solitary or 2-3, 30-40 mm long. Bracteoles leafy. Corolla tube 22-27 mm long. Capsule 12-16 mm long, glabrous.

Distribution: Tropical Africa, Arabia, Pakistan, south west India, Sri Lanka.

Stony or sandy ground particularly lower Sindh and Balochistan. A desert plant. September-March.

8. **Pteracanthus** 1 Sp

A genus sometimes included in *Strobilanthes,* characterised by 4 fertile stamens, curved and inflated corolla tube. Bracts soon falling off.

497. **P. urticifolius** (Kuntze) Bremek Pl. 80
Syn.: *Strobilanthes alatus* Nees

Erect shrub up to 1.5 m tall with sticky parts. Leaves elliptic-ovate, to ovate-cordate, toothed, narrowing into a winged petiole. Flowers deep blue to purple, in paniculate spikes. Bracts shorter than the calyx.

Distribution: Afghanistan, Pakistan, Kashmir, India to Nepal.

Common in shady places in forest undergrowth from 1800-3300 m in Chitral, Dir, Swat, Hazara and Murree hills. July-September.

9. **Eranthemum** 1 sp.

Shrubs or perennial herbs with simple petiolate and entire leaves. Flowers in terminal or axillary spikes. Bracts leafy. Corolla tube long, cylindrical; limb of 5 spreading overlapping lobes. Stamens 4, only two fertile.

498. **E. pulchellum** Andrews Pl. 80
Syn.: *Daedalacanthus nervosus* T. Anders.

Branched shrub up to 120 cm tall. Leaves large, elliptic-ovate to elliptic. Flowers deep blue, in spikes 20-50 cm long. Bracts quite conspicuous, elliptic-ovate, 10-12 x 5-8 mm.

Corolla tube 15-20 mm long; lobes obtuse.

Distribution: Tropical Asia.

Often cultivated for its pretty flowers. Wild in sub-Himalayan tracts in moist shady places up to 900 m. February-March.

10. Strobilanthes c. 5 spp.

Shrubs or herbs with opposite, entire to crenate or toothed leaves. Flowers solitary or in axillary or terminal spikes, panicles or heads. Bracts solitary. Capsule oblong, 4-8 seeded.

499. **S. glutinosus** Nees Pl. 80

An erect branched shrub up to 1.5 m tall with glandular-sticky parts. Leaves petiolate, elliptic to elliptic-ovate, 4-7 x 2-3.5 cm, hairy on both surfaces. Flowers showy, purplish-mauve, 30-50 mm long. Bracts and bracteoles glandular-hairy. Calyx deeply 5-lobed, glandular-pubescent. Capsule up to 2 cm long, 4-seeded.

Distribution: Pakistan, Kashmir to west Nepal.

Found in the sub-Himalayan tracts up to 1200 m. The plant parts especially the flower is strong smelling. Easily recognized by its winter flowering habit. November-March.

VERBENACEAE

Verbena Family

A large tropical and subtropical family of trees, shrubs or herbs with opposite leaves. Flowers irregular, bisexual. Corolla 2-lipped or subequally 5-lobed. Stamens 4, in two groups. Fruit a drupe or a capsule.

1.	+	Leaves palmately compound.	**7. Vitex**
	–	Leaves entire or dissected.	2
2.	+	Trailing herbs of wet places, rooting at the node.	**1. Phyla**
	–	Erect herbs or shrubs of drier situations.	3
3.	+	Fruit dry, splitting, of 4-seeded parts.	**2. Verbena** **3. Caryopteris**
	–	Fruit a drupe.	4
4.	+	Fruiting calyx much enlarged, saucer like.	**4. Holmskioldia**
	–	Fruiting calyx not so.	5

5. + Flowers small, in terminal or axillary heads or spikes. **5. Lantana**

– Flowers generally large and showy, paniculate or in corymbs. **6. Clerodendron**

1. Phyla 1 sp.

Prostrate perennial herbs with opposite leaves. Flowers very small, in ovoid, spike-like heads. Corolla 2-lipped, lower lip 3-lobed. Stamens exserted.

500. **P. nodiflora** (L.) Greene *Makra* Pl. 80
Syn.: *Lippia nodiflora* (L.) L. C. Rich.

Leaves somewhat fleshy, 5-30 x 5-15 mm, toothed on upper half. Spikes up to 4 cm long, solitary axillary. Flowers pale white, c. 3 mm long.

Distribution: Tropics and subtropics of the Old World.

Common and gregarious in damp places in the plains to 1600 m in all provinces of country. Flowering mostly throughout the year.

2. Verbena 3 spp.

Leaves deeply divided or toothed. Flowers in terminal simple or branched spikes. Corolla 2-lipped, 5-lobed, the lobes subequal. Stamens 4, in two groups. Fruit splitting into four 1-seeded parts (nutlets).

501. **V. tenuisecta** Briq. Pl. 80

A perennial herb up to 70 cm tall with pretty pink-mauve flowers. Spikes 6-10 cm long, terminal. Corolla tube 8 mm long, limb 5-lobed.

Distribution: Native to south America. Naturalized elsewhere.

Open places in the plains and up to 1200 m. Flowering mostly throughout the year.

3. Caryopteris 2 spp.

Shrubs with medium to small flowers in terminal or axillary clusters. Corolla tube short, cylindric; limb 2-lipped, 5-lobed. Stamens 4, exserted. Fruit a capsule.

502. **C. odorata** (D. Don) B. L. Robinson Pl. 80

A branched shrub up to 3 m tall with pretty purplish-blue flowers 10-12 mm long. Stamens exserted. Capsule 5 mm broad, globose.

Distribution: Pakistan, Kashmir to Bhutan, India.

Both cultivated in gardens and wild from plains to 1200 m in Swat, Hazara and Margalla hills. February-April.

4. Holmskioldia 1 sp.

Calyx enlarging in fruit, saucer-like. Corolla tube cylindric, curved; limb 2-lipped. Fruit a drupe, enclosed by the persistent red-scarlet calyx.

503. **H. sanguinea** Retz. *Turk's Turban* Pl. 81

A woody shrub with angled stems, brick-red flowers and prominent calyx. Leaves elliptic-ovate to ovate, 5-10 x 3-7 cm, toothed or subentire.

Distribution: Pakistan, Kashmir to Bhutan, India.

Both cultivated and wild in sub-Himalayan tracts to 1400 m especially Margalla hills. October-February.

5. Lantana 2 spp.

Shrubs or undershrubs with 4-angled stems. Leaves opposite, simple, usually crenate or toothed. Flowers small, congested in short spike-like or capitate clusters. Stamens 4, fruit a drupe.

+	Stems with short hooked spines. Flowers in capitate heads, yellow, red, orange or white.	504. **L. camara**
–	Stems unarmed. Flowers spicate, white.	505. **L. indica**

504. **L. camara** L. *Panch Phuli* Pl. 81

A vigorous rambling or straggling shrub with rough stems and branches. Leaves ovate to ovate-oblong. Flowering heads 20-30 mm broad. Flowers about 6 mm broad, yellow or orange sometimes pink but fading darker with age. Drupes 4 mm broad, shiny, black.

Distribution: Native of tropical America. Widely naturalized.

Both cultivated and wild and apparently spreading in the plains and foothills. Flowering mostly throughout the year.

505. **L. indica** Roxb. *Ghaneri* Pl. 81

A smaller shrub with white flowers in cylindrical to ovoid spikes.

Distribution: Pakistan, India, Thailand.

Common in the foothill zone to 900 m. March-September.

6. Clerodendron 2 spp.

Shrubs or trees with opposite leaves. Flowers often showy, in axillary or terminal clusters. Corolla limb subequaly 5-lobed. Stamens 4, exserted. Fruit a drupe.

506. **C. philippinum** Schau. Pl. 81
Syn.: *C. fragrans* Willd.

An erect leafy shrub up to 1.2 m tall. Leaves subcordate to broadly ovate, 8-20 x 7-15 cm. Flowers fragrant, white or pink, in dense terminal clusters.

Distribution: Native to China. Naturalized elsewhere.

Gregarious. Both planted as hedge and wild in waste places in sub-Himalayan tracts. Spreads by root suckers. Flowers fragrant. Evil smelling when crushed. September-October.

7. Vitex 1-2 spp.

Shrub or small trees with 3-5 (-7) palmately divided leaves. Flowers in terminal or axillary panicles or racemes, small, fragrant or not. Corolla tube cylindrical; limb 2-lipped. Stamens 4, exserted. Fruit a drupe.

507. **V. negundo** L. *Nirgud* Pl. 81

Shrubs 1-2.5 m tall. Stem and branches with white tomentose hairs. Leaflets 3-5, 5-12 x 15-40 mm, the middle one being the largest. Flowers pale violet to pink-mauve, small, c. 3 mm long. Corolla with 5 subequal lobes. Drupe c. 5 mm broad, subglobose.

Distribution: North Africa, Pakistan, India. Widely naturalized.

Cultivated and as an escape along water channels etc. from plains to 1600 m mainly in sub-Himalayan zone. Flowering mostly throughout the year.

AVICENNIACEAE

A monogeneric family of small trees or shrubs. Sometimes included under *Verbenaceae*. Forms an important component of mangrove vegetation in tropical coastal areas of world.

Avicennia 1 sp. *Salt Water Mangrove Family*

Shrubs or small trees with aerial roots (pneumatophores) projecting out of mud. Leaves simple, opposite, often greyish or yellowish-tomentose. Flowers in dense clusters. Fruit a broad compressed, one seeded capsule. Seed germinate in the fruit.

508. **A. marina** (Forssk.) Vierh. *Tivar, Timir* Pl. 81

Small shrubby trees with aerial roots. Leaves acute to acuminate, entire, shiny green above. Flowers yellowish in axillary or terminal clusters. Seeds often germinate on the parent plant before falling.

Distribution: Tidal swamps of Red Sea to Arabian Sea.

A dominant component of our mangrove vegetation along coastal areas of Balochistan and Sindh especially Indus delta area. The bark has tanning properties and leaves are used as fodder for camel and goats. Wood is used for fuel by local population. February-June.

LABIATAE *Mint Family*

Herbs or undershrubs usually with gland dotted leaves and 4-angled stems. Leaves opposite or whorled. Flowers irregular, bisexual. Corolla 4-5-lobed, limb usually distinctly 2-lipped. Stamens 4, in two groups, fertile or only 2 fertile. Ovary distinctly 4-lobed. Fruit of four 1-seeded nutlets.

1.	+	Plants spiny.	**3. Otostegia** **4. Lagochilus**
	–	Plants not spiny.	2
2.	+	Inflorescence densely hairy and velvet-like. Calyx lobes linear, plumosely hairy.	**1. Colebrookia**
	–	Inflorescence and calyx not so.	3
3.	+	Fertile stamens 2, the upper pair reduced.	**2. Salvia** **5. Perovskia**
	–	Fertile stamens 4.	4

4.	+	Upper lip of corolla entire or 2-5-lobed and reduced.	**6. Ajuga** **7. Teucrium** **8. Micromeria**
	–	Upper lip of corolla well developed.	5
5.	+	Plants with stellate, forked or scale-like hairs (sometimes simple in **Stachys**).	**9. Phlomis** **10. Stachys**
	–	Plants with simple hairs or glabrous parts.	6
6.	+	Fruiting calyx membranous, enlarged and narrow.	**11. Eremostachys** **12. Hymenocrater**
	–	Fruiting calyx not so.	7
7.	+	Calyx 15-ribbed. Upper stamen pair longer than lower.	**13. Nepeta**
	–	Calyx 5-15-ribbed. Upper stamen pair shorter than lower.	8
8.	+	Upper lip of corolla hooded.	**14. Prunella** **15. Scutellaria** **16. Leucas** **17. Lamium**
	–	Upper lip of corolla flat to somewhat curved.	**18. Thymus** **19. Origanum** **20. Mentha** **21. Anisomeles**

1. Colebrookia 1 sp.

Densely hairy or woolly shrubs with opposite, petiolate, crenate leaves. Flowers small, in branched woolly reddish-brown spikes. Corolla lobes 4. Stamens 4, included.

509. **C. oppositifolia** Smith *Shakardana* Pl. 81

A branched shrub up to 2.5 m tall. Stem and branches pale coloured. Leaves 8-18 x 10-70 mm, softly pubescent on both sides. The conspicuous fruiting spikes are 5-10 cm long, cylindrical. Corolla white.

Distribution: Pakistan, Kashmir to Bhutan, India, Burma, south west China, Indo-China.

Foothills up to 1300 m. Quite common in the lower Murree hills adjacent to the river Jhelum, also lower Swat. January-April.

2. Salvia 16 spp. *Sage*

Herbs or shrubs. Flowers small or large. Fertile stamens 2, the upper pair small and not fertile. Corolla 2-lipped.

1.	+	Flowers yellow.	510. **S. nubicola**
	–	Flowers pink, purple or shades of it.	2
2.	+	Flowers 5-6 mm long, white to pale pink, speckled with purple or blue.	511. **S. aegyptiaca**
	–	Flowers 20-30 mm long, pale pink or lilac, red or magenta.	3
3.	+	Plants densely white-canescent or woolly.	512. **S. lanata**
	–	Plants glabrous.	4
4.	+	Flowers red or magenta. Leaves pinnately divided.	513. **S. bucharica**
	–	Flowers pale pink or lilac. Leaves sub-orbicular, crenate.	514. **S. cabulica**

510. **S. nubicola** Wall. ex Sweet *Kharnar* Pl. 82

Perennial herb 100 cm or more tall, with sticky, strong scented parts. Flowers yellow, 24-26 mm long with brown markings on lower lip. Corolla tube much exceeding the calyx. Upper lip somewhat curved. Leaves ovate with spreading basal lobes.

Distribution: East Afghanistan, Pakistan, Kashmir to Bhutan.

Forest clearings, field edges, water courses or open places etc. from 2000-3700 m including Chitral, Swat, Hazara. July-August.

511. **S. aegyptiaca** L. *Tak Malanga* Pl. 82

A low branched undershrub up to 40 cm tall with rigid branches. Leaves ovate-oblong or narrow, 12-30 x 4-6 mm, crenate or toothed. Corolla white to pale pink, throat mottled; upper lip notched.

Distribution: Canary and Cape Verde Isls., north and north west Africa, Iran, Arabian Peninsula, Afghanistan, Pakistan, India.

A desert plant from plains and low dry hills in all provinces up to 1700 m. February-April.

512. **S. lanata** Roxb. Pl. 82

Perennial herb up to 30 cm tall, with dense softly woolly parts. Leaves 8-16 x 2-4 cm, oblanceolate or oblong, crenulate, densely white on undersurface. Flowers violet or pale blue-pink, 12-14 mm broad, disposed in distant whorls on the flowering axis. Corolla with upper lip well developed.

Distribution: Pakistan, Kashmir to west Nepal.

A handsome plant of open areas, slopes from 1200-3100 m in Murree hills, Swat, and Indus Kohistan. April-May.

513. **S. bucharica** M. Popov *Sursandah* Pl. 82

Shrub up to 60 cm tall. Leaf segments narrow. Flowers showy red or magenta, 20-30 mm long.

Distribution: Afghanistan, Pakistan, central Asia.

Open slopes, rocky places. Gregarious from 1700-2600 m, especially north Balochistan. April-May.

514. **S. cabulica** Benth. *Ghara Butai* Pl. 82

Shrubs 40-80 cm tall with whitish stems and ovate-oblong to suborbicular, crenate leaves. Flowers pale pink to lilac, 20-25 mm long.

Distribution: Afghanistan, Pakistan.

Dry hills from 1600-2400 m in Balochistan. Medicinal, locally used to cure colds and fevers. April-May.

3. **Otostegia** 3 spp.

Shrubs with spiny bracts and entire, toothed or crenate leaves. Corolla 2-lipped; upper lip oblong, hairy to the outside. Stamens 4, the lower pair longer.

515. **O. limbata** (Benth.) Boiss. *Chiti Buti, Marjoram* Pl. 82

Branched slender plants with whitish stems and branches. Leaves oblanceolate, 10-30 x 5-8 mm. Flowers in axillary clusters, pale yellow with orange throat, up to 22 mm long.

Distribution: Pakistan, Kashmir.

Dry places in the plains and lower hills up to 1000 m, especially Punjab, Salt Range and N.W.F.P. April-May.

4. Lagochilus 2 spp.

Resembling *Otostegia,* but with calyx equally 5-lobed, not enlarging in fruit.

516. **L. cabulicus** Benth. Pl. 82

Much branched perennial, with long linear leaf segments. Flowers pale yellow with a distinct persistent calyx.

Distribution: Afghanistan, Pakistan, Iran.

Stony slopes at Shahi Mahar (Gilgit) at 2500 m, also in Chitral. July.

5. Perovskia 2 spp.

Shrub with branches arising mainly from the base; leaves subentire to pinnately divided. Flowers pale violet. Corolla 2-lipped, exceeding calyx length; upper lip subequally 4-lobed. Stamens 4, the upper pair sterile.

517. **P. abrotanoides** Karel. *Tirk* Pl. 83

Aromatic shrub up to 100 cm tall with greyish hairy parts. Leaves pinnatisect, segments up to 5 mm long. Flowers in clusters of 2-4 on the nodes of the flowering axis. Corolla densely hairy on the outside, 8-10 mm long.

Distribution: Afghanistan, Pakistan to Kashmir, central Asia.

Stony slopes or dry valleys from 2100-4000 m. Common in Balochistan, Gilgit and Hunza. June-August.

6. Ajuga 2 spp.

Annual or perennial herbs with entire, toothed or divided leaves. Flowers in clusters of 2-5. Corolla 2-lipped, upper lip 2-lobed and reduced, the lower 3-lobed and longer. Stamens 4, fertile.

518. **A. bracteosa** Wall. ex Benth. *Kauri Booti* Pl. 83

Perennial pubescent herb. Leaves mostly radical, 15-40 x 10-20 mm, oblanceolate to elliptic or subspathulate, crenulate or subentire. Bracts leafy. Corolla pale white with pinkish streaks, 5-7 mm long.

Distribution: East Afghanistan, Pakistan, Kashmir to Bhutan, India, China, south Japan, Burma, Malaysia.

Open dry places especially northern Balochistan, N.W.F.P. from plains to 2200 m. March-December.

7. Teucrium 4-5 spp.

Resembling *Ajuga,* but corolla 1-lipped and 5-lobed.

519. **T. quadrifarium** Buch.-Ham. ex D. Don Pl. 83

Perennial herb up to 100 cm tall with hairy parts. Leaves ovate to oblong-ovate, base cordate, 25-70 mm long, toothed. Flowers reddish-purple, in stout racemes up to 14 cm long. Bracts leaf-like, broadly ovate. Corolla tube short, barely exceeding the calyx.

Distribution: Pakistan, Kashmir to Sikkim, India, Burma, China, Indo-China, Malaysia.

Rare. Found in the chir *(Pinus roxburghii)* zone, up to 1200 m. In Pakistan as yet recorded only from Margalla Hills above Islamabad. September.

8. Micromeria 1 sp.

Perennial herbs. Calyx generally 13-ribbed with narrow lobes. Flowers small. Corolla 2-lipped; lower lip 3-lobed; tube straight. Stamens 4.

520. **M. biflora** (Buch.-Ham. ex D. Don) Benth. Pl. 83

A straggling herb. Leaves sessile, 3-6 x 2-4 mm, ovate. Flowers in axillary, subsessile clusters of 2-4, small, pale pink. Upper lip of corolla somewhat flat. Stamen pair unequal.

Distribution: East Afghanistan, Pakistan, Kashmir to Bhutan, India, Burma and China.

Rock crevices, gravel slopes from 700-2200 m. in N.W.F.P., Chitral, Swat and Murree hills. April-October.

9. Phlomis 4 spp.

Perennial tomentose herbs with stellate or scale-like hairs. Leaves entire or subentire. Calyx 5-10-ribbed. Corolla 2-lipped. Upper lip hooded, tube about calyx length. Stamens 4.

1.	+	Leaves lanceolate, mostly cauline and shortly petiolate.	521. **P. stewartii**
	–	Leaves cordate or ovate-cordate, mostly basal, long petioled.	2

2. + Stems and branches angled. Plants 30-60 cm tall. Upper lip of corolla pilose on the outside. 522. **P. bracteosa**

– Stem and branches terete. Plants stouter, up to 120 cm tall. Upper lip of corolla densely hairy on the outside. 523. **P. spectabilis**

521. **P. stewartii** Hook. f. Pl. 83

Shrubby plant up to 50 cm tall, with floccose-tomentose parts. Leaves 9 cm, lanceolate. Bracts subulate, densely hairy. Corolla rosy-pink, upper lip densely tomentose.

Distribution: East Afghanistan, Pakistan.

Fairly common in dry places, stream beds in northern Balochistan and Waziristan from 900-2000 m. April-May.

522. **P. bracteosa** Royle ex Benth. Pl. 83

A low growing perennial, with tomentose parts. Leaves petiolate, cordate-ovate, crenate. Flowers bluish-purple or pinkish-purple, in axillary whorls, 25-35 mm broad. Corolla 15-20 mm long; upper lip erect, hood-like.

Distribution: East Afghanistan, Pakistan, Kashmir to Bhutan, India.

Alpine and sub-alpine meadows. Quite common on dry rocky slopes from 2900-3800 m from Chitral eastwards to Deosai plateau and Galis. June-August.

523. **P. spectabilis** Falc. ex Benth. *Gar* Pl. 84

A much more attractive and stouter plant than *P. bracteosa,* with long petioled cordate leaves, and stems and branches which are not angled. Flowers rose-purple.

Distribution: East Afghanistan, Pakistan, Kashmir to west Nepal.

Forest slopes, clearings from 900-2600 m in juniper zone, Balochistan, N.W.F.P. and Murree hills. July-August.

10. Stachys 6 spp.

Annual or perennial herbs with simple, stellate or scale-like hairs. Leaves simple. Flowers in whorls of up to 20. Calyx 5-10-ribbed. Corolla 2-lipped; upper lip entire or notched. Stamens 4, exserted.

+ Plants densely tomentose. Corolla c. 5 mm long, red. Stamens included in corolla tube. 524. **S. parviflora**

– Plants villous, with longer hairs. Corolla 12-15 mm long, pink-purple. Stamens excluded from the corolla tube. 525. **S. sericea**

524. **S. parviflora** Benth. *Bui* Pl. 84

A white hoary, erect perennial herb up to 45 cm tall. Leaves lanceolate, 40-60 x 10-20 mm, upper surface dark green, white hoary on the undersurface. Corolla almost included in the calyx.

Distribution: North east Iran, Afghanistan, Pakistan, Pamir-Alaj.

Often gregarious. Field edges etc. Plains to 1600 m. Evil smelling when crushed. April-May.

525. **S. emodi** Hedge Pl. 84
Syn.: *S. sericea* Wall. ex Benth.

Erect hairy herb up to 100 cm tall. Stems usually simple. Leaves cordate-ovate, 4-9 x 2-5 cm, crenate or toothed. Flowers pink, spotted purplish within, 12-18 mm long. Corolla well exserted from the calyx, 2-lipped; upper lip erect, entire.

Distribution: Afghanistan, Pakistan, Kashmir to Bhutan, India.

Common in forests. 1200-3100 m. July-August.

11. Eremostachys 8 spp.

Robust perennial herbs. Flowers 2-6 per whorl, showy. Corolla 2-lipped, upper lip hooded; tube hairy at the throat. Stamens 4.

1. + Calyx prominent, lobes triangular subulate, up to 7 mm long. Inflorescence broad pyramidal. 526. **E. thyrsiflora**

– Calyx not so prominent. Inflorescence not broad pyramidal. 2

2. + Flowers yellow. 527. **E. superba**

– Flowers white. 528. **E. vicaryi**

526. **E. thyrsiflora** Benth. Pl. 84

An unbranched plant up to 50 cm tall, with an attractive inflorescence of bracts, prominent calyces and white corolla suffused with yellow or pale purple.

Distribution: Afghanistan, Pakistan.

In dry stony places from 1600-2100 m in northern Balochistan. April-May.

527. **E. superba** Royle ex Benth. *Gajar Mula* Pl. 84

Spectacular perennial herb up to 100 cm tall, with simple stems. Leaves basal, 20-30 x 10-16 cm, broadly ovate, pinnately lobed into lanceolate or ovate-lanceolate segments. Flowers pale yellow, in axillary whorls of 5-11, on a stout axis.

Distribution: Afghanistan, Pakistan, Kashmir to Uttar Pradesh (India).

Edge of fields, open places etc. 300-1200 m especially sub-Himalayan tracts. March-April.

528. **E. vicaryi** Benth. ex Hook. f. *Gurganna* Pl. 84

Similar to *E. superba* but with white flowers.

Distribution: South east Iran, Pakistan and north west India.

Roadsides and bare plains up to 1800 m from southern Balochistan and N.W.F.P., extending to Rawalpindi and Attock districts. March-April.

12. Hymenocrater 1 sp.

Perennial herbs. Fruiting calyx enlarged, venose, often coloured. Corolla 2-lipped, lower 3-lipped. Stamens 4.

529. **H. sessilifolius** Benth. Pl. 85

A low growing branched shrub up to 40 cm tall. Leaves ovate-cordate, toothed. Flowers reddish-purple. Calyx 17-20 mm long, purplish suffused. Corolla tube 12-15 mm long.

Distribution: East Iran, Afghanistan, Pakistan.

Gregarious on dry slopes often making a spectacular display from 1600-2800 m in Balochistan. April-May.

13. Nepeta c. 48 spp. *Catmint*

Perennial herbs with erect to suberect stems. Leaves toothed or crenate. Flowers in dense or interrupted spikes, forming verticillasters or not. Calyx 15-ribbed. Corolla 2-lipped; upper lip 2-fid; the lower 3-lobed, with the median lobe longer than the two laterals; tube straight or curved. Stamens 4, the upper pair longer than lower.

+ Flowers spicate, blue-purple. 530. **N. laevigata**

– Flowers in distant whorls, blue or pale blue. 531. **N. clarkei**

530. **N. laevigata** (D. Don) Hand. -Mazz. Pl. 85
Syn.: *N. spicata* Wall. ex Benth.

Plants up to 40 cm tall. Leaves stalked, 20-40 cm, ovate, toothed. Flowers blue-purple, in spicate racemes. Upper lip of corolla 2-lobed.

Distribution: Afghanistan, Pakistan, Kashmir to south west China.

Common in meadows and forest clearings from 1800-3800 m in Chitral, Swat, Kaghan and Murree hills. July-September.

531. **N. clarkei** Hook. f. Pl. 85

Differing from *N. laevigata* in the lighter coloured pale blue or mauve flowers which are in interrupted spikes. Corolla tube 12-15 mm long, upper lip by lobed, lower lip white with blue spots.

Distribution: Pakistan, Kashmir and Tibet.

Open places and stony slopes. Common from 2100-3400 m from Chitral to Gilgit. July-August.

14. Prunella 1 sp.

Perennial herbs with entire to divided leaves. Flowers in dense, short, broad spikes. Calyx 2-lipped, 10-veined. Corolla 2-lipped, upper lip hooded; throat hairy. Stamens 4, exserted.

532. **P. vulgaris** L. *Self-Heal* Pl. 85
Syn.: *Brunella vulgaris* Moench

An erect hairy perennial 8-20 cm long. Leaves elliptic-ovate to elliptic-lanceolate, 20-40 cm long, usually toothed. Flowers in compact oblong spikes up to 8 cm long, blue-violet.

Distribution: Afghanistan, Pakistan, Kashmir to Bhutan. Temperate Europe and north America. Naturalized in south Africa and Australia.

Common in meadows, moist slopes and irrigation channels from 1800-3300 m in the Himalaya and the Karakorams. June-August.

15. Scutellaria 16 spp.

Perennial herbs with flowers in racemes or spikes. Calyx 2-lipped. Corolla 2-lipped; lower lip broad, somewhat recurved; tube long. Stamens 4, the lower pair longer.

533. **S. linearis** Benth. *Mastiari* Pl. 85

Stems prostrate to suberect, branched. Leaves 8-15 x 3-7 mm, lanceolate or narrowly so. Flowers c. 20 mm long, in short racemes, rose pink with the lower lip of corolla pale yellow.

Distribution: East Afghanistan, Pakistan, Kashmir to Uttar Pradesh (India).

Open gravelly slopes, banks from 700-1800 m in Balochistan, N.W.F.P. and Murree hills at lower levels. March-April.

16. Leucas 7 spp.

Erect to suberect hairy herbs. Flowers in axillary or terminal whorls. Calyx tubular, 10-toothed. Corolla 2-lipped, upper lip hook-like, hairy, lower 3-lobed; tube more or less included in calyx. Stamens 4, in unequal pairs.

+ Flower heads 3-5 cm broad. Leaves lanceolate, crenate or toothed. 534. **L. cephalotes**

– Flower heads up to 2 cm broad. Leaves ovate, toothed. 535. **L. urticaefolia**

534. **L. cephalotes** (Roth) Spreng. Pl. 85

Erect annual up to 35 cm tall. Leaves 3-5 x 1.5-2 cm, lanceolate, crenate-serrate. Flowers white, many, in heads 3-5 cm broad.

Distribution: East Afghanistan, Pakistan, Kashmir, India.

Common field weed in and around cultivation. Plains to 1600 m from Peshawar, Swabi, lower Swat and Rawalpindi. July-September.

535. **L. urticifolia** (Vahl) R. Br. Pl. 86

Similar to *L. cephalotes,* but with broader leaves and smaller flower heads.

Distribution: East Africa, Arabian peninsula, Pakistan, India.

Weed from plains to 900 m in Sindh and Punjab. July-September.

17. Lamium 2 spp.

Annual decumbent or ascending herb with hairy parts. Flowers in axillary whorls, the upper whorls crowded. Calyx 5-toothed. Corolla 2-lipped; upper hood-like, lower 3-lobed, the laterals very small, tube narrowing towards the base. Stamens 4, in unequal pairs.

536. **L. amplexicaule** L. Pl. 86

A prostrate branched herb with stems and branches up to 30 cm long. Leaves suborbicular to reniform-cordate, 8-30 x 5-10 mm. Flowers in whorls of 6-14, purplish-red, 12-16 mm long. Floral leaves larger than the stem leaves. Upper lip of corolla 4-5 mm long, hairy.

Distribution: Temperate Eurasia, introduced elsewhere.

Common weed of fields from plains to 2500 m in Balochistan, N.W.F.P., and Salt Range. March-May.

18. Thymus 1 sp. *Thyme*

Aromatic perennial herbs or undershrubs with entire leaves. Flowers usually in terminal heads, small. Calyx 2-lipped, 10-13-veined. Corolla 2-lipped, upper lip 2-lobed, lower 3-lobed; tube straight. Stamens 4.

537. **T. linearis** Benth. Pl. 86
Syn.: *T. serpyllum* auct. non L.

A prostrate spreading often tufted herb. Leaves elliptic-ovate to ovate-lanceolate, 4-10 x 4-5 mm, gland dotted. Flowers 5-6 mm long, pink, in terminal heads.

Distribution: Afghanistan, Pakistan, Kashmir, India, west Nepal, Tibet, China, Japan.

A variable species widespread in the Himalaya from 1600-3500 m. Two subspecies are recognized in our area; the subspecies *hedgei* is confined to Balochistan and east Afghanistan. Leaves are dried and used as tea and to flavour meat dishes. June-August.

19. Origanum 2 spp.

Perennial herbs often with a woody base. Stems several, ascending or erect. Flowers in dense spike-like branched inflorescences. Bracts yellowish-green. Calyx 5-toothed, 10-13-veined. Corolla 2-lipped; upper lip 2-lobed. Stamens 4.

538. **O. vulgare** L. *Banjawain, Wild Marjoram* Pl. 86

An erect aromatic pubescent herb up to 50 cm tall. Leaves stalked, ovate, 15-20 x 10-14 mm. Flowers small, white to pale pink, 5-6 mm long, on spikes 6-20 mm long. Floral leaves with a purplish tinge.

Distribution: Temperate Europe, The Mediterranean, Pakistan, Kashmir to Bhutan, China and Taiwan.

Common and gregarious in the forest zone especially Swat, Murree hills. Open places and dry banks from 900-3000 m often as a pioneer on eroded slopes. Leaves used as a herb to flavour food. July-September.

20. Mentha c. 6 spp. Mint

Perennial herbs, often stoloniferous with aromatic herbage. Flowers in whorls on axillary or terminal spikes. Calyx 5-toothed, 10-13-veined. Corolla subequally 4-lobed; upper broader; tube shorter than calyx. Stamens 4.

539. **M. longifolia** (L.) L. *Horse Mint* Pl. 86
Syn.: *M. sylvestris* L.

Erect rhizomatous herb up to 70 cm tall. Stems 4-angled. Leaves 20-60 x 10-30 mm, elliptic-oblong to oblanceolate, toothed, densely hairy like the stems. Flowering spikes 20-90 cm long. Flowers pink to mauve or violet, small, stamens exserted.

Distribution: North and west Asia, Europe, Pakistan, Kashmir to central Nepal, south Africa.

Leaves used for flavouring food. Wet places, stream sides. Very common and gregarious. Plains to 3300 m from Chitral east to Baltistan and higher hills of Hazara. July-August.

21. Anisomeles 1 sp.

Erect herbs. Leaves toothed. Flowers in dense terminal spikes. Corolla 2-lipped; upper lip erect, entire; lower 3-lobed. Stamens 4.

540. **A. indica** (L.) O. Kuntze Pl. 86
Syn.: *A. ovata* R. Br.

A branched herb up to 120 cm tall with red flowers. Leaves ovate, 20-25 mm long, crenate. Flowers reddish-purple, tinged white, 10-12 mm long. Seeds shiny black.

Distribution: Pakistan, Kashmir to Sikkim, India, Sri Lanka, China, Malaysia.

Evil smelling. Found from the plains to 1600 m in Salt Range, Attock, Hazara and Rawalpindi districts. July-September.

NYCTAGINACEAE *Four O'Clock Family*

Trees, shrubs or herbs with alternate or opposite leaves. Flowers in cymose clusters. Bracts sometimes petaloid. Perianth 5-lobed. Stamens as many as the lobes and alternating with them.

1.	+	Flowers large, showy. Bracts prominent.	**1. Mirabilis**
	–	Flowers small. Bracts not prominent.	2
2.	+	Fruit 5-ribbed, glabrous or not, but without wart-like glands.	**2. Boerhavia**
	–	Fruit 10-ribbed, with large wart-like glands.	**3. Commicarpus**

1. Mirabilis 1 sp.

Perennial herbs with opposite leaves. Flowers in axillary cymes, surrounded by 5 bracts. Perianth tube funnel form. Stamens 5-6, exserted. Fruit ribbed.

541. **M. jalapa** L. *Gul-e-Abbasi, Four O'Clock Plant* Pl. 87

Herb up to 1.5 m tall with leafy stems that are reddish and swollen at the nodes. Leaves ovate, 5-10 x 2-5 cm, acuminate. Flowers red or sometimes yellow or white, in clusters of 4-6. Perianth tube 20-30 mm long. Fruit ovoid, 8-9 mm long, black, tubercled.

Distribution: A native of south America. Cultivated and naturalized elsewhere.

Near waste places, in and around cultivation. Naturalized. Plains and foothills up to 1200 m. Flowers open in the late afternoon. September-October.

2. Boerhavia 4-5 spp.

Flowers small, in panicles or capitate heads. Stamens usually 5, exserted. Fruit club-like, 5-ribbed, glabrous or glandular.

542. **B. procumbens** Banks ex Roxb. *Itsit* Pl. 87

A prostrate slender glabrous spreading herb, often with reddish stems and branches. Leaves ovate or oblong, 10-40 x 5-30 mm, margin somewhat wavy. Flowers pink c. 3 mm long, in axillary panicles. Stamens 2-3, exserted. Fruit club-like, 5-ribbed, glandular and papillose.

Distribution: South west Asia, Pakistan, India.

A common tropical herb. From plains to the foothills up to 2200 m in all four provinces. Leaves can be used as a pot herb. Roots medicinal. September-August.

3. Commicarpus 4 spp.

Like *Boerhavia,* but fruit 10-ribbed, with large wart-like glands.

543. **C. boissieri** (Heimerl) Cufod. Pl. 87
Syn.: *Boerhavia verticillata* auct. non Poir.

A straggling or decumbent branched perennial. Leaves ovate, subcordate, 20-50 x 10-45 mm, sometimes repand or sinuate, more or less fleshy. Flowers in superposed umbels, pink, 3-4 mm long. Fruit 5-8 mm long, nodding, club-like, with a ring of wart-like glands around the apex.

Distribution: South Iran, Pakistan, India.

A straggler on bushes. Sandy places, low hills up to 700 m. Sindh, Punjab and Balochistan. September-October.

CHENOPODIACEAE *Goosefoot Family*

Annual or perennial salt loving herbs with usually succulent leaves. Flowers small, greenish. Sepals 3-5. Petals absent. Stamens as many as or fewer than sepals.

Chenopodium c. 13 spp.

Annual with erect or prostrate stems. Leaves alternate, entire or lobed. Flowers in spiked panicles. Stamens 5, filaments free. Styles 3.

544. **C. foliolosum** (Moench) Asch. *Strawberry Goosefoot* Pl. 87
Syn.: *C. blitum* Hook. f.

An attractive prostrate herb with bright red, axillary clusters of fruit. Leaves triangular, coarsely toothed, petiolate.

Distribution: Temperate Europe, north Africa, Pakistan, Kashmir to central Nepal.

Waste places, cultivated areas from 1500-3600 m, very common in Gilgit. Leaves used as pot herb. June-August.

POLYGONACEAE *Bistort and Sorrel Family*

Herbs or shrubs often with swollen stem nodes. Leaves alternate, simple. Stipules ochreate. Flowers bisexual, small, regular. Perianth 3-6, greenish or coloured, in two whorls. Stamens 6-9.

1. + Shrubs. Stamens 10-16. Fruit bristly. **1. Calligonum**

– Herbs, sometimes with a woody base. Stamens less than 10. Fruit not bristly. 2

2.	+	Perianth segments 4.	**2. Oxyria**
	–	Perianth segments 5 or 6.	3
3.	+	Perianth segments 5.	**3. Bistorta** **4. Polygonum**
	–	Perianth segments 6.	**6. Rheum** **5. Rumex**

1. Calligonum 1 sp.

545. **C. polygonoides** L. Pl. 87

Branched shrubs 2-3 m tall with small leaves and silvery white woody branches. Stipules ochreate. Flowers whitish, c. 5 mm broad, in dense axillary clusters. Stamens 10-16. Fruit covered with bristly hairs.

Distribution: Palestine, Syrian Desert, Soviet Armenia, north west and central Iran, Sinai, Pakistan, India.

Desert tracts particularly where shifting sand dunes occur, characteristic of Cholistan, Thal and Balochistan. March-April.

2. Oxyria 1 sp.

Perennial herb with basal kidney shaped leaves. Perianth segments 4. Stamens 6.

546. **O. digyna** (L.) Hill Pl. 87

A low growing tufted herb up to 25 cm tall. Leaves basal, long stalked, reniform, more or less fleshy. Flowers greenish. Fruit winged, 4-5 mm broad, reddish.

Distribution: Arctic or subarctic Siberia, Europe, north America, west and central Asia, Pakistan, Kashmir to Bhutan, Tibet, west China, Japan.

Common. Alpine and sub-alpine slopes particularly in wet conditions throughout northern region of country from 2200-4600 m. May-June.

3. Bistorta 7 spp.

Like *Polygonum,* but flowers in spike-like clusters.

+	Leaves linear or linear lanceolate. Flowering stems up to 30 cm tall.	547. **B. affinis**

– Leaves ovate-lanceolate, with a sheathing subcordate base. Flowering stems 35-50 cm long. 548. **B. amplexicaulis**

547. **B. affinis** (D. Don) Green Pl. 88
Syn.: *Polygonum affine* D. Don

Low growing carpet-like herbs with a woody base. Leaves mostly radical, 4-10 cm long, linear or linear-lanceolate, margin inrolled. Flowers red, in dense short spikes 6-8 cm long.

Distribution: Afghanistan, Pakistan, Kashmir to Nepal.

Alpine. Gregarious on scree slopes in wet places in Himalaya from 3000-4800 m. June-August.

548. **B. amplexicaulis** (D. Don) Green *Masloon* Pl. 88
Syn.: *Polygonum amplexicaule* D. Don

A taller plant with a single flowering stalk up to 60 cm tall. Flowers white, pink or red. Leaves ovate-lanceolate, base sheathing.

Distribution: East Afghanistan, Pakistan, Kashmir to Bhutan, west and central China.

Gregarious on forest slopes, clearings from 1800-3100 m in Himalaya. June-August.

4. Polygonum

Annual or perennial herbs. Flowers small, in axillary clusters. Stipules ochreate, membranous.

549. **P. plebeium** R. Br. Pl. 88

A prostrate mat forming herb. Leaves linear to oblong. Stipules hyaline. Flowers red or pink, 2-3 mm long, in axillary clusters.

Distribution: Africa, west Asia, Himalaya, Pakistan, India, China, Ryukyu Isls., Malaysia, Australia.

Common in moist places in the plains to 1500 m. Also found in cultivated fields. February-May.

5. Rumex c. 10 spp. *Sorrel*

Annual or perennial herbs. Flowers in axillary clusters in simple or branched racemes. Perianth segments 6, in two whorls of three each; the inner much larger than the outer. Stamens 6. Fruit a nutlet.

1.	+	Plant 30-90 cm tall, leafy. Leaves 5-18 cm long, oblong obtuse, margin crisp.	550. **R. dentatus**
	–	Low growing or bushy plant. Leaves up to 5 cm, hastate or ovate.	2
2.	+	Leaves hastate. Calyx not inflated in fruit.	551. **R. hastatus**
	–	Leaves ovate. Calyx inflated in fruit.	552. **R. vesicarius**

550. **R. dentatus** L. Pl. 88
subsp. **klotzschianus** (Meissn.) Rech. f.

An erect annual up to 90 cm tall. Leaves mostly radical; pale green, with a crisp margin. Flowers in terminal branched racemes, small. Fruiting sepals enlarged, ovate or obovate, winged; wings nervose, toothed.

Distribution: Iran, Afghanistan, Pakistan, Kashmir to Sikkim, India, China.

Common weed of plains and hills up to 3300 m in all four provinces. March-May.

551. **R. hastatus** D. Don Pl. 88

A bushy perennial up to 90 cm tall. Leaves hastate, pale green. Flowers in terminal paniculate clusters, pinkish, small. Nutlet pinkish.

Distribution: Afghanistan, Kashmir to Bhutan, west China.

A colonizer of disturbed slopes, banks etc. Low hills to 2500 m. generally in Murree hill areas, lower Swat, N.W.F.P., Chitral and Gilgit. March-August.

552. **R. vesicarius** L. Pl. 88

An annual up to 30 cm tall. Leaves ovate, 8-25 x 5-18 mm; obtuse; calyx inflated in fruit, greenish, with a pinkish tinge.

Distribution: North Africa, west Asia, Pakistan, Kashmir to Nepal.

Plains. Dry slopes and desert areas from Makran to N.W.F.P., Salt Range and Attock district. The inflated calyx is characteristic. March-April.

6. Rheum c. 5 spp. *Rhubarb*

Perennial herbs with a woody base and large palmate, prominently nerved leaves. Flowers bisexual, paniculate. Perianth segments 6, free, not enlarging in fruit. Stamens 9.

553. **R. webbianum** Royle Pl. 89

A stout perennial 50-120 cm tall. Leaves long stalked, 10-40 cm broad. Stipules leafy. Flowers pale yellowish, in axillary branched terminal spikes. Nutlets broadly winged.

Distribution: Pakistan, Kashmir to west Nepal.

Open rocky places etc. from 2200-3400 m. In Balochistan the stalks are used as a vegetable. June-August.

THYMELEACEAE *Daphne Family*

Perennial herbs with entire leaves. Flowers in racemes, spikes or fascicles, bisexual, regular. Petals small or absent. Fruit a berry or drupe.

+	Calyx without scale-like hairs.	**1. Daphne**
–	Calyx with scale-like hairs.	**2. Wikstroemia**

1. Daphne 3 spp.

Undershrubs with entire somewhat leathery leaves. Flowers in heads or short racemes. Calyx 4-lobed, lobes spreading, petals absent. Fruit a berry.

+	Involucral bracts absent. Leaves elliptic-oblong, 30-50 mm long.	554. **D. mucronata**
–	Involucral bracts present. Leaves elliptic-lanceolate, 60-140 mm long.	555. **D. papyracea**

554. **D. mucronata** Royle Pl. 89

Plants up to 2 m tall. Flowers white, in axillary or terminal clusters. Corolla tube 6-8 mm long. Berries globose, 8-10 mm broad, bright red.

Distribution: North Africa, south Europe, Iran, Afghanistan, Pakistan, Kashmir to Uttar Pradesh (India).

Common in dry areas from Balochistan and throughout northern areas, also Hazara district. River banks, shrubberies, from 800-3000 m. Plant parts are medicinal, poisonous to livestock. April-September.

555. **D. papyracea** Wall. ex Steud. Pl. 89

Evergreen plants 1-2 m tall. Stem and branches glabrous. Flowers white, in terminal clusters of 4-10, 8-10 mm long. Berry ovoid, 8-10 mm long, scarlet.

Distribution: Pakistan, Kashmir to west Nepal.

Forest undergrowth from 1700-3300 m. March-April.

2. **Wikstroemia** 1 sp.

Trees or shrubs with opposite leaves. Flowers in short terminal heads, racemes or spikes. Calyx 4-lobed. Stamens 8, in two series. Fruit a berry.

556. **W. canescens** Meissn. Fig. between p. 102-103

Shrub up to 1 m tall. Leaves elliptic, oblong or lanceolate, 25-70 mm long. Flowers yellow, axillary or extra-axillary. Berry 4-5 mm long, ovoid.

Distribution: Afghanistan, Pakistan, Kashmir to Nepal, Khasia, Sri Lanka, China.

In forests. 2100-3100 m. Bark of plants used in fibre and paper making. June-September.

ELAEAGNACEAE *Oleaster Family*

Trees or shrubs, often spinose. Leaves simple, often covered with stellate or scale-like hairs. Flowers in racemes, spikes or clusters, regular. Perianth 2-4-lobed. Stamens as many or twice the number of perianth lobes. Fruit drupe-like.

Hippophae 1 sp.

Deciduous shrubs or small trees with narrow leaves. Flowers unisexual. Male flowers in axillary clusters. Perianth lobes 2, opposite. Stamens 4.

557. **H. rhamnoides** L. *Sea Buckthorn* Pl. 89
subsp. **turkestanica** Rousi

A thorny shrub up to 2.5 m tall. Leaves 15-50 cm, elliptic-lanceolate to oblong. Male flowers c. 2.5 mm long, in clusters at the base of the shoot; calyx lobes suborbicular. Fruit subglobose, 5-6 mm broad, in clusters, succulent, orange.

Distribution: Afghanistan, Pakistan, Kashmir to Himachal Pradesh.

Dry areas from 1800-3600 m, particularly Gilgit, Hunza, and Baltistan. Often planted as a hedge in northern areas of Pakistan. April-May.

LORANTHACEAE *Mistletoe Family*

Evergreen semiparasitic shrubs with opposite, simple and entire leaves. Flowers in clusters or racemes, rarely solitary, regular. Perianth of 4-8 free or united parts. Stamens attached to and as many as the perianth parts. Fruit a berry or a drupe.

Loranthus 4 spp.

Flowers bisexual, in axillary racemes or fascicles. Perianth lobes 4-5, free or united into a tube. Stamens opposite the lobes. Fruit a berry.

558. **L. longiflorus** Desr. *Parand* Pl. 89

A leafy shrub semi-parasitic on various trees. Leaves 4-10 cm long, elliptic-ovate to elliptic-lanceolate or oblaceolate, opposite, somewhat leathery. Flowers orange-red, c. 2.5 cm long, 5-lobed; lobes narrow. Berry 8-10 mm broad, oblong.

Distribution: Tropical and subtropical Himalaya, Sumatra, New Guinea, Assam, Sri Lanka.

Conspicuous when in flower. Semi-parasitic on various trees of the foothill zone up to 800 m, such as siris *(Albizzia lebbek)*, acacia *(Acacia modesta, A. catechu)*, silk cotton tree *(Bombax ceiba)* etc. The nectariferous flowers and mucilagenous berries are attractive to birds which are responsible for its propagation. December-February.

SANTALACEAE *Sandal Wood Family*

Trees, shrubs or semiparasitic herbs. Leaves simple. Flowers small, regular, in racemes, spikes or heads. Perianth 4-5-lobed. Stamens as many as perianth lobes. Ovary inferior. Fruit a drupe or a nut.

Thesium 2 spp.

Perennial root parasites. Stem simple or branched. Leaves alternate, sessile, narrow, 1-3-nerved. Flowers in racemes or panicles. Perianth lobes 5, persistent. Fruit a nut.

559. **T. himalense** Royle ex Edgew. Pl. 89

Slender decumbent herbs up to 30 cm tall. Leaves 12-30 mm, linear to linear-lanceolate, 1-nerved, glabrous. Flowers white, racemose, 2.5-3.5 mm long, funnel-form. Bracts exceeding flower length. Perianth persistent in fruit and overlapping it. Lobes obtuse. Nut subglobose, 2.5-3 mm long, nervose.

Distribution: Pakistan, Kashmir to Sikkim, India, west China.

Uncommon. Foothills from 600-2200 m. April-June.

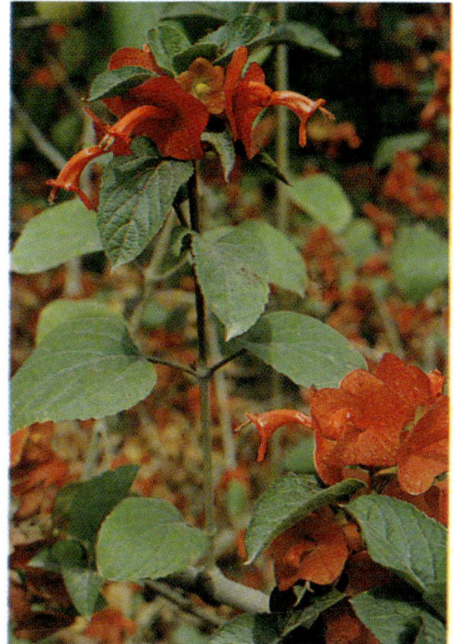

503. *Holmskioldia sanguinea* Margalla hills, Y. J. Nasir

507. *Vitex negundo* Margalla hills, S. A. Sultan

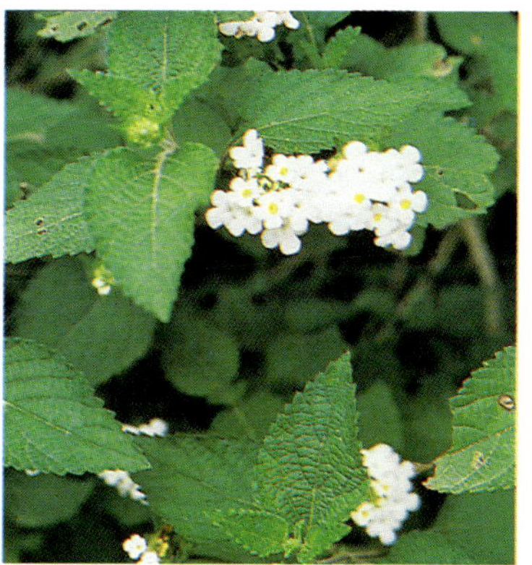

505. *Lantana indica* Margalla hills, S. A. Sultan

508. *Avicennia marina* Karachi, Rubina A. Rafiq

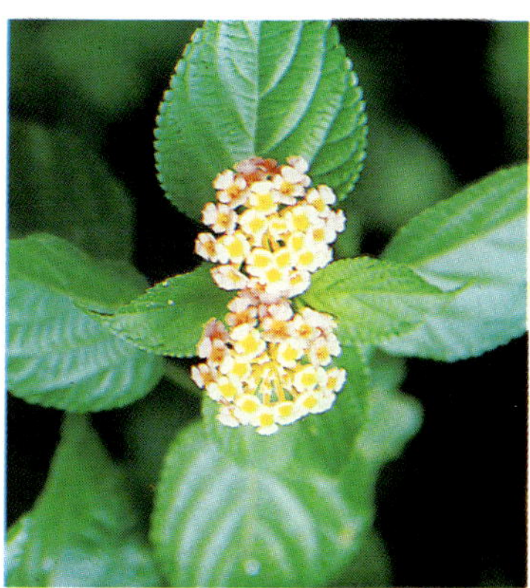

504. *Lantana camara* Margalla hills, S. A. Sultan

509. *Colebrookia oppositifolia* Margalla hills, Y. J. Nasir

506. *Clerodendron philippinum* Islamabad, Y. J. Nasir

515. *Otostegia Limbata* Margalla hills, Rubina A. Rafiq

511. *Salvia aegyptiaca* Karachi, Y. J. Nasir

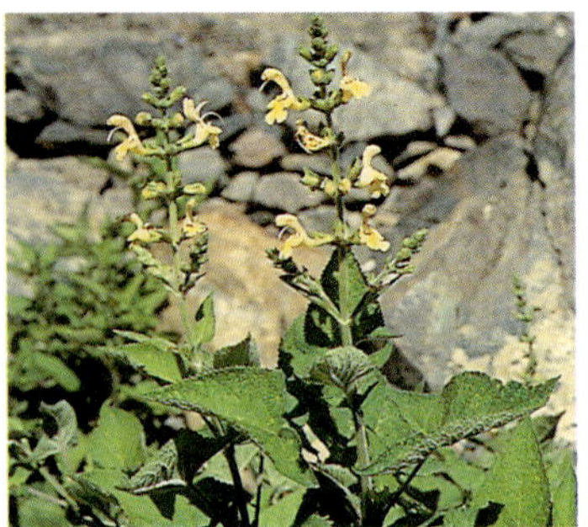

510. *Salvia nubicola* Chitral, T. J. Roberts

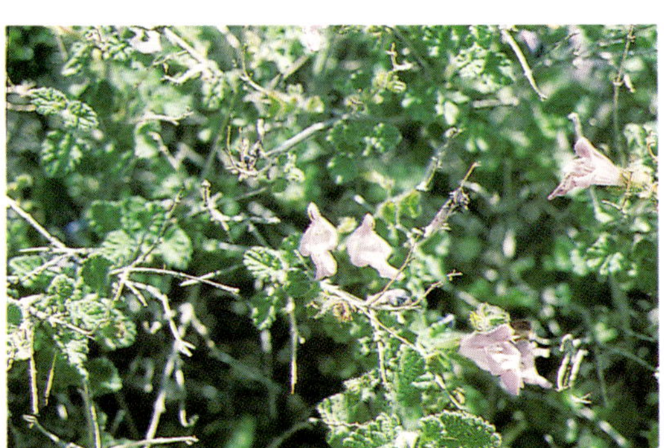

514. *Salvia cabulica* Chiltan hills, Rubina A. Rafiq

516. *Lagochilus cabulicus* Yasin, Gilgit, T. J. Roberts

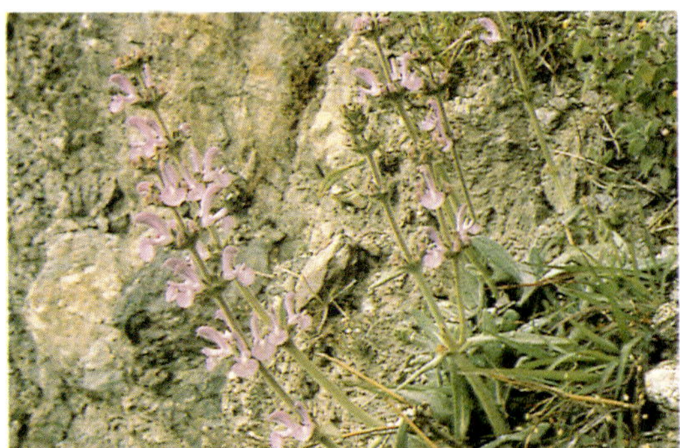

512. *Salvia lanata* Shangla pass, T. J. Roberts

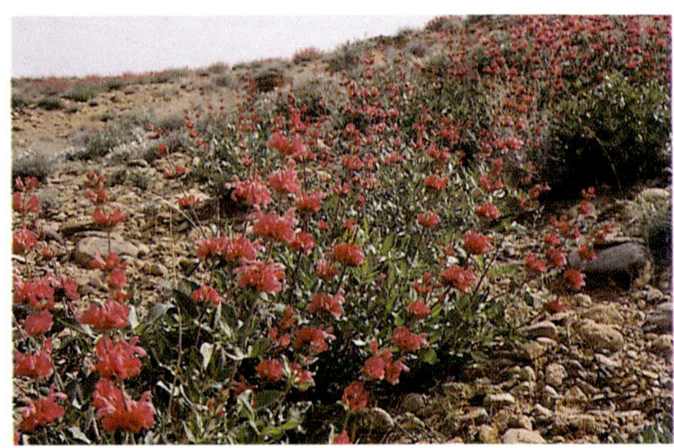

513. *Salvia bucharica* Surkhab, Pishin, Rubina A. Rafiq

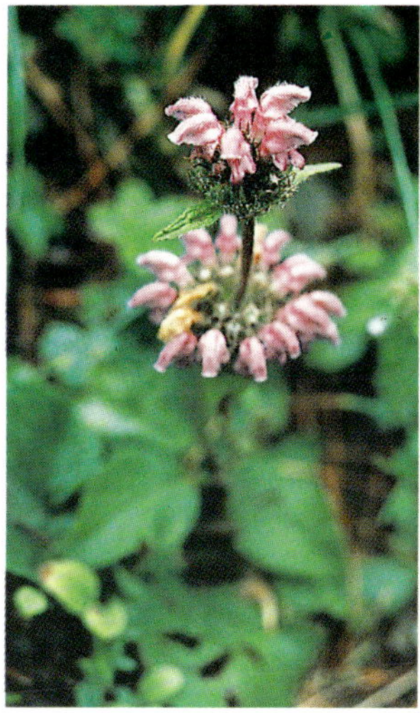

522. *Phlomis bracteosa* Kaghan, Rubina A. Rafiq

521. *Phlomis stewartii* Sanjavi, Loralai, Rubina A. Rafiq

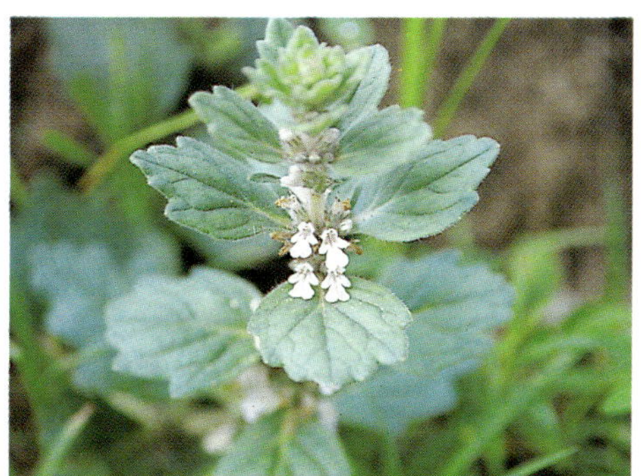

518. *Ajuga bracteosa* Islamabad, S. A. Sultan

519. *Teurcrium quadrifarium* Margalla hills, S. A. Sultan

517. *Perovskia abrotanoides* Skardu, Fritz Berger

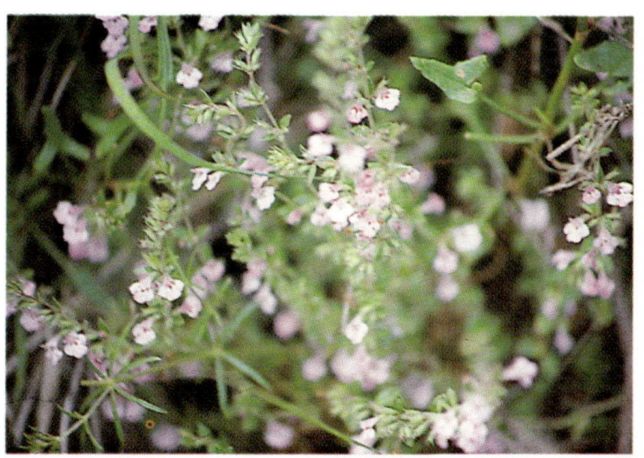

520. *Micromeria biflora* Murree hills, S. A. Sultan

524. *Stachys parviflora* Margalla hills, S. A. Sultan

526. *Eremostachys thyrsiflora* Chiltan hills, Rubina A. Rafiq

528. *Eremostachys vicaryi* Islamabad, Y. J. Nasir

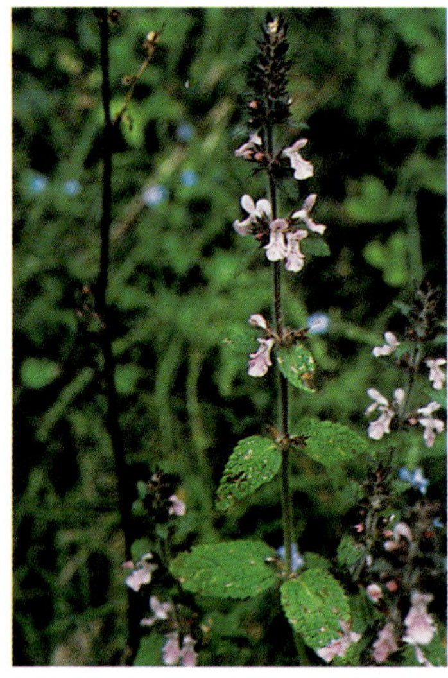

525. *Stachys emodi* Thandiani, S. A. Sultan

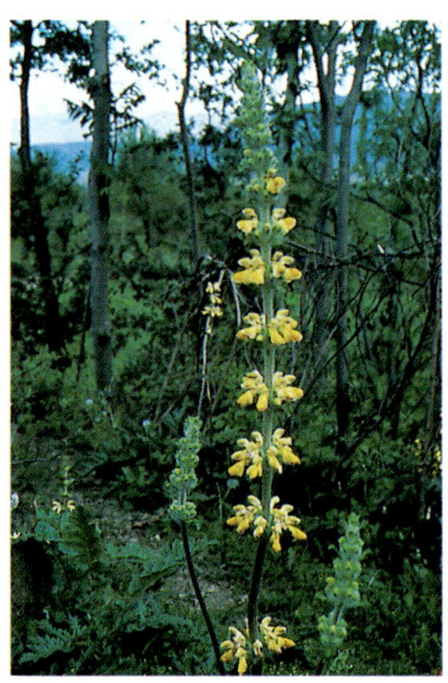

527. *Eremostachys superba* Swat, S. A. Sultan

523. *Phlomis spectabilis* Margalla hills, S. A. Sultan

531. *Nepeta clarkei* Lalazar, Kaghan, Y. J. Nasir

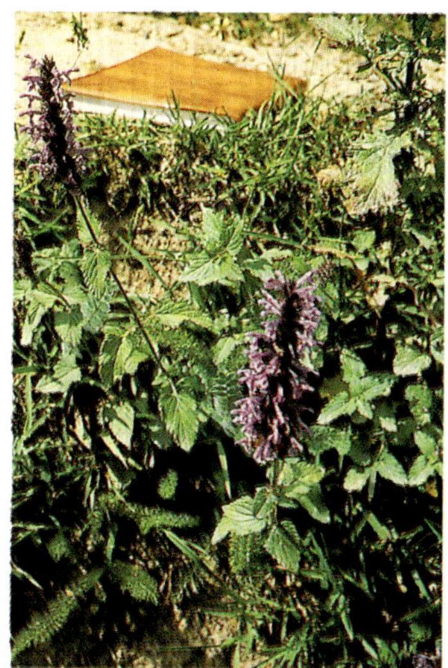

530. *Nepeta laevigata* Lalazar, Kaghan, Y. J. Nasir

534. *Leucas cephalotes* Margalla hills, S. A. Sultan

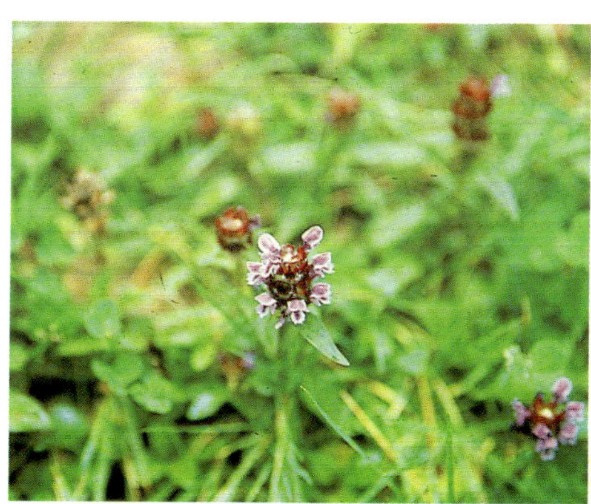

532. *Prunella vulgaris* Nathia Gali, S. A. Sultan

533. *Scutellaria linearis* Lower Kaghan, S. A. Sultan

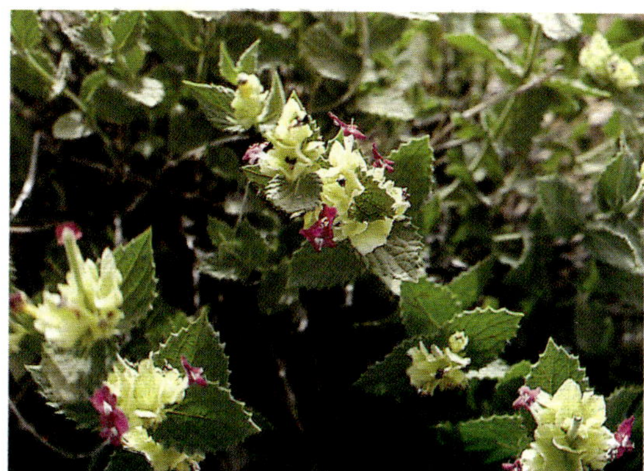

529. *Hymenocrater sessilifolius* Surkhab, Pishin, Rubina A. Rafiq

540. *Anisomeles indica* Margalla hills, S. A. Sultan

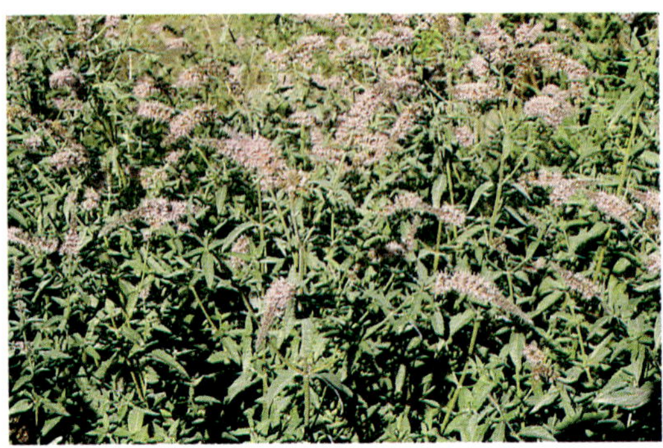

539. *Mentha longifolia* Lowari pass, Y. J. Nasir

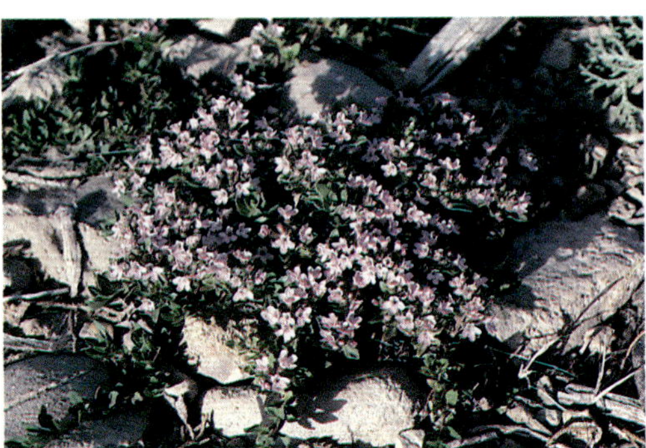

537. *Thymus linearis* Ziarat, Balochistan, Rubina A. Rafiq

538. *Origanum vulgare* Swat, Y. J. Nasir

536. *Lamium amplexicaule* Margalla hills, Rubina A. Rafiq

535. *Leucas urticifolia* Karachi, Y. J. Nasir

541. *Mirabilis jalapa* Murree hills, Rubina A. Rafiq

546. *Oxyria digyna* Kalam, Fritz Berger

543. *Commicarpus boissieri* Sindh, Khan Muhammad Khan

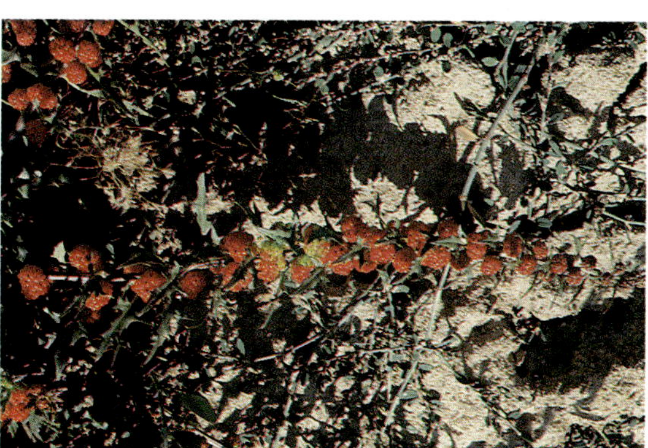

544. *Chenopodium foliolosum* Lalazar, Kaghan, Y. J. Nasir

542. *Boerhavia procumbens* Islamabad, S. A. Sultan

545. *Calligonum polygonoides* Thal desert, Y. J. Nasir

552. *Rumex vesicarius* Attock, S. A. Sultan

547. *Persicaria affinis* Kalam, Fritz Berger

551. *Rumex hastatus* Lower Swat, Fritz Berger

549. *Polygonum plebeium* Islamabad, Y. J. Nasir

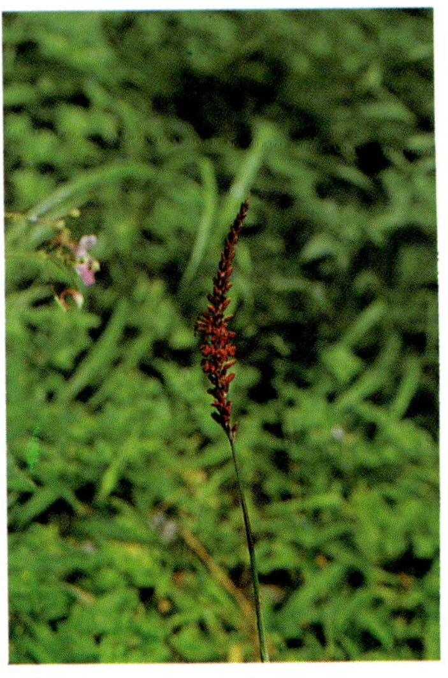

548. *Persicaria amplexicaulis* Kaghan, Y. J. Nasir

550. *Rumex dentatus* Islamabad, Tahira Z. Mahmood

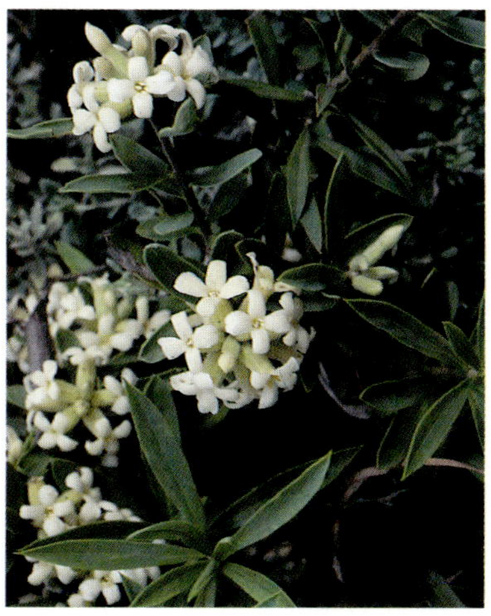

554. *Daphne mucronata* Swat, S. A. Sultan

555. *Daphne papyracea* Murree hills, S. A. Sultan

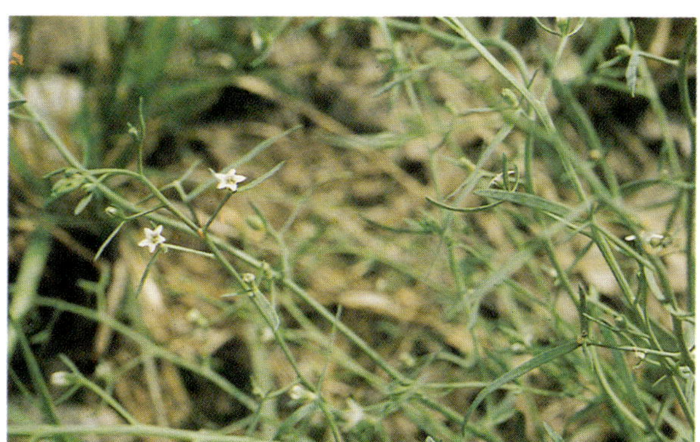

559. *Thesium himalense* Margalla hills, Y. J. Nasir

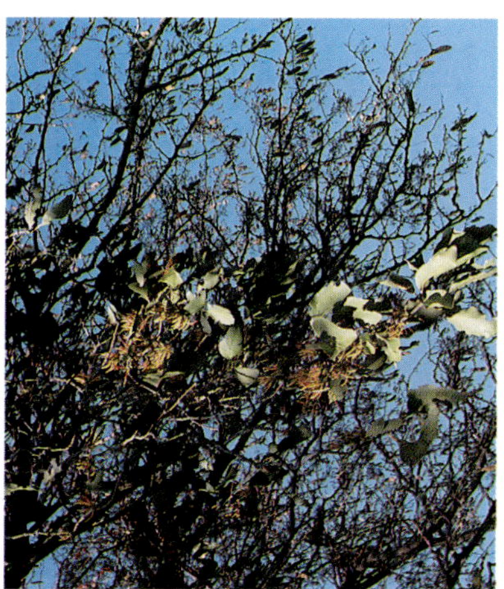

558. *Loranthus longiflorus* Margalla hills, Rubina A. Rafiq

557. *Hippophae rhamnoides subsp. turkestanica* Chitral, T. J. Roberts

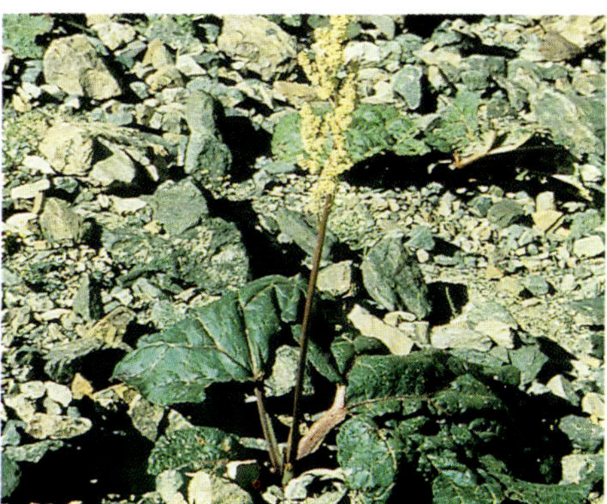

553. *Rheum webbianum* Northern Chitral, T. J. Roberts

564. *Euphorbia prolifera* Islamabad, Y. J. Nasir

560. *Euphorbia caducifolia* Bela, Makran, Y. J. Nasir

563. *Euphorbia osyroidea* Chiltan hills, Y. J. Nasir

562. *Euphorbia wallichii* Mokshpuri, Y. J. Nasir

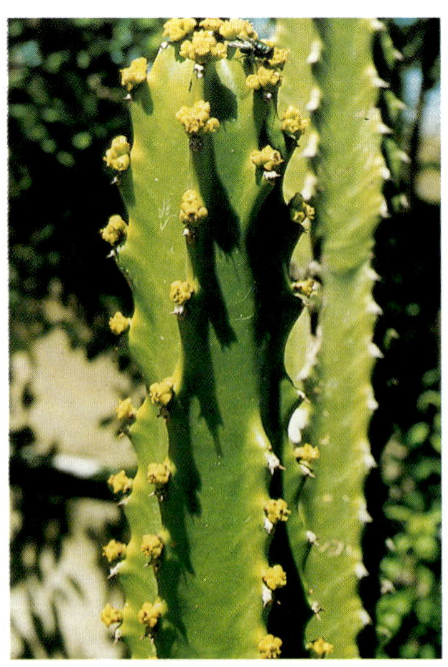

561. *Euphorbia royleana* Margalla hills, Y. J. Nasir

567. *Mallotus philippensis* Islamabad, Rubina A. Rafiq

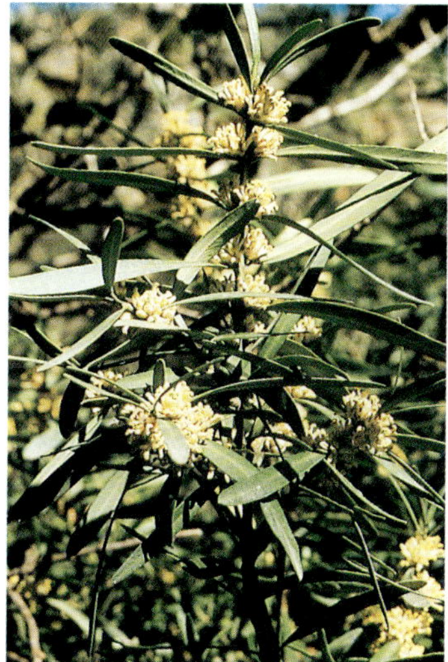

568. *Buxus papillosus* Margalla hills, Tim Holmes

566. *Phyllanthus emblica* Margalla hills, Y. J. Nasir

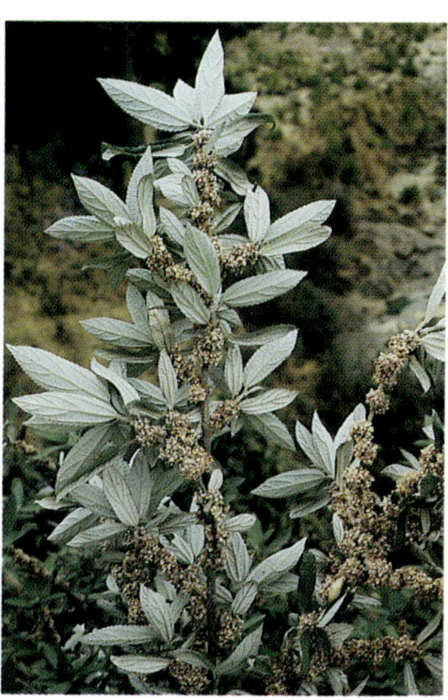

571. *Debregeasia saeneb* Lower Murree hills, Rubina A. Rafiq

569. *Juglans regia* Shogran, Kaghan, S. A. Sultan

570. *Girardinia palmata* Nathiagali, Rubina A Rafiq

576. *Quercus leucotrichophora* Margalla hills, Y. J. Nasir

577. *Salix denticulata* Islamabad, Y. J. Nasir

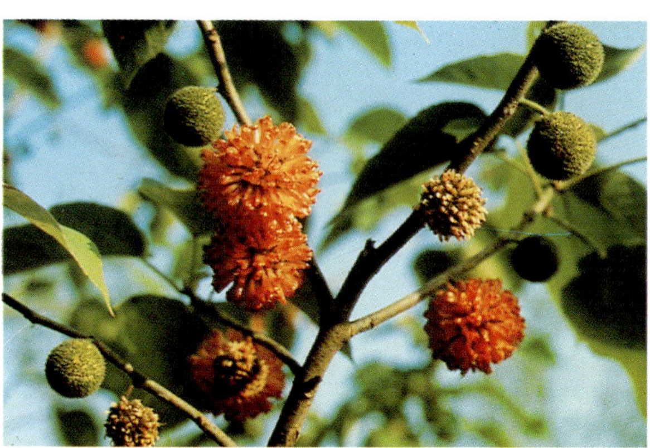

574. *Broussonetia papyrifera* Islamabad, T. J. Roberts

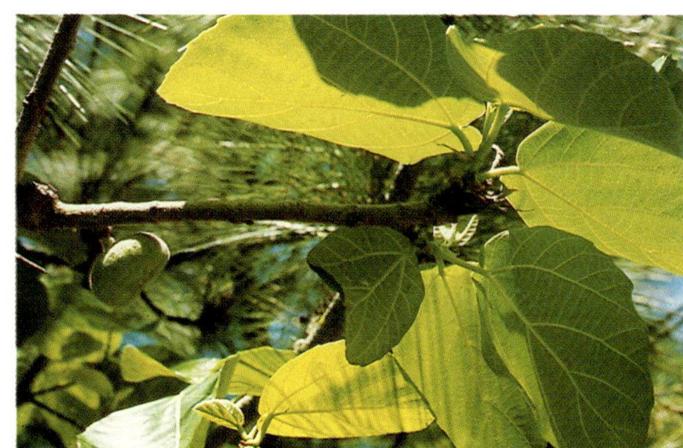

573. *Ficus virgata* Margalla hills, Y. J. Nasir

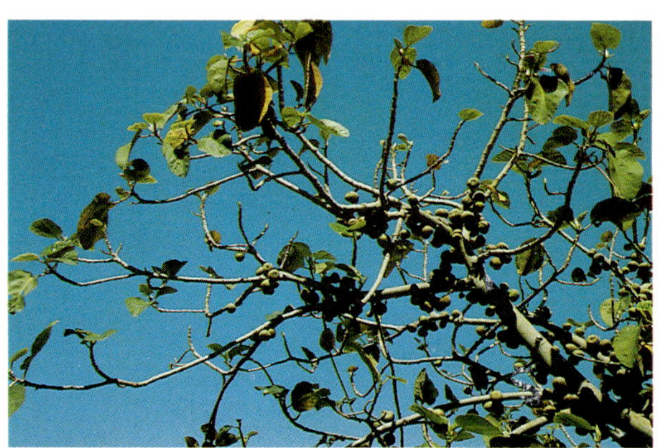

572. *Ficus auriculata* Margalla hills, Rubina A. Rafiq

575. *Betula utilis* Upper Swat, Fritz Berger

582. *Nervilea gammieana* Margalla hills, Y. J. Nasir

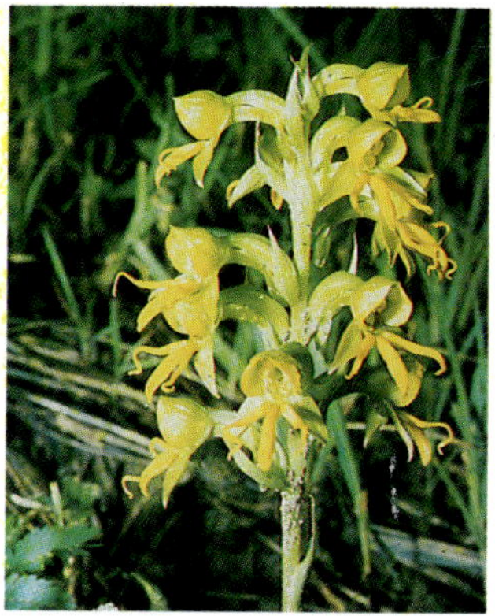

584. *Habenaria marginata* Swat, Jenny Renz

580. *Cymbidium macrorhizon* Murree hills, Y. J. Nasir

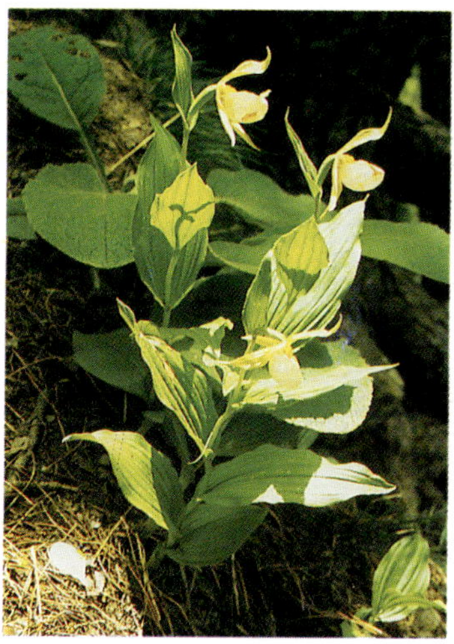

579. *Cypripedium cordigerum* Kaghan, Y. J. Nasir

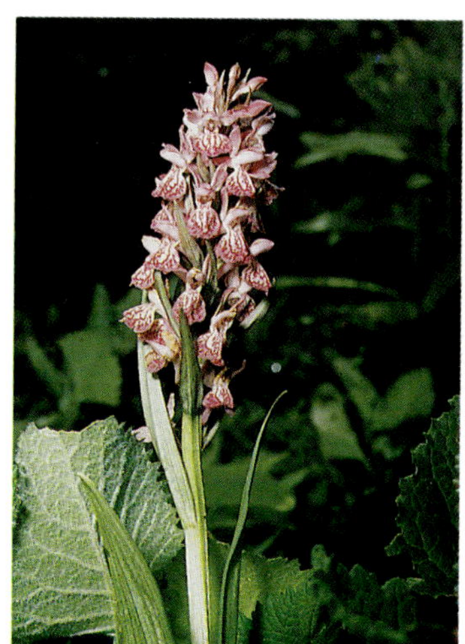

581. *Dactylorhiza hatagirea* Swat, Fritz Berger

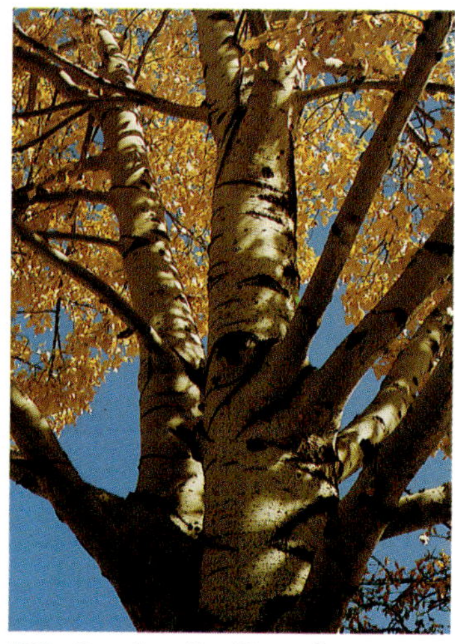

578. *Populus alba* Utror, Swat, Y. J. Nasir

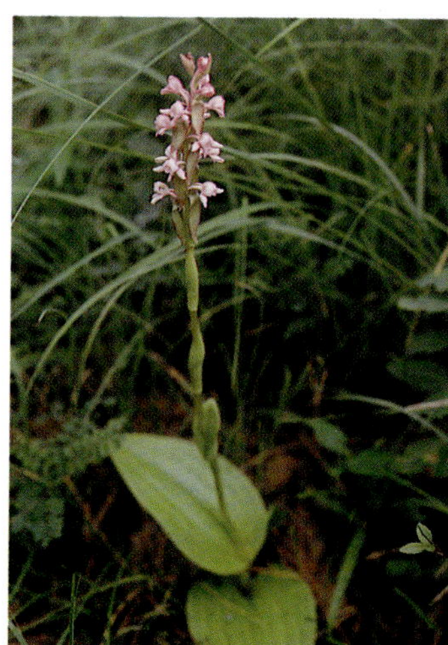

583. *Satyrium nepalense* Murree hills, Rubina A. Rafiq

589. *Cephalanthera longifolia* Thandiani, Y. J. Nasir

585. *Habenaria digitata* Murree hills, Rubina A. Rafiq

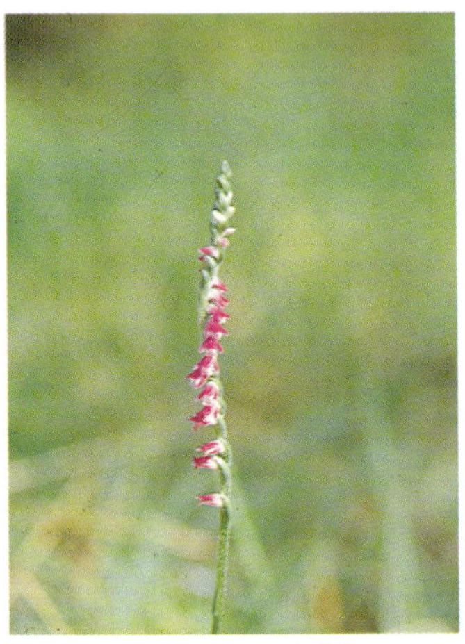

590. *Spiranthes sinensis* Murree hills, S. A. Sultan

◀a b▲

586. *Habenaria latilabris* a, b: Murree hills, Rubina A. Rafiq

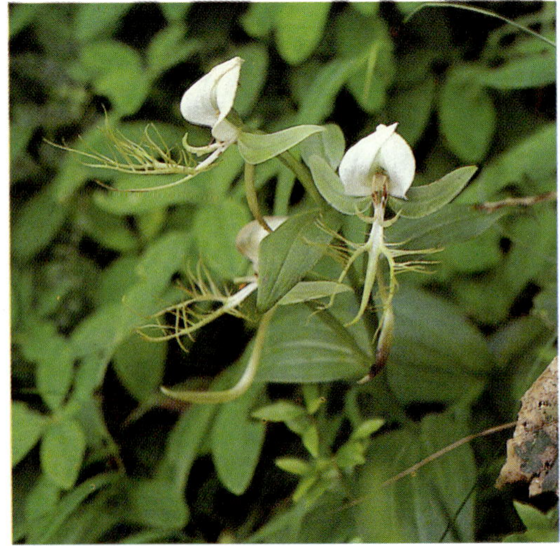

587. *Habenaria intermedia* Murree hills, Rubina A. Rafiq

b ▶

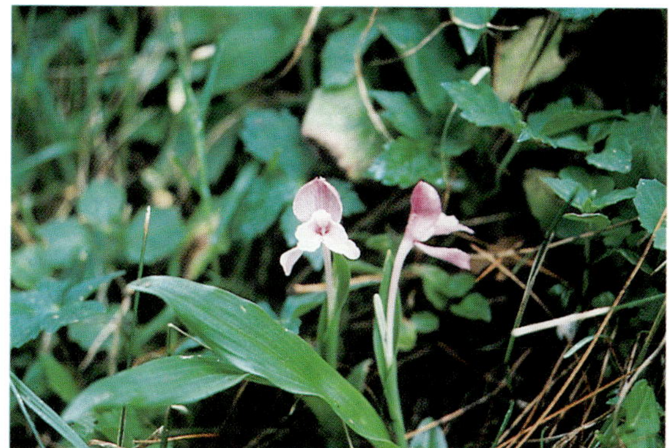

593. *Roscoea alpina* Thandiani, Rubina A. Rafiq

▼ a

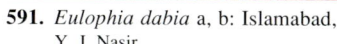

591. *Eulophia dabia* a, b: Islamabad, Y. J. Nasir

594. *Iris aitchisonii* Swat, Fritz Berger

592. *Zeuxine strateumatica* Islamabad, Y. J. Nasir

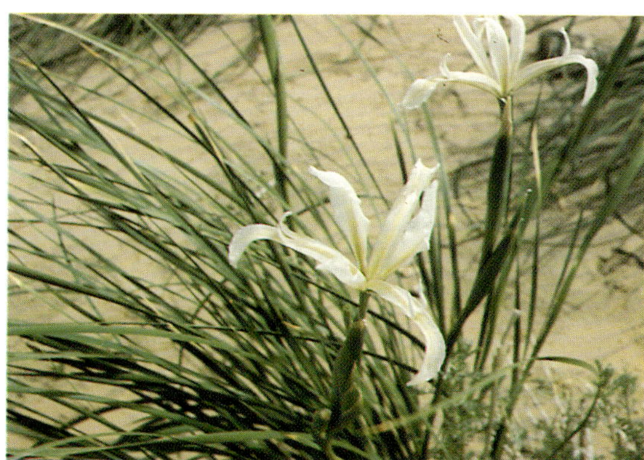

596. *Iris songarica* Khanozai, Muslimbagh, Rubina A. Rafiq

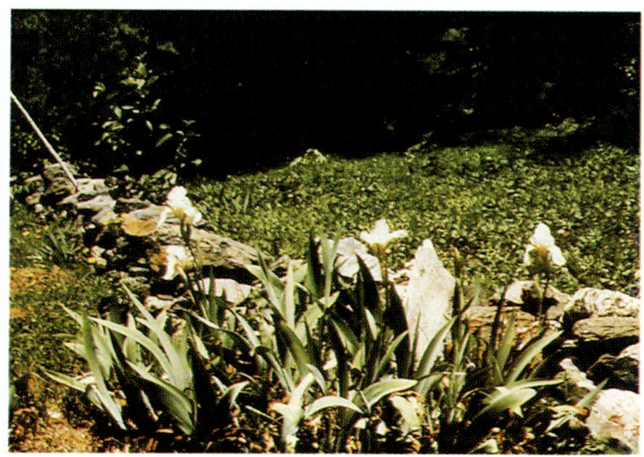

595. *Iris germanica* Shahran, Kaghan, T. J. Roberts

598. *Iris loczyi* Ziarat, Balochistan, Rubina A. Rafiq

599. Gynandrisis sisyrinchium Lower Swat, Rubina A. Rafiq

601. *Allium humile* Shahran, Kaghan, T. J. Roberts

603. *Allium caspium* Quetta, Rubina A. Rafiq

600. *Allium fedtschenkoanum* Deosai, Fritz Berger

597. *Iris hookeriana* Kaghan, S. A. Sultan

602. *Allium neopolitanum* Chaprot, Hunza, Y. J. Nasir

EUPHORBIACEAE

Spurge or Castor Oil Family

A large tropical and subtropical family of trees, shrubs or herbs, generally exuding a milky juice. Leaves simple or compound, stipulate. Flowers unisexual, regular, small, 3-6-merous, variously disposed in axillary, terminal cymes, panicles, racemes or sometimes solitary. Calyx or corolla sometimes absent. Stamens 3 to many. Styles 3. Fruit splitting into 3 parts or indehiscent or a drupe.

1.	+	Inflorescence a cyathium. Stamen 1.	**1. Euphorbia**
	–	Inflorescence various, but not as above. Stamens more than 1.	2
2.	+	Trees 2-6 m tall.	**4. Mallotus**
	–	Generally annual or perennial herbs or shrubs up to 1.2 m tall (Trees in *Phyllanthus* also).	3
3.	+	Hairs if present simple. Fruit baccate or a drupe.	**3. Phyllanthus**
	–	Hairs stellate. Fruit schizocarpic.	**2. Chrozophora**

1. Euphorbia 48 spp.

Herbs, shrubs or trees, with a milky juice. Leaves mostly sessile, simple. Male and female flowers much reduced and enclosed in a cup-like involucre. Styles 3. Fruit 3-lobed, dehiscent.

1.	+	Shrubs or small trees, 3-5 m tall cactus-like, spiny.	2
	–	Perennial herbs up to 1.2 m tall, not spiny.	3
2.	+	Shrubs, often tree-like, but not forming a trunk.	560. **E. caducifolia**
	–	Tree-like, with distinct trunk.	561. **E. royleana**
3.	+	Bracts leafy, yellow.	562. **E. wallichii**
	–	Bracts not so.	4
4.	+	Uppermost leaves bract-like, suborbicular or ovate-rhombic.	563. **E. osyroidea**
	–	Uppermost leaves ovate to ovate-lanceolate.	564. **E. prolifera**

560. **E. caducifolia** Haines Pl. 90

Tree-like shrubs, branched from the base, up to 3.5 m high. Branches green. Leaves 20-50 mm long, thick, ovate or broadly so and sprouting only during the monsoon. Stipules spiny, paired, stout.

Distribution: Pakistan, India.

Common in sandy and stony desert subject to monsoon influence. Found up to 800 m particularly Thar desert and Las Bela. February-March.

561. **E. royleana** Boiss. *Danda Thor* Pl. 90

Similar to *E. caducifolia,* but plants with a distinct stem forming a candelbra shape. Stipules 3-5 mm long.

Distribution: Sub-Himalayan tracts from Indus eastward to Kashmir, Uttar Pradesh (India).

Common from 700-1800 m in sub-Himalayan tract and Salt Range. April-May.

562. **E. wallichii** Hook. f. Pl. 90

Perennial tufted herbs up to 70 cm tall. Leaves 30-100 x 8-25 mm, elliptic-lanceolate to ovate-lanceolate, bright green above. Flowers in umbellate heads, surrounded by leaf-like, attractive yellow bracts.

Distribution: Afghanistan, Pakistan, Kashmir to Nepal, south west China.

The most attractive of our herbaceous euphorbias. Open places, slopes in the forest zone from 2100-3200 m. Quite common in the Murree hills. July-August.

563. **E. osyroidea** Boiss. Pl. 90

Perennial up to 1 m tall; stems many, arising from a rootstock. Leaves linear-lanceolate, 8-50 x 8-35 mm. Uppermost leaves suborbicular to rhombic-ovate, 4-10 mm long, cuspidate.

Distribution: Iran, Afghanistan, Pakistan.

Common in dry rocky places, stream sides etc., from 1500-2200 m in Balochistan and Sindh. March-April.

564. **E. prolifera** Buch.-Ham. Pl. 90

Perennial herb up to 60 cm tall. Stems several, arising from a rootstock. Leaves linear-lanceolate or oblanceolate, 12-50 x 3-10 mm. Uppermost leaves ovate to ovate-lanceolate.

Distribution: Pakistan, Kashmir to Bhutan, India, north Burma, south west China.

Field borders. The plains to 1600 m in northern Punjab and N.W.F.P. February-September.

2. Chrozophora 4 spp.

Annual or perennial herbs, stellately hairy. Leaves alternate, simple, petiolate, margin often wavy. Flowers unisexual, in spikes, racemes or subpanicles. Female flowers long peduncled in fruit. Fruit splitting by valves.

565. **C. tinctoria** (L.) Rafin. Fig. between p. 102-103

Perennial herbs up to 80 cm tall, covered with silvery scales. Lamina 25-60 x 10-50 mm, rhombic-ovate or broadly ovate, 3-nerved, margin wavy. Male flowers 3-4 mm long, yellowish-green. Female flowers with peduncles 20-30 mm in fruit. Fruit subglobose, 3-lobed, 5 x 8 mm, covered with silvery scales.

Distribution: North west Africa, Spain, Arabia, Pakistan, India.

Weed in harvested fields, roadsides from 800-2200 m in Punjab, Sindh and Balochistan. January-September.

3. Phyllanthus 10 spp.

Trees, shrubs or herbs. Leaves alternate, simple, entire. Flowers unisexual, small, solitary or clustered; female flowers with stouter pedicels. Ovary 3-lobed. Fruit a berry or a drupe.

566. **P. emblica** L. *Amla* Pl. 91

A small to medium sized tree up to 15 m tall, with pendant branches and attractive foliage. Leaves simple (but appearance as compound), 60-170 per shoot, 5-12 x 3-4 mm, linear-oblong. Male flowers yellowish-green. Fruit subglobose, c. 2.5 cm broad, pale green.

Distribution: Pakistan, Kashmir to Bhutan, Assam, north Burma, south China, Indo-China, Malaysia.

Fairly common and wild in sub-Himalayan tract but often planted around houses. Source of vitamin C. Fruit used in pickles. Cultivated in plains, 900-1000 m. March-May.

4. Mallotus 1 sp.

Trees and shrubs. Flowers appearing before the leaves. Female flowers with ovary 3-lobed. Fruit globose, compressed, 3-lobed.

567. **M. philippensis** (Lam.) Muell. -Arg. *Kamila* Pl. 91

A small tree or shrub up to 4 m tall. Leaves petiolate; lamina ovate or broadly so, 4-12 x 2-8 cm, subacuminate, 3-nerved, upper surface dark green. Male flowers in erect short cluster of spikes. Fruit 3-lobed, subglobose, 6-8 mm broad, brick red, with a powdery covering.

Distribution: Pakistan, Kashmir to Bhutan, India, Sri Lanka, Indo-China, Malaysia, Australia, Polynesia.

Common in the foothills up to 1200 m. The red powder on the fruit is medicinal and is also used as a dye. February-November.

BUXACEAE *Box Wood Family*

Evergreen monoecious shrubs or trees, closely related to the *Euphorbiaceae,* but differing in the absence of milky juice and styles free from the base.

Buxus 2 spp.

Branched shrubs. Leaves opposite. Flowers small, greenish-yellow, in dense racemose clusters. Fruit a 3-beaked capsule, dehiscing by 3 valves.

568. **B. papillosa** C.K. Schneid. *Shamshad* Pl. 91

A leafy shrub up to 2.5 m tall. Branches ascending from the base. Leaves lanceolate or elliptic-oblong, 10-50 x 5-10 mm. Flowering racemes 8-14 mm long, with a terminal female flower. Stamens of male flowers exserted. Capsule 3-angled, 7-9 mm long, ovoid, horns 2, erect.

Distribution: Endemic to Pakistan.

Dryer foothill zone between 600-3000 m especially northern Balochistan and N.W.F.P. and Margalla hills. February-March.

JUGLANDACEAE *Walnut Family*

Monoecious deciduous trees with alternate, compound leaves. Flowers small; male flowers in catkins, female 1-5, in terminal short spikes or racemes. Ovary inferior. Fruit a drupe or nut.

Juglans 1 sp. *Walnut*

Bark and foliage resinous. Male catkins lateral. Stamens 8-40. Stigmas 2, recurved or spreading, feathery. Drupe large with a green spotted skin, shell woody.

569. **J. regia** L. *Akhrot, Himalayan Walnut* Pl. 91

Trees up to 25 m tall. Leaves 17-40 cm long; leaflets 5-9, opposite or subopposite, ovate to elliptic-ovate, 6-18 x 2.5-7 cm. Male catkins up to 12 cm long, greenish-yellow. Female flowers 1-3, terminal on short spikes. Drupe 3-5 cm broad, globose to subglobose, skin glandular, shell 2-valved. Seeds 2-4-lobed, deeply rugose longitudinally.

Distribution: Central America, south west Europe, Caucasus, Syria, north Iran, Afghanistan, Pakistan, Kashmir to Nepal, Tibet, north Burma, west China.

Both cultivated and wild from 1000-3200 m. In the drier Himalayan valleys, specimens often grow to magnificent proportions, especially in Chitral and Swat Kohistan. Valuable for its wood which is used in making furniture. The thin shelled variety is valued for its fruit. The bark of young twigs is good for the gums. February-April.

URTICACEAE *Nettle Family*

Herbs or shrubs, often with stinging hairs. Leaves simple, alternate or opposite. Flowers small, greenish, regular, 4-5-merous, unisexual, in axillary clusters. Fruit an achene or a drupe.

\+ Plants with stinging hairs.
Leaves lobed or toothed. **1. Girardinia**

– Plants without stinging hairs.
Leaves entire, undersurface white. **2. Debregeasia**

1. Girardinia 1 sp.

Herbs or subshrubs with large 3-ribbed, toothed and petiolate leaves. Flowers in panicles. Stems 4-5. Fruit an achene, enclosed by the persistent enlarged calyx.

570. **G. palmata** (Forssk.) Gaudich. Pl. 91
Syn.: *G. diversifolia* (Link) Fries

Vigorous herbaceous perennial herb up to 1.2 m tall, bristly hairy with stinging hairs up to 5 mm long. The handsome leaves are 6-27 x 4-17 cm, broadly ovate, lobed. Flowers 2-3 mm broad; female flowers in dense spikes.

Distribution: Yemen, Pakistan, Kashmir to Bhutan, India, Sri Lanka, Burma, central China, Malaysia.

Forest undergrowth from 2200-3100 m in moist forest in Murree hills. The largest and most painful of our nettles. It is fortunately not very common. July-August.

2. Debregeasia 1 sp.

Shrubs with leaves silvery-white on undersurface. Flowers unisexual, in globose heads arranged in panicles or racemes.

571. **D. saeneb** (Forssk.) Hepper & Wood Pl. 91
Syn.: *D. salicifolia* (D. Don) Rendle

Dioecious shrub up to 2 m tall. Leaves 3-12 x 0.5-2.5 cm, oblanceolate or lanceolate, serrate. Female flowers in smaller clusters than male. Calyx 4-lobed. Stamens 3-5, exserted. Achenes yellow.

Distribution: Abyssinia, Iran, Afghanistan, Pakistan, Kashmir to Nepal.

Evergreen shrub with willow shaped leaves common in foothills to 1900 m, especially in lower Murree hills and Hazara area. March-May.

MORACEAE *Mulberry Family*

A tropical or sub-tropical family of trees or shrubs, often with a milky juice. Leaves simple, entire or not, stipulate, petiolate. Flowers small, unisexual, regular, in axillary racemes, catkins or heads. Sepals 4. Petals absent. Fruit a drupe.

+ Inflorescence a hypanthodium. 1. **Ficus**

– Male and female flowers borne externally, in catkins or heads. 2. **Broussonetia**

1. Ficus *Fig*

Trees, shrubs, sometimes of a twining or lianous habit. Leaves often palmately lobed. Stipules united or free, enclosing the terminal bud, soon falling off and leaving a ring-like scar. Flowers many, unisexual, borne in an urn-like structure with a terminal opening that is guarded by scales. Fruit fleshy.

+ Leaves 8-32 cm long, broadly ovate to oblong-ovate, glabrous above, 5-7-nerved. 572. **F. auriculata**

– Leaves 5-9 cm long, ovate-cordate to suborbicular, softly tomentose on both surfaces. 3-5-nerved. 573. **F. virgata**

572. **F. auriculata** Lour. *Trembal* Pl. 92
Syn.: *F. roxburghii* Wall. ex Miq.

A medium sized evergreen tree up to 7 m tall with spreading branches. Trunk pale grey to brown. Leaves large, 5-7-nerved, base cordate. Fruit borne in clusters on the branches, top shaped or depressed, globose, 3-7 cm broad, purplish-brown, pubescent.

Distribution: Pakistan, Kashmir to Nepal, north Burma, north east India, China, Indo-China.

Fairly common in the foothills especially in the Chir Pine zone up to 1600 m in lower Murree and Hazara hills. Fruit is edible. August-November.

573. **F. virgata** Wall. ex Roxb. *Phagwara* Pl. 92
Syn.: *F. palmata* Forssk.

Deciduous trees up to 8 m tall. Trunk smooth, greyish-brown. Leaves 5-9 cm long, softly pubescent on the upper surface, toothed, ovate or palmately divided, 3-5-nerved. Fruit axillary solitary or paired, top or pear shaped, 20-30 mm long, pale yellowish to purplish.

Distribution: East Africa, Arabia, Peninsula, south Iran, north west Afghanistan, Pakistan, Kashmir to Nepal, India.

Variable in leaf shape and size. Common especially in the foothill zone and occuring in all four provinces, from plains to 2100 m. Fruit is edible. May-November.

2. Broussonetia 1 sp.

Unarmed trees or shrubs with a milky juice. Leaves usually alternate, simple, entire or not, stipulate. Male flowers in catkins. Female flowers in globose heads. Fruit a globose head of small juicy drupes.

574. **B. papyrifera** (L.) L'Herit ex Vent. *Paper Mulberry* Pl. 92

A medium sized and very quick growing dioecious tree up to 16 m tall. Trunk whitish-grey. Leaves 4-18 x 3-14 cm, broadly ovate to elliptic-ovate, sometimes palmately lobed, subcordate, scabrid above, softly hairy on undersurface. Male catkins appearing before the leaves, 3-5 cm long. Fruit juicy, orange-red, in globose heads, 3 cm broad.

Distribution: Native to Japan, China, south west Asia. Introduced and naturalized elsewhere.

Self-sown and now considered as a pest around the capital area. The attractive fruit are produced during monsoon. Plains to 1400 m. February-March.

BETULACEAE *Birch Family*

Deciduous, monoecious trees or shrubs. Flowers small, 2-4-merous. Male flowers small, in catkins, the female in erect spikes or strobili. Fruit a small compressed nut.

Betula 1 sp. *Birch*

Male catkins in solitary axillary clusters of 2-4. Bracteoles 2. Female flowers with the bracteole pair united to bract to form a 3-lobed fruiting scale. Nut with a membranous wing.

575. **B. utilis** D. Don *Bhojpattra* Pl. 92

Small trees up to 8 m tall. The attractive bark white or reddish-white, peeling horizontally. Leaves ovate to rhomboid, 20-70 x 15-50 mm, base subcordate or tapering. Male catkins 30-40 x 4-5 mm long. Female strobili 20-40 x 10-12 mm, erect. Nut elliptic-ovate, 2-3 mm long.

Distribution: Afghanistan, Pakistan, Kashmir to Bhutan, Tibet, west China.

Found in the sub-alpine zone at the upper limit of trees (3000-) 3200-4600 m. Gregarious on slopes. Bark used for roofing purposes and substitute for writing paper; said to have medicinal properties. May-June.

FAGACEAE *Chestnut Family*

Monoecious, evergreen or deciduous shrubs or trees. Leaves simple, alternate. Male flowers in slender catkins or spikes. Female flowers 1-3. Carpels 3, united. Ovary inferior. Fruit an acorn (nut), partially enclosed in a cupule, formed by the fused involucral bracts.

Quercus 5 spp. *Oak*

Trees rarely shrubs, with toothed or lobed leaves. Male flowers in solitary or few catkins. Perianth 3-6-lobed. Stamens 6. Female flowers on erect spikes. Seeds solitary.

576. **Q. leucotrichophora** A. Camus *Silver or White Oak* Pl. 92
Syn.: *Q. incana* Roxb.

Trees up to 18 m tall. Leaves elliptic-lanceolate to ovate-lanceolate, 5-12 x 1.5-4 cm, toothed, undersurface white tomentose, upper dark green. Catkin 5-14 cm long. Stamens 4-6. Cupule 10-12 mm broad, half enclosing the oblong-ovoid nut.

Distribution: Pakistan, Kashmir to Nepal, upper Burma.

Common and gregarious in sub-Himalayan tract from 700-2000 m. The trees are much damaged by lopping for browse and the wood used for construction purposes. The finest specimens survive in Kao forest, Dunga Gali. April-May.

SALICACEAE *Willow Family*

Dioecious trees or shrubs with simple, alternate leaves. Flowers small, in catkins, which elongate in fruit. Male flowers: Stamens 2 or 5-10 or more. Fruit capsular, dehiscing by 2-4 valves. Seeds with long white hairs for dispersal.

+	Petioles up to 10 mm long. Stamens 2-10 in number. Capsule dehiscing by 2 valves.	**1. Salix**
–	Petioles 40-80 mm long. Stamens many. Capsule opening by 2-4 valves.	**2. Populus**

1. Salix *Willow*

Leaf stalks short. Catkins solitary in the axil of the bracts. Filaments long, free or partially united, exserted. Capsule conical, opening by 2 valves.

577. **S. denticulata** Anderson Pl. 92

Bushy shrub up to 2 m tall. Leaves elliptic to elliptic-lanceolate, 20-40 mm long, minutely toothed, undersurface lighter coloured. Male catkins 30-40 mm long; female catkins 70-80 mm in fruit, on leafy shoots.

Distribution: Afghanistan, Pakistan, Kashmir to central Nepal.

Gregarious on slopes in the Himalayan forest zone from 2100-4500 m, and along stream beds in alpine areas especially Chitral, Gilgit and Deosai plains. March-May.

2. Populus *Poplar, Aspen*

Leaves including stalk well developed, 3-5-nerved. Buds often sticky. Stamens usually many. Capsule 2-4-valved.

578. **P. alba** L. *Safeda, White Poplar* Pl. 93

Tree up to 16 m tall. Bark light grey or whitish. Undersurface of leaves dense cottony below. Leaves 4-10 cm long, ovate, margin sinuate, base 3-5-nerved. Male catkins 4-9 cm long, female shorter.

Distribution: The Himalaya.

A variable species as regards the leaf lobes. Found both wild and cultivated from 1000-3100 m. Popular as a roadside plantation. The timber is valued in the match industry. April.

ORCHIDACEAE *Orchid Family*

Perennial herbs with rhizomes or underground tubers, often of a saprophytic habit. Leaves radical or cauline. Flowers irregular, bisexual, generally of an elaborate nature and striking appearance. Perianth 6, in two whorls, the inner whorl (petals) with two laterals and one medium, the labellum (lip). Fertile stamens 1 or 2. Capsule ovoid or cylindrical, generally ribbed.

1.	+	Fertile anthers 2. Lip boat-like.	**1. Cypripedium**
	–	Fertile anther 1. Lip not boat-like.	2
2.	+	Plants saprophytic, weak stemmed, whitish, leafless.	**2. Cymbidium**
	–	Plants not so. Leaves well developed, green.	3
3.	+	Underground tubers palm-like.	**3. Dactylorhiza**
	–	Underground tubers or corms entire, rhizomes often present.	4
4.	+	Plants with tubers or corms.	**4. Nervilia** **5. Satyrium** **6. Habenaria**
	–	Plants with rhizomes.	5
5.	+	Lip of flower divided.	**7. Epipactis** **8. Cephalanthera**
	–	Lip of flower entire or lobed.	6
6.	+	Flowers spirally arranged on the flowering axis.	**9. Spiranthes**
	–	Flowers not so.	7
7.	+	Leaves appearing after flowering. Rhizome tuberous.	**10. Eulophia**
	–	Leaves appearing with flowering. Rhizome tuberous.	**11. Zeuxine**

1. Cypripedium 1 sp.

Herbs with leafy stems. Flowers showy. Lip boat-like. Fertile stamens 2. Stigmas 3.

579. **C. cordigerum** D. Don *Ladies Slipper* Pl. 93

Perennial up to 70 cm tall with large broadly elliptic to elliptic-ovate leaves. Flowers terminal, 1-3. Bract leaf-like. Flowers showy, pale yellow with a white lip.

Distribution: Pakistan, Kashmir to Bhutan, south east Tibet.

Forest clearings from 2100-3200 m from Indus Kohistan to Hazara. Uncommon. In the Murree hills it has suffered from over-collecting. June-July.

2. Cymbidium 1 sp.

Leafless saprophytic or epiphytic herb. Flowers racemose, showy. Lip entire or 3-lobed. Stigma entire.

580. **C. macrorhizon** Lindley Pl. 93

A delicate herb up to 18 cm tall. Scape whitish. Flowers 4-8, pink-white. Bracts small, 5-10 mm, lanceolate. Lip ovate, 3-lobed, whitish, with purplish dots.

Distribution: Pakistan, Kashmir to Burma, Khasia hills, Thailand.

In forests from 800-1600 m mainly eastern part of Murree hill range. Rare. September-November.

3. Dactylorhiza 3 spp.

Underground tubers palm-like. Leaves both basal and cauline. Flowers in spikes or racemes. Lip prolonged into a spur, entire or 3-lobed.

581. **D. hatagirea** (D. Don) Soo *Marsh Orchid* Pl. 93
Syn.: *Orchis latifolia* auct. non L.

Plants rather stout, up to 50 cm tall with a leafy hollow stem. Leaves elliptic to broadly lanceolate, 5-16 x 2-6 cm. Flowers many rather variable on stout spikes up to 16 cm long, purplish-pink to rose. Lip broader than long, 8-10 x 11-13 mm. Spur short.

Distribution: Pakistan, Kashmir to Bhutan, south east Tibet.

Not uncommon in marshy places in the Himalayan forest zone and sub-alpine areas from 2200-3600 m. June-July.

4. Nervilia 2 spp.

Plants with a globose corm. Leaf solitary, plicate, long petioled, appearing after flowering. Flowers showy, solitary or many. Lip entire or 3-lobed.

582. **N. gammieana** (Hook.f.) Pfitzer Pl. 93

Leaf stalk 8-15 cm long. Lamina cordate, fan-like, 6-12 cm broad, pale green, nerves prominent. Flowers appearing before the leaves, in racemes up to 17 cm long, pink, with a pale yellow throat. Lip distinctly divided into a upper and lower half.

Distribution: Pakistan, Kashmir to Sikkim.

Forest humus. 500-1200 m in lower Murree hills. Rare. A singularly attractive plant with pretty pink flowers and usual striking foliage. July.

5. Satyrium 1 sp.

Plants with underground tubers. Leaves few, basal and cauline, flowers with paired spurs arranged in spikes.

583. **S. nepalense** D. Don Pl. 93

Robust tuberous plants up to 100 cm tall. The broad leaves are usually basal. Flowers small, pale pink in dense spikes.

Distribution: West Himalaya to Nepal, Sikkim, Bhutan, China from 3000-4000 m.

On moist slopes under Chir Pine Forest to 2500 m in Murree hills. July-November.

6. Habenaria 8 spp.

Plants with fleshy tubers. Leaves few, basal or cauline. Flowers usually on a well developed axis, greenish, white or yellow. Lateral sepals spreading or incurved. Petals entire or split. Lip spurred, entire or 3-lobed or partite, the side lobes usually further divided.

1.	+	Leaves basal.	584. **H. marginata**
	–	Leaves cauline.	2
2.	+	Petals divided into two.	585. **H. digitata**
	–	Petals entire.	3
3.	+	Lip entire.	586. **H. latilabris**
	–	Lip 3-lobed, the side lobes fimbriate.	587. **H. intermedia**

584. **H. marginata** Colebr. Pl. 93

Plants up to 25 cm tall with fleshy tubers. Radical leaves 3-5, appressed to the ground, oblong to ovate. The delicately shaped flowers are green and yellow, in cylindrical spikes up to 14 cm long. Petals greenish-yellow, 8-10 x 5 mm. Lip yellow, 3-lobed.

Distribution: Pakistan, Kashmir to Bhutan, Burma.

Not common. In forests from 600-2000 m in eastern part of the Murree hills. August-September.

585. **H. digitata** Lindl. Pl. 94

Plants 20-40 cm tall with stalked tubers. Stem leafy. Leaves 5-7, orbicular to oblanceolate, 8-11 cm long. Flowers many, greenish-white. Petals 2-parted to the base. Lip 3-parted, the mid lobe 10-12 mm long, longer than the two laterals.

Distribution: Pakistan, Kashmir, India.

In pine forests from 500-700 m in lower Murree hills. Uncommon. Flowers fragrant at night. July-August.

586. **H. latilabris** (Lindl.) Hook. f. Pl. 94

Plants up to 40 cm tall, with sessile more or less oblong tubers. Leaves oblong, 6-12 x 3-5 cm. Flowers green-yellow, 7-8 mm long, in spikes up to 17 cm long. Lip greenish-yellow, entire, with a deflexed spur 10-12 mm long.

Distribution: Pakistan, Kashmir to Bhutan.

In forest clearings, slopes. More widespread than the previous one, Swat Kohistan to Murree hills from 1300-3000 m. July-August.

587. **H. intermedia** D. Don *Reindeer Orchid* Pl. 94

Plants rather stout, 25-40 cm tall with sessile oblong tubers. Leaves 3-5, ovate-oblong, up to 10 cm long, sheathing. Flowers 4-5, white, on a stout axis. Bracts leafy. Lip white, 3-lobed, side lobes fimbriate. Spur c. 5 cm long, flexuose.

Distribution: Pakistan, Kashmir to Nepal.

A rare orchid of striking beauty, found on forest slopes and clearings from 1300-2500 m in Chir pine zone late in the monsoon. August-September.

7. **Epipactis** 4 spp.

Plants rhizomatous. Leaves sessile. Flowers more or less nodding, racemose. Lip without a spur, 2-parted.

588. **E. veratrifolia** Boiss. & Hohen. Fig. between p. 102-103

Robust plants up to 120 cm tall, leafy. Leaves ovate-lanceolate to lanceolate. Flowers pale green, tinged with brownish-red. Lower part of lip 3-lobed.

Distribution: Somalia, Turkey, Cyprus, Syria, Sinai Peninsula, Iran, Afghanistan, Pakistan, Kashmir to Nepal.

Moist places, stream banks up to 2500 m from Balochistan, N.W.F.P., Chitral and Gilgit. April-July.

8. Cephalanthera 1 sp.

Plants rhizomatous. Stems leafy. Leaves sessile. Flowers generally many. Lip 2-parted, lower part shortly spurred.

589. **C. longifolia** (L.) Fritsch Pl. 94

Plants up to 50 cm tall, leafy. Leaves linear-lanceolate, spreading. Flowers 1-12, white, nodding, 10-15 mm long. Lip 2-parted, lower part orange-yellow.

Distribution: Euro-Siberian, Pakistan, Kashmir to Bhutan, south east Tibet, Burma.

In moist forest from 1500-2800 m in Chitral, Dir, Swat, Murree and Hazara hills. May-June.

9. Spiranthes 1 sp.

Slender plants with somewhat fleshy roots. Leaves stalked. Flowers pink (in our species) on a spirally twisted spike. Lip spurless, entire, apex with a wavy margin.

590. **S. sinensis** (Pers.) Ames. *Ladies Tresses* Pl. 94

Plants 15-30 cm tall. Leaves erect spreading, lanceolate. Flowers arranged on a spiral twisted axis. Lip 4-5 mm long, ovate.

Distribution: South east Asia, China, Japan, Australia, New Zealand.

Uncommon. In moist places, forest clearings, grassy slopes from 2000-3500 m in Chitral, Swat, Gilgit, Baltistan, Hazara and Murree hills. May-September.

10. Eulophia 2 spp.

Plants usually with a tuberous corm. Leaves appearing before or after flowering. Lip sac-like, entire or 3-lobed; the upper part crested.

591. **E. dabia** (D. Don) Hoch Pl. 95

Leaves 2, linear, appearing after flowering. Flowers nodding, somewhat one sided on the axis. Sepals yellowish-green, purplish striped, 10-12 mm long. Petals spreading. Lip

3-lobed, obovate, shortly compressed, 3-lobed; laterals yellowish-green, mid-lobe purple, with wavy margin..

Distribution: Pakistan, Kashmir to Sikkim, Burma, west China.

In damp places, field edges, in the plains in northern Punjab. Uncommon. March-April.

11. Zeuxine 1 sp.

Small rhizomatous plants with leafy stem. Flowers many, spicate. Perianth segments unequal, inner (petals) forming a hook. Labellum (lip) fleshy. Stigmas 2, free.

592. **Z. strateumatica** (L.) Schlechter Pl. 95
Syn.: *Z. sulcata* (Roxb.) Lindl.

A delicate plant 4-20 cm tall. Leaves sessile, linear-lanceolate. Flowers many, white or pale pink, spicate. Lip fleshy, sac-like. Lateral lobes oval, yellowish.

Distribution: South Iran, Pakistan, Kashmir to Burma, Assam, Nepal, China, Japan.

A relatively small orchid, uncommon in moist grassy places, stream edges etc up to 1000 m from Sindh, N.W.F.P., lower Swat and northern Punjab. March-April.

ZINGIBERACEAE *Ginger Family*

Perennial aromatic and rhizomatous herbs. Flowers showy and of unusual structure, in spikes or solitary in the axil of bracts. Corolla subequally 3-lobed, tube slender. Fertile stamens 1. Lateral staminodes petaloid. Ovary inferior.

Roscoea 1 sp.

Leaves lanceolate or elliptic. Flowers in terminal spikes. Corolla tube as long as or longer than calyx. Upper lobe erect, hood-like. Capsule membranous, 3-valved.

593. **R. alpina** Royle Pl. 95

Plants up to 20 cm tall. Leaves 60-120 x 10-15 mm, sheathing, appearing after flowering. The beautifully shaped flowers are rose-purple, 1-6. Corolla tube longer than calyx. Upper lobe suborbicular, 15-20 mm long, bilobed.

Distribution: Pakistan, Kashmir to Bhutan, Tibet.

Uncommon on damper forest slopes, mainly Hazara district and Galis from 2100-2700 m. July-August.

IRIDACEAE *Iris Family*

Perennial rhizomatous herbs with narrow plicate leaves. Flowers showy, bisexual. Perianth of sepals in two whorls (3 + 3), petaloid. Stamens 3. Ovary inferior, 3-locular. Capsule 3-valved.

+	Rootstock not a corm. Perianth tube present. Leaves usually terete or flat.	**1. Iris**
–	Rootstock a corm with fibrous covering. Perianth tube absent. Leaves more or less folded longitudinally.	**2. Gynandrisis**

1. Iris 13-14 spp.

Flowers showy, in the axil of spathes. Perianth segments 6; the inner (standard) 3, erect and united to form a tube. The outer (falls) spreading or recurved, hairy, crested or smooth.

1.	+	Flowers yellow.	594. **I. aitchisonii**
	–	Flowers white, mauve or purplish.	2
2.	+	Plants robust, with leaves up to 5 cm broad. Falls 3-5 cm broad.	595. **I. germanica**
	–	Plants less robust, with leaves up to 2 cm broad. Falls 3 cm broad or less.	3
3.	+	Falls not bearded. Flowers pale violet to white or creamish.	4
	–	Falls bearded. Flowers purple-blue, subsessile, present amidst foliage.	597. **I. hookeriana**
4.	+	Flowering stems well developed. Perianth tube 4-6 cm long. Flowers white or pale violet.	596. **I. songarica**
	–	Flowering stems obsolete. Perianth tube 9-11 cm long. Flowers creamish with deep brown or purplish streaks.	598. **I. loczyi**

594. **I. aitchisonii** (Baker) Boiss. *Yellow Iris* Pl. 95
Syn.: *I. aitchisonii* var. *chrysantha* Baker

Plants 30-50 cm tall, tufted. Leaves well developed at flowering time, linear, 25-35 cm long. Flowers yellow. Perianth tube c. 3 cm long. Falls 4-5 cm long, tongue-like. Standards reflexed, 1.2-2.5 cm long, 3-lobed, the mid-lobe longer than laterals.

Distribution: Afghanistan, Pakistan.

Locally common in Swat valley. Also found in Hazara area particularly Mansehra district. Dry places, edge of fields. Plains to 900 m. March-April.

595. **I. germanica** L. Pl. 95

Plants 60 cm or more tall, with a stout rhizome. Leaves 2-5 cm broad, sheathing. Flowers 4-6, deep blue-purple or white. Falls 6-8 cm long, obovate, with a yellowish beard. Standard similar but smaller. Capsule 4-5 cm long.

Distribution: Widely distributed in west Asia and Europe.

Often planted around graveyards and occasionally as an escape. Plains to 3000 m. July-August.

596. **I. songarica** Schrenk Pl. 95

Tufted plants up to 60 cm tall, leaves linear, flat, sheathing. Flowering stems equalling or exceeding foliage. Flowers white to pale violet with darker veins. Falls 4-6 cm long elliptic-oblong. Standard erect, 8-11 mm long, erect. Capsule 4-6 cm long, cylindric-ellipsoid.

Distribution: Iran, Afghanistan, Pakistan, Turkestan, Uzbekistan, Kazakhstan, Mongolia.

In dry places from 1800-2100 m. Found in Balochistan and N.W.F.P. April-May.

597. **I. hookeriana** Foster Pl. 96

Tufted plants up to 50 cm tall. Leaves broadly linear. The beautiful flowers are purple-blue, solitary or 2-3, on short stems up to 12 cm long. Perianth tube 2-3 cm long. Falls obovate, bearded. Capsule cylindric-ellipsoid.

Distribution: Pakistan, Kashmir.

Common often gregarious in sub-alpine meadows in Chitral, Swat, Baltistan, Kaghan, Hazara and Galis from 2300-3300 m. June-July.

598. **I. loczyi** Kanitz Pl. 96

Similar to *I. songarica,* but with flowers subsessile and perianth tube 9-11 cm long.

Distribution: Pakistan, Kashmir, Iran, Afghanistan, Dsungaro-Alatau, Tien Shan, Pamir Alaj.

In Juniper tract from 2000-2700 m in Balochistan. April-May.

2. Gynandrisis 1 sp.

Perennial herbs related to *Iris,* but differing in the characters mentioned in the key. Flowers blue to white. Outer perianth segments (falls) longer than the inner (standards).

599. **G. sisyrinchium** (L.) Parl. Pl. 96
Syn.: *Iris sisyrinchium* L.

Plants 30-40 cm tall, with a round underground corm up to 32 mm broad. Flowers 2-6, deep blue to mauve or violet, with white-yellow to the inside. Falls obovate-elliptic with yellow stripe, up to 37 mm long, spreading. Standards erect.

Distribution: Mediterranean, south west Asia.

In and around cultivated fields from 900-1300 m in Balochistan and Swat valley. March-April.

ALLIACEAE *Onion Family*

Allium 40 spp. *Onion*

Perennial bulbous scapigerous herbs. Underground bulbs solitary or in clusters. Leaves mostly basal or radical, cylindrical, fistular or not, sheathing. Flowers umbellate. Perianth of 6 tepals, in two whorls, free or connate. Tepals 1-nerved of various colours. Stamens 6, exserted or not. Ovary superior, 3 locular. Fruit capsular.

1.	+	Flowers yellow.	600. **A. fedtschenkoanum**
	–	Flowers white, lilac to purplish-blue.	2
2.	+	Flowers white.	601. **A. humile**
	–	Flowers white-pink or deeper shades.	3
3.	+	Plants generally cultivated, but occasionally found as an escape.	602. **A. neopolitanum**
	–	Plants native.	4
4.	+	Leaves 1-4 cm broad. Filaments exserted. Scape stout. Pedicels 3-10 cm long.	603. **A. caspium**
	–	Leaves less than 5 mm broad. Scape slender. Pedicels 1-2 cm long.	604. **A. griffithianum** 605. **A. jacquemontii**

600. **A. fedtschenkoanum** Regel Pl. 96

Leaves 1-3, cylindrical, hollow, sheathing. Scapes 15-48 cm tall. Umbels 2-2.5 cm broad, subglobose. Flowers yellow. Tepals ovate to elliptic-oblong, c. 8 mm long. Capsule oblong, 5-6 mm long.

Distribution: Central Asia, Afghanistan, Pakistan.

Alpine meadows, marshy places in northern areas from 3000-4200 m. June-July.

601. **A. humile** Kunth Pl. 96

Plants with cylindrical bulbs. Leaves 4-6 in number, narrow. Scape 20-30 cm tall. Flowers white, umbellate. Tepals lanceolate, 7-9 mm long.

Distribution: Himalaya.

One of the smaller alliums of alpine and sub-alpine regions up to 4200 m in Kaghan valley, Swat and Chitral. June-July.

602. **A. neopolitanum** Cyr. Pl. 96

Plants up to 60 cm tall. Bulbs globose. Leaves linear, flat, not fistular, shorter than the scape. Flowers white, tinged with pink. Pedicels up to 3 cm long. Tepals 7-9 mm long, elliptic-ovate.

Distribution: South Europe.

Cultivated and as an escape. In hilly regions up to 3000 m. July-August.

603. **A. caspium** (Pall.) M. Bieb. Pl. 96

Striking plants with ovoid bulbs. Leaves 3, up to 6 cm broad, lanceolate. Scape stout, 20 x 1 cm. Umbels large, globose. Flowers purplish. Pedicels up to 10 cm long. Tepals 6-8 mm long, oval. Filaments exserted.

Distribution: Central Asia, Iran, Afghanistan, Pakistan.

An attractive plant in dry open places from 1600-2000 m in Quetta valley. April-May.

604. **A. griffithianum** Boiss. *Jangli Piaz* Pl. 97

Plants with ovoid bulbs. Leaves 2-3, linear, cylindrical, fistular, grooved. Umbels globose. Flowers many, purplish-pink. Tepals ovate to broadly lanceolate, 6-8 mm long.

Distribution: Central Asia, Afghanistan, Pakistan, India.

Common locally in the lower hills up to 1800 m from Balochistan, N.W.F.P., Chitral, Swat, Salt Range and lower Hazara district. March-April.

605. **A. jacquemontii** Kunth Pl. 97
Syn.: *A. rubellum* auct. non Bieb.

Similar to *A. griffithianum,* but with smaller size and paler flowers and stamens more than half the tepal length.

Distribution: Central Asia, Afghanistan, Pakistan, India.

Common in stony areas in lower hills up to 2200 m. March-April.

DIOSCOREACEAE *Yam Family*

Dioecious perennial herbs often with a twining habit and underground tubers. Leaves alternate, simple or compound. Flowers small, regular, in axillary spikes, racemes or panicles. Perianth segments 6. Stamens 6, in two whorls. Ovary inferior, 3 locular. Fruit a capsule or a berry.

Dioscorea 4 spp.

Tuberous herbs. Leaves often with bulbils in their axil, petiolate. Male flowers with perianth segments (tepals) in two whorls (3 + 3). Tepals in female flowers free. Capsule 3-winged.

606. **D. melanophyma** Burkill & Prain Pl. 97

Twining herbs with palmately compound leaves. Tubers globose. Leaflets 3-5, 4-14 x 1-3 cm, lanceolate or elliptic-lanceolate. Bulbils black. Male racemes 20-50 mm long, 1 or 2. Female spikes solitary, 3-4 cm long. Capsule 12-15 x 5 mm, rounded at the base.

Distribution: Pakistan, Kashmir to Nepal, Assam, Yunnan.

In outer sub-Himalayan tracts from 900-2400 m. August-September.

LILIACEAE *Lily Family*

Perennial herbs, sometimes climbers, often with underground bulbs, corms or tubers. Flowers in panicles, racemes or solitary. Perianth segments 6, tepals often petaloid. Stamens 6. Fruit a capsule or berry.

1. + Plants climbing by leaf tendrils. Flowers yellow to orange-red, fading to red. **1. Gloriosa**

   – Plants without tendrils. Flowers variously coloured, but not as above. 2

2.	+	Flowers 5-10 cm long, showy.	**2. Notholirion** **3. Tulipa** **4. Lilium** **5. Fritillaria**
	–	Flowers not exceeding 2 cm long.	3
3.	+	Flowers yellow or greenish-white.	**6. Gagea** **7. Eremurus**
	–	Flowers white, blue to purplish or green-white.	4
4.	+	Bushy herbs often stragglers, with a woody base, with stem modified into leaf-like structure (cladodes). Leaves much reduced.	**8. Asparagus**
	–	Erect, low growing or scandent herbs with simple stems. Leaves well developed.	5
5.	+	Leaves cauline, in numerous whorls or alternate. Flowers in axillary drooping racemes.	**11. Polygonatum**
	–	Leaves basal. Flowers in racemes or spikes.	6
6.	+	Flowers greenish-white.	**9. Dipcadi**
	–	Flowers white, blue to purple.	7
7.	+	Flowers white.	**10. Asphodelus**
	–	Flowers blue to purple.	**12. Scilla** **13. Muscari**

1. Gloriosa 1 sp.

Herbs with tuberous roots. Stem simple, weak, climbing by means of leaf tendrils. Leaves alternate, the apex modified into a simple tendril. Flowers solitary axillary. Perianth segments erose and wavy, free. Stamens spreading. Fruit capsular.

607. **G. superba** L. *Climbing lily* Pl. 97

A week climber. Leaves elliptic-lanceolate to lanceolate, 6-10 x 1-4 cm, pale green. Pedicels up to 7 cm long, reflexed. The beautiful flowers are, c. 7-9 cm broad, red and yellow, becoming all red with age, with spreading crinkly tepals. Capsule oblong, glabrous.

Distribution: Tropical Africa, Asia.

Sub-Himalayan tracts up to 1200 m. Uncommon. Tubers are medicinal. Flowering late monsoon. July-September.

2. Notholirion 1 sp.

Perennial herbs with underground bulbs with a brown scaly covering. Leaves narrow. Stamens not spreading. Stigma 3-lobed; the lobes short and recurved.

608. **N. thomsonianum** (Royle) Stapf *Hazara or Thomson's Lily* Pl. 97

Stems simple up to 100 cm tall. Leaves 16-35 cm long, lanceolate. Flowers large, racemose, many, campanulate, pale lilac to white, 5-8 cm long, fragrant. Capsule 20-25 mm long. Tepals c. 5 mm broad, oblanceolate.

Distribution: Afghanistan, Pakistan, Kashmir to west Nepal.

Rocky places in the lower hills up to 1200 m especially in the Hazara, Swat and lower Neelam valley. One of the most beautiful plants of Pakistan worth cultivating. March-April.

3. Tulipa c. 3 spp. *Tulip*

Perennial herbs. Bulbs with several layers of leathery coverings. Leaves 1-3, narrow. Flowers showy, solitary. Perianth segments (tepals) 6, free, in two whorls, variously coloured. Fruit an erect capsule.

+ Flowers red or scarlet. 609. **T. lehmanniana**

– Flowers yellow or white. 610. **T. stellata**

609. **T. lehmanniana** Merckl. Pl. 97
Syn.: *T. montana* auct. non Lindl.

Perennial herb. Tunic of bulbs densely hairy within. Stem glabrous. Flowering scape not exceeding the flower. Tepals abruptly acuminate.

Distribution: Afghanistan, Pakistan.

In dry stony areas from 1100-1600 m in Balochistan. March-April.

610. **T. stellata** Hook. f. *Star Tulip* Pl. 98

A polymorphic species. The typical plant (var. *stellata*) has white flowers with red on the back (outside).

Distribution: Pakistan, Kashmir to Uttar Pradesh (India).

Very common as a weed in wheat fields, hill slopes from plains to 3000 m in northern Punjab and N.W.F.P. The yellow form (var. *chrysantha*) with backs of tepals red, tends to grow in drier regions at higher elevation (i.e. generally above 900 (-1400 m) in Balochistan . March-April.

4. Lilium 1 sp. *Lily*

Tall perennial herbs with leafy stem. Bulbs scaly, without a tunic. Flowers showy, in terminal racemes. Tepals free.

611. **L. polyphyllum** D. Don Pl. 98

Plant up to 1.2 m tall. Leaves 8-14 cm long, linear-lanceolate. Pedicels long. Flowers 7-9 cm long, fragrant, drooping pale yellowish-green on the outside, mottled purplish within. Stamens prominent. Anthers red.

Distribution: Afghanistan, Pakistan, Kashmir to Uttar Pradesh (India).

In forests from 2300-3200 m from Swat Kohistan to Hazara. Uncommon. June-August.

5. Fritillaria 3 spp.

Perennials with underground bulbs covered with thick scale-like leaves. Flowers showy, drooping. Fruit capsular.

612. **F. roylei** Hook. f. Pl. 98

Spectacular plants up to 50 cm tall. Flowers solitary, broadly bell-shaped, 4-5 cm long, yellowish-green to brownish-purple streaked with dull purple. Leaves narrow, 2-6 per node.

Distribution: Pakistan to Uttar Pradesh (India).

In forest shrubberies and clearings. Uncommon from 2300-3200 m in Swat Kohistan and Kaghan valley. June-July.

6. Gagea c. 14 spp. *Yellow Star of Bethelehem*

Low growing bulbous herbs with 1-2 basal glossy leaves, rarely cauline. Flowers small, yellow. Perianth 6-8, star-like, with spreading tepals. Fruit capsular.

\+ Leaves 3-5 mm broad, linear. 613. **G. pseudo-reticulata**

– Leaves 8-20 mm broad, broadly lanceolate. 614. **G. elegans**

613. **G. pseudo-reticulata** Vved. Pl. 98

Leaves very narrow, 8-20 cm. Stamens very short, slender. Flowers subumbellate. Pedicel pubescent. Tepals 10-15 mm long, acute. Capsule as long as the perianth.

Distribution: North Africa, Greece, Afghanistan, Pakistan, Kashmir.

Common. Plains to 1500 m from Balochistan, N.W.F.P., Hazara and Gilgit. March-April.

614. **G. elegans** Wall. ex D. Don Pl. 98
Syn.: *G. lutea* auct. non Schultz f.

Similar but with broader glossy leaves.

Distribution: Pakistan, Kashmir to central Nepal.

Alpine. Gregarious, often near melting snow from 2200-3300 m. From juniper zone in Balochistan to Chitral, Swat, Gilgit, Kaghan and Galis. The leaves are used by hill people as a vegetable. May-June.

7. Eremurus 7 spp. *Foxtail Lily*

Erect perennial herbs with a short, thick rhizome and cluster of tuberous roots. Leaves basal, in rosettes, linear. Flowers many, in elongated racemes. Capsule globose, dehiscing by 3 valves.

1.	+	Flowers yellow.	615. **E. stenophyllus**
	–	Flowers white or creamy.	2
2.	+	Leaves 12-45 mm broad. Capsule 12-14 mm broad. Plants 70-110 cm tall.	616. **E. himalaicus**
	–	Leaves up to 10 mm broad. Capsule 18-22 mm broad. Plants up to 70 cm tall.	617. **E. persicus**

615. **E. stenophyllus** (Boiss. & Buhse) Baker *Desert Candle* Pl. 98
Syn.: *E. aurantiacus* Baker

Plants up to 120 cm tall. Flowers yellow in dense racemes. Pedicels 10-18 mm long, suberect. Tepals 9-13 mm long, turning brownish with age. Capsules 5-8 mm broad, globose.

Distribution: Central Asia, Afghanistan, Pakistan.

Gregarious in dry areas, edge of fields, stony hills from 1500-2700 m. Mainly in juniper zone in Balochistan and Chitral. April-May.

616. **E. himalaicus** Baker *Mountain Candle* Pl. 98

Flowers white, in dense racemes. Pedicels up to 26 mm long, suberect in fruit. Tepals 15-20 mm long. Capsule 12-14 mm broad, globose.

Distribution: Central Asia, Afghanistan, Pakistan, Kashmir to Himachal Pradesh (India).

Gregarious on open hillsides, forest clearings and alpine slopes from 2500-3300 m. Young leaves used as greens. May-June.

617. **E. persicus** (Jaub. & Spach) Boiss. *Hoal, Hoad* Pl. 99
Syn.: *E. sikesarus* O. Fedtsch.

A much smaller plant than *E. himalaicus* with narrower leaves and broader capsules. Flowers white.

Distribution: Iran, Afghanistan, Pakistan, Kashmir.

Locally common in Balochistan. Uncommon in upper Punjab, Salt Range and Hazara. Lower hills on limestone, up to 1400 m. Leaves used as vegetable. March-April.

8. Asparagus c. 11 spp.

Bushy plants, often stragglers. Leaves reduced to scales. Stems modified into leaf-like structures (cladodes). Cladodes in clusters of 2-20 at the nodes. Flowers small. Fruit a berry.

\+ Flowers axillary, in clusters of 3-5. Plants erect. 618. **A. gracilis**

– Flowers in short racemes, many.
Plants procumbent. 619. **A. adscendens**

618. **A. gracilis** Royle Pl. 99
Syn.: *A. capitatus* Baker

Plants straggling. Lower stems and branches prickly. Cladodes 2-5, 3-6 mm long, straight. Flowers white, subsessile. Tepals spreading.

Distribution: Pakistan, India, Kashmir.

In scrub forest, from plains to 1000 m from Salt Range, lower Swat to Hazara. March-April.

619. **A. adscendens** Roxb. *Safed Musli* Pl. 99

Stems erect, up to 100 cm tall, spiny. Spines straight. Cladodes 6-20 , 8-12 mm long. Racemes 20-60 mm long. Flowers white, subsessile.

Distribution: Sub-Himalayan tract and outer Himalaya up to 1500 m.

Common in the Punjab and lower Murree hills. October-November.

9. Dipcadi 2 spp.

Scapigerous herbs with underground tubers. Flowers racemose. Perianth cylindric. Tepals 6, erect. Stamens included. Capsule short and broad.

+	Scapes up to 50 cm tall. Bracts shorter than the pedicel.	620. **D. hysudricum**
–	Scapes up to 18 cm tall. Bracts equalling the pedicel.	621. **D. erythraeum**

620. **D. hysudricum** (Edgew.) Baker Pl. 99

Leaves narrow, 8-14 cm long. Scape up to 50 cm long. Flowers greenish-white, greenish to the outside. Tepals 12-15 mm long.

Distribution: Pakistan, Kashmir to west Nepal.

Plains and lower foothills up to 800 m especially Punjab, Salt Range and lower foothill areas. March-April.

621. **D. erythraeum** Webb & Berth. Pl. 99

A smaller plant than *D. hysudricum,* with bracts about equalling the pedicel.

Distribution: North Africa, Arabia, Pakistan and adjacent Rajasthan (India).

Sandy dry areas, low hills especially lower Sindh. February-March.

10. Asphodelus 1 sp. *Asphodel*

Annual or perennial scapigerous herbs with fleshy or fibrous roots. Leaves narrow, radical, fistular. Flower racemose. Perianth segments 6, united basally to form a tube. Stamens 6. Capsule few-seeded.

622. **A. tenuifolius** Cavan. *Piazi* Pl. 99

An erect glabrous annual, up to 50 cm tall, branched from the base. Stems hollow. Flowers white; tepals with dark brown midrib. Capsule globose, 4-5 mm broad, valves wrinkled.

Distribution: Canary Isls., Pakistan, India.

A pernicious weed of plains and lower hills up to 2000 m in Sindh, Balochistan, N.W.F.P., Punjab and Kashmir. March-April.

11. Polygonatum 3 spp. *Solomons Seal*

Perennial herbs with a thick creeping rootstock. Stem erect or scandent, simple, leafy. Leaves whorled or alternate. Flowers drooping, axillary. Perianth tubular, 6-lobed. Stamens 6. Ovary 3-locular. Fruit a berry.

623. **P. verticillatum** (L.) All. Pl. 99

Stems erect, up to 1.3 m tall. Leaves in whorls of 3-8, linear-lanceolate, 8-17 cm long, acute, lower surface glabrous. Racemes 2-3-flowered, axillary. Flowers 6-8 mm long, greenish-white, pendulous. Berries globose, 5-6 mm broad, purple when mature.

Distribution: Europe, Turkey, central Asia, Pakistan, Kashmir to south east Tibet.

In forests from 1600-3200 m in Murree and Hazara hills. Not common. July-August.

12. Scilla 2 spp.

Bulbous plants. Bulbs covered with a dark tunic. Bracts basal. Flowers racemose. Tepals 6, free or connate basally. Capsule dehiscing by 3 valves.

624. **S. griffithii** Hochr. Pl. 100
Syn.: *S. hohenackeri* Hook. f. non Fisch. & Mey.

Leaves 3-5, 70-180 x 4-6 mm, linear. Scape 8-17 cm long, many-flowered. Flowers purple-blue or blue. Tepals 15-20 mm long. Anthers blue.

Distribution: Iran, Afghanistan, Pakistan, Kashmir.

Gregarious. Main valleys and lower hills up to 1200 m in N.W.F.P., Chitral and Hazara. Often a weed in cultivated fields in Swat and Margalla hills. March-April.

13. Muscari 1 sp.

Perianth urn-like or cylindrical. Lobes very short. Plants bulbous. Bulbs covered by brown to blackish tunic. Leaves basal, 2-6, narrow. Flowers spicate. Capsule 3-angled and thin walled. Seeds wrinkled.

625. **M. neglectum** Guss. Pl. 100
Syn.: *M. racemosum* Lam. & DC. non Miller

Bulb ovoid. Leaves narrow, 5-10 cm long. Scape up to 12 cm long, many-flowered.

Flowers purplish to deep bluish-purple, drooping, sessile, perianth urn-like, 7-8 mm long; lobes white, recurved.

Distribution: North Africa, Europe to south east England, central Russia, west Syria, Cyprus, Caucasia, Iran, Afghanistan, Pakistan, central Asia.

Dry places, meadows from 1400-1900 m in northern Balochistan. April-May.

AMARYLLIDACEAE *Amaryllis Family*

Perennial herbs with underground bulbs, corms or rhizomes. Leaves radical or cauline, simple. Flowers solitary, umbellate, paniculate or in clusters. Flowering axis (scape) well developed. Tepals 6, free or slightly united at the base. Stamens 6. Ovary inferior. Fruit capsular.

Ixiolirion 2 spp.

Bulbous plants. Stems with narrow leaves. Flowers few to several, showy. Tepals acute.

626. **I. tataricum** (Pall.) Herb. Pl. 100

Plants up to 100 cm tall. Stem simple. Bulbs 2-3 cm broad, globose. Leaves 4-5 mm broad. Flowers 4-8, in umbellate clusters, purplish-blue. Capsule oblong.

Distribution: North Africa, central Asia, Iraq, Iran, Afghanistan, Pakistan.

Though a weed of grain fields, it is a handsome plant worth cultivating. Plains to 2500 m. widespread in dryer mountainous tracts of Balochistan, Swat and Kala Chita hills. March-April.

TRILLIACEAE *Trillidium Family*

Related and sometimes included in the *Liliaceae,* but differing in the terminal whorled leaves, 4-12 stamens and 3-10 carpels.

Trillidium 1 sp.

Rhizomatous perennial herb. Leaves in whorls of 3. Leaf base sheathing the stem base. Leaves 3, overlapping the stem. Flowers solitary, 3-merous. Tepals linear-lanceolate, the outer 3 broader than the inner 3. Fruit a berry.

627. **T. govanianum** (Wall. ex D. Don) Kunth Pl. 100
Syn.: *Trillium govanianum* Wall. ex. D. Don

Plants up to 20 cm tall. The three broad petiolate leaves make a pleasing pattern. Lamina broadly ovate, 30-11 x 10-20 cm, glabrous. Tepals lanceolate, spreading, reflexed in fruit, dark purple, outer 3 broader. Stamens 6, in two whorls. Styles 3, purplish. Berry red, 10-20 mm broad.

Distribution: Pakistan, Kashmir to Bhutan.

Shade loving. Forest undergrowth from 2200-3300 m in Hazara and Murree hills. May-June.

BUTOMACEAE *Flowering Rush Family*

Perennial scapigerous herbs with a short rhizome. Leaves narrow, sheathing at the base. Flowers in terminal umbellate cymes, 3-merous, bisexual, regular. Perianth 6, in two whorls (3 + 3), the inner whorl petal-like. Stamens 6 or more. Fruit of 6 to many-seeded follicles.

Butomus 1 sp.

628. **B. umbellatus** L. *Flowering Rush* Pl. 100

An aquatic, glabrous erect herb. Root fibrous. Flowering stalk 40-100 cm long. Leaves radical, about as long as the scape. Flowers pale pink, on stalks 4-10 cm long. Petal-like inner sepals obovate, 10-16 mm long. Stamens c. 9. Follicles 8-10 mm long, obovate.

Distribution: Temperate and sub-tropical Eurasia.

Marshy places in the plains and lower hills to 2000 m in Punjab, particularly in rice growing tracts. May-July.

COMMELINACEAE *Spiderwort Family*

Erect to prostrate perennial herbs with succulent jointed stems, often swollen at the nodes. Leaves alternate. Flowers in condensed lateral or axillary cymes, subtended by bracts or spathes, 3-merous. Perianth of 6 tepals, the inner 3 petal-like. Stamens 6, in two whorls. Fruit capsular.

Commelina 3 spp.

Prostrate to suberect plants. Leaves sheathing at base. Flowers 2, subtended by a hooded, funnel-shaped spathe. Inner tepals 3, the middle one larger than the other two. Fertile stamens 3.

1. + Leaves ovate to ovate-oblong.
   Flowers deep blue. 629. **C. benghalensis**

– Leaves lanceolate. Flowers sky blue. 2

2. + Spathe 10-13 mm long. Seeds black, spotted with dull yellow. 630. **C. albescens**

– Spathe 18-35 mm long. Seeds leadish, greyish-white, minutely dotted. 631. **C. paludosa**

629. **C. benghalensis** L. Pl. 100

A perennial prostrate herb with ovate to ovate-oblong leaves, 20-40 x 15-35 mm. Flowers pale blue, subtended by a spathe.

Distribution: Tropical and subtropical Asia and Africa.

Moist places. Plains to 1200 m in Punjab, N.W.F.P. and Azad Kashmir. Medicinal. July-September.

630. **C. albescens** Haussk. Pl. 100

Prostrate branched perennial herb with thickened nodes. Leaves lanceolate, 5-8 cm long. Spathe 10-13 mm long, enclosing 3-5 flowers. Flowers sky-blue to blue. Seeds black, spotted yellow.

Distribution: Tropical Africa, Arabia, Pakistan.

Plains and low hills in rocky or sandy places up to 400 m in Sindh. Flowering mostly throughout the year.

631. **C. paludosa** Blume *Kanjuna* Pl. 101

Similar but with longer lanceolate leaves, 40-122 x 20-80 mm. Flowers pale bluish.

Distribution: Pakistan, India, Sri Lanka east to Indonesia, Philippines.

Foothills and adjacent plains, up to 1500 m in Punjab and N.W.F.P. Plant medicinal. August-October.

COLCHICACEAE *Crocus or Colchicum Family*

Related and sometimes included in the *Liliaceae,* but differing in the distinct styles, the extrorse dehiscence of anthers and nature of the splitting of the capsule.

Colchicum 3 spp.

Perennial herbs with underground corms. Corms with a brown tunic. Flowers 1-3, on a very small scape. Perianth with 6 tepals, united at the base to form a short tube. Stamens 6. Styles 3.

+ Flowers yellow. 632. **C. luteum**

– Flowers white to pale lilac. 633. **C. aitchisonii**

632. **C. luteum** Baker *Surinjan Talakh* Pl. 101

Corm ovoid to oblong. Leave 3-6, linear to lanceolate, 80-200 x 5-18 mm. Flowers appearing before the leaves (or when leaves are young), 3-4 cm broad. Perianth tube 7-9 cm long. Tepals lanceolate to oblanceolate, 2-3 cm long. Anthers 10-15 mm long, yellow. Capsule 2-3 cm long, ovoid.

Distribution: Central Asia, Afghanistan, Pakistan, Kashmir to Himachal Pradesh (India).

Hills and adjacent plains from 600-3100 m in Waziristan, Chitral, Dir, Gilgit, Baltistan, Murree hills and Hazara. Open places, forest shrubbery. The alkaloid colchicine is extracted from the corm and seed. January-April.

633. **C. aitchisonii** (Hook.f.) E. Nasir Pl. 101
Syn.: *Merendera aitchisonii* Hook. f.

Similar to *C. luteum,* but perianth split to the base and not forming a tube. Flowers white or pale lilac.

Distribution: Afghanistan, Pakistan.

Low hills and rocky places from 600-2200 m in Balochistan, N.W.F.P., Chitral, Dir, Karakoram, Kaghan, Margalla, Salt Range. Also a source of colchicine. March-April.

ARACEAE *Arum Family*

Monoecious or dioecious perennial herbs or shrubs, sometimes aquatic or epiphytic, with underground rhizomes, corms or tubers. Leaves simple or compound. Flowers spicate, the inflorescence subtended by a large bract-like structure, the spathe. Perianth generally absent. Stamens 6, free or united. Fruit a berry.

Arisaema 6 spp.

Tuberous herbs with digitately compound leaves. The inflorescence (spadix) included or excluded from the spathe. Male flowers with 2-5 stamens. Style short or absent in female flowers. Berries red when mature.

1. + Leaflets 3, rhombic to suborbicular.
Spathe brown-purple with whitish stripes. 634. **A. utile**

– Leaflets 5-14, elliptic-ovate to lanceolate or linear. Spathe green or greenish-yellow. 2

2. + Leaflets 7-13. Spathe tube shorter than the limb. 3

– Leaflets 5-7. Spathe tube longer than the limb 635. **A. jacquemontii**

3. + Leaves linear-lanceolate. Spathe tube cylindrical, 25-40 mm. 636. **A. tortuosum**

– Leaves elliptic-ovate to lanceolate. Spathe tube ovoid, 10-20 mm long. 637. **A. flavum**

634. **A. utile** Hook. f. ex Schott Pl. 101

Leaves prominent. Stalk 20-35 cm long. Leaflets 3, 8-20 x 4-18 cm, middle leaflet the largest. Spadix present below the foliage and resembling the hood of a cobra, purple-brown with whitish ribs, apex ending in an appendage 10-28 mm long. Berries ovoid to subglobose, c. 5 mm broad.

Distribution: Pakistan, Kashmir to Bhutan.

Forest floor in humus. Conspicuous during monsoon season. From 2300-3300 m in Himalayan moist forest. June-July.

635. **A. jacquemontii** Blume Pl. 101

Bulb subglobose, 2-3 cm broad. Leaves 1 or 2, pale green. Leaflets 4-14 x 2-4 cm, acuminate, cuneate. Petiole up to 20 cm. Spathe green, equalling or exserted above foliage; tube cylindric, 35-55 mm long, limb oblong-ovate, shorter than tube, tip prolonged into a tail, 4-8 cm long. In male, spadix 3-5.5 cm, in female longer. Berries 4-5 mm broad.

Distribution: Afghanistan, Pakistan, Kashmir to Sikkim, Tibet, north Assam.

Common and more widespread than *A. utile.* Forest openings from 2100-3400 m. June-July.

636. **A. tortuosum** (Wall.) Schott var. **curvatum** (Roxb.) Engl. *Kiriki Kukri* Fig. p. 267

Monoecious or dioecious herbs. Leaves 2, pedate. Leaflets 11-13, linear-lanceolate, 8-19 x 0.5-2 cm. Spathe exserted above foliage, green. Spadix appendage curled upwards beyond the spathe. Berries 6-7 mm broad.

Distribution: Pakistan, Kashmir to west China, north Burma.

609. *Tulipa lehmanniana* a,b: Chiltan hills, Rubina A. Rafiq

606. *Dioscorea melanophyma* Margalla hills, S. A. Sultan

608. *Notholirion thomsonianum* Lower Hazara hills, S. A. Sultan

607. *Gloriosa superba* Murree foothills, Rubina A. Rafiq

605. *Allium jacquemontii* Margalla hills, S. A. Sultan

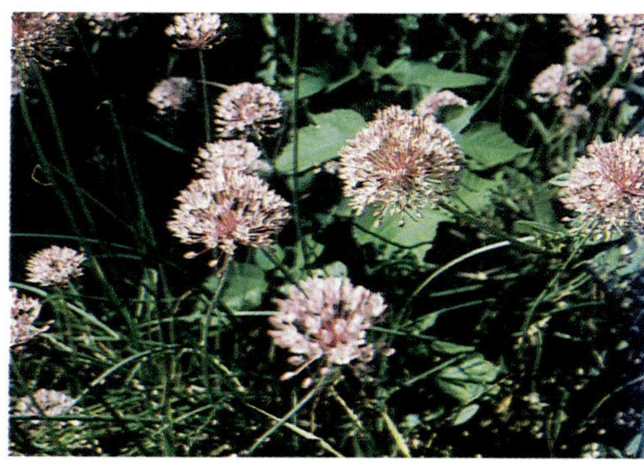

604. *Allium griffithianum* Rawal lake, Y. J. Nasir

610. *Tulipa stellata* a. Margalla hills, Y. J. Nasir
b. Balochistan, T. J. Roberts

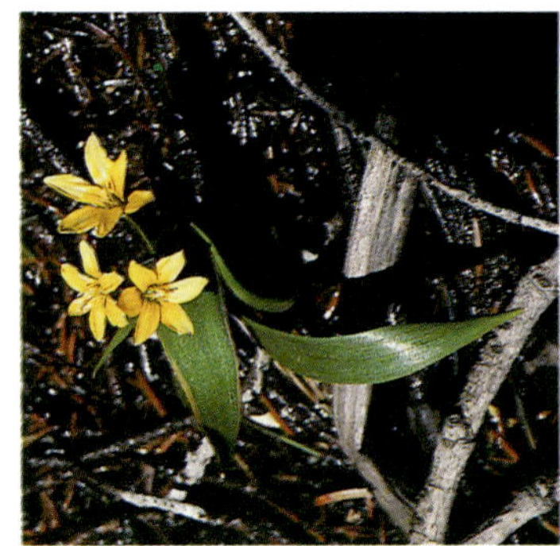

614. *Gagea elegans* Kaghan, Y. J. Nasir

611. *Lilium polyphyllum* Mahodand, Upper Swat, Fritz Berger

613. *Gagea pseudoreticulata* Islamabad, S. A. Sultan

612. *Frittilaria roylei* NW Himalaya, Chris Chadwell

616. *Eremurus himalaicus* Deosai, Fritz Berger

615. *Eremurus stenophyllus* Balochistan, T. J. Roberts

621. *Dipcadi erythraeam* Karachi, Y. J. Nasir

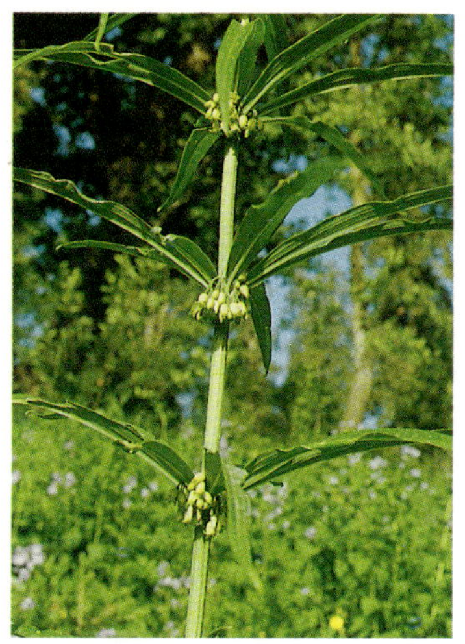

623. *Polygonatum verticillatum* Kaghan, Y. J. Nasir

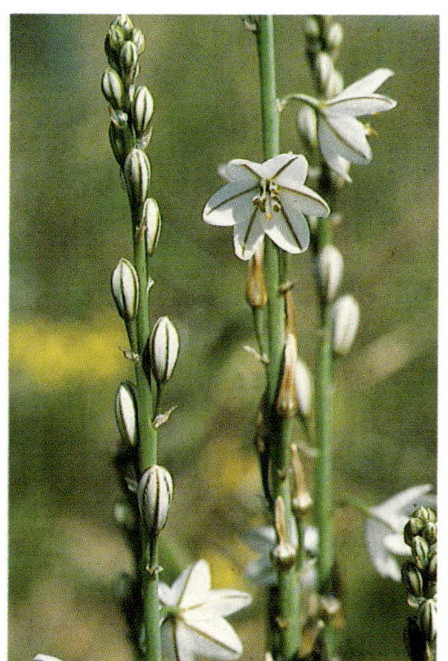

622. *Asphodelus tenuifolius* Margalla hills, Y. J. Nasir

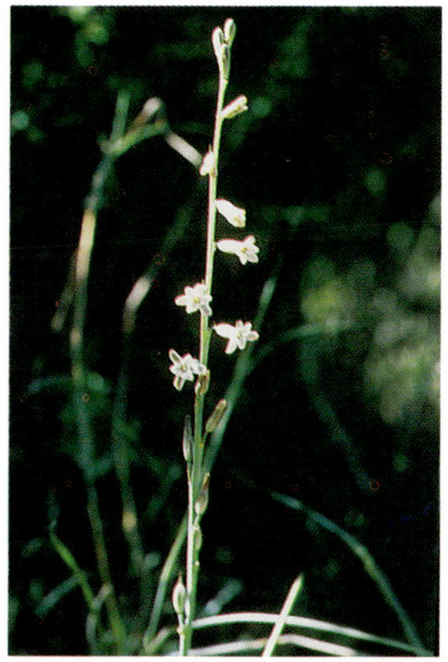

620. *Dipcadi hysudricum* Islamabad, Y. J. Nasir

617. *Eremurus persicus* Zarchi, Kalat, Rubina A. Rafiq

618. *Asparagus gracilis* Margalla hills, S. A. Sultan

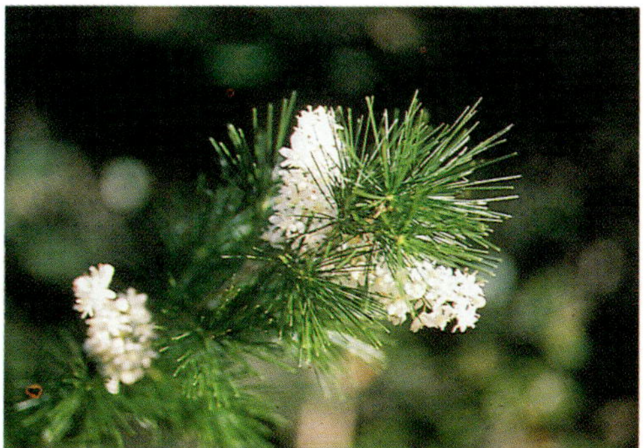

619. *Asparagus adscendens* Margalla hills, David Warr

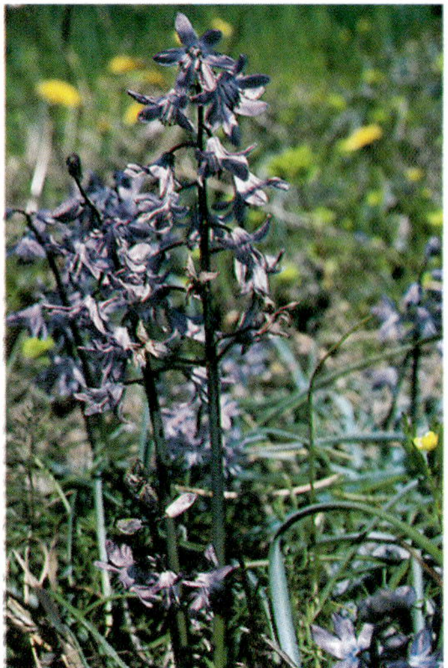

624. *Scilla griffithii* Margalla hills, S. A. Sultan

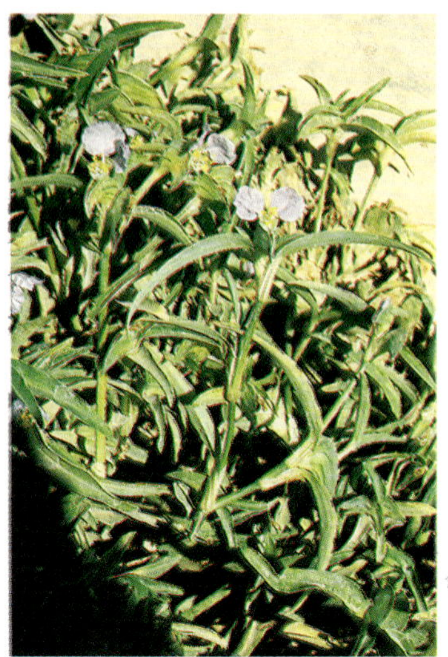

630. *Commelina albescens* Karachi, Y. J. Nasir

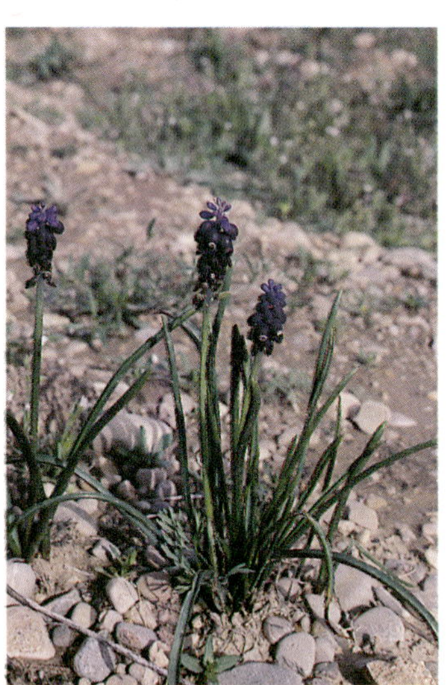

625. *Muscari neglectum* Quetta, Y. J. Nasir

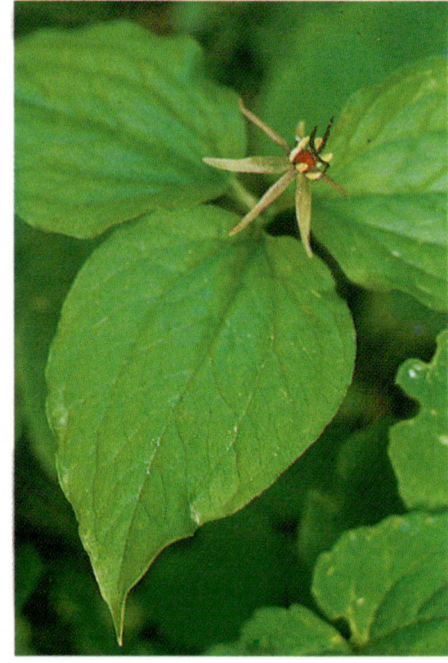

627. *Trillidium govanianum* Kaghan, Y. J. Nasir

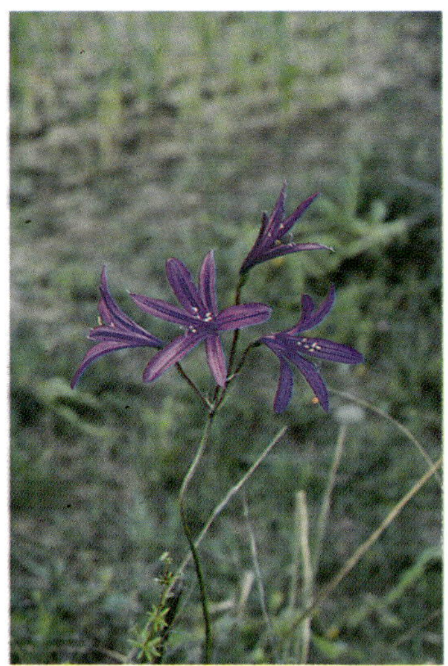

626. *Ixiolirion tataricum* Attock, S. A. Sultan

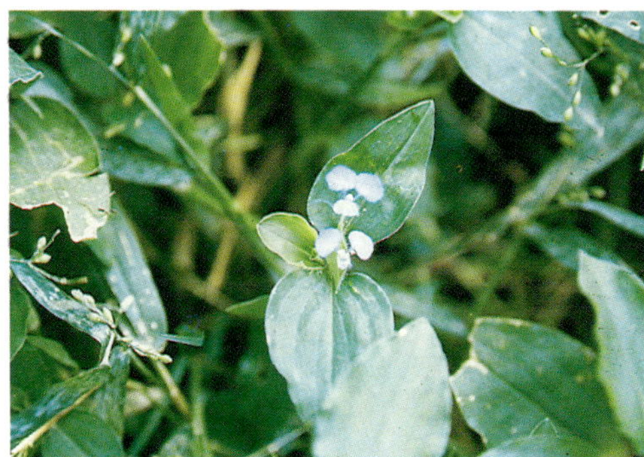

629. *Commelina bengalensis* Margalla hills, S. A. Sultan

628. *Butomus umbellatus* Lahore, Rubina A. Rafiq

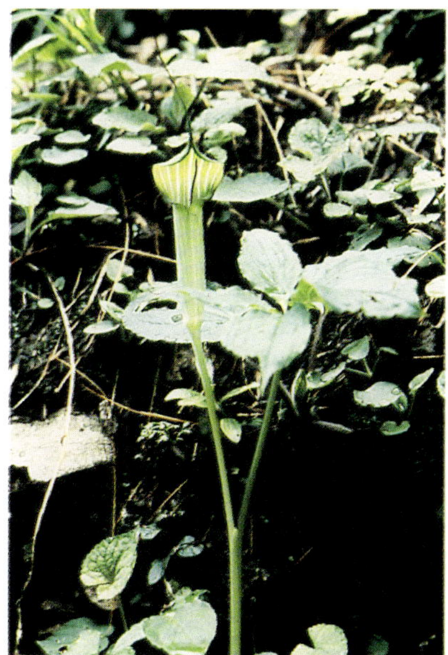

635. *Arisaema jacquemontii* Dunga gali, T. J. Roberts

631. *Commelian paludosa* Islamabad, S. A. Sultan

634. *Arisaema utile* Changla Gali, S. A. Sultan

633. *Colchicum aitchisonii* Khairi Murat Range, Rubina A. Rafiq

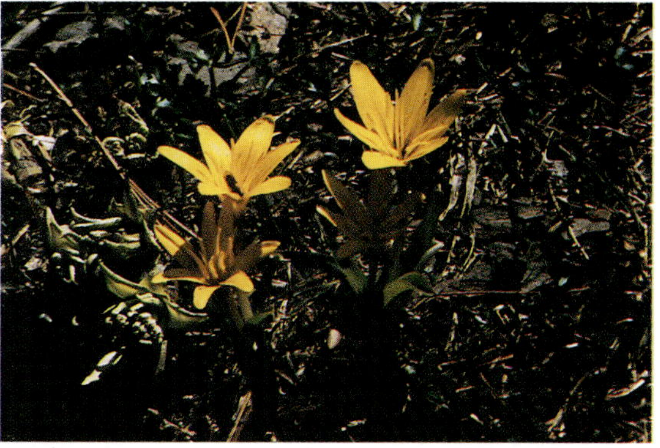

632. *Colchicum luteum* Batrassi pass, T. J. Roberts

639. *Juncus membranaceus* Deosai, Fritz Berger

641. *Phoenix sylvestris* Margalla hills, Rubina A. Rafiq

642. *Sagittaria guayanensis subsp. lappula* Wah, Y. J. Nasir

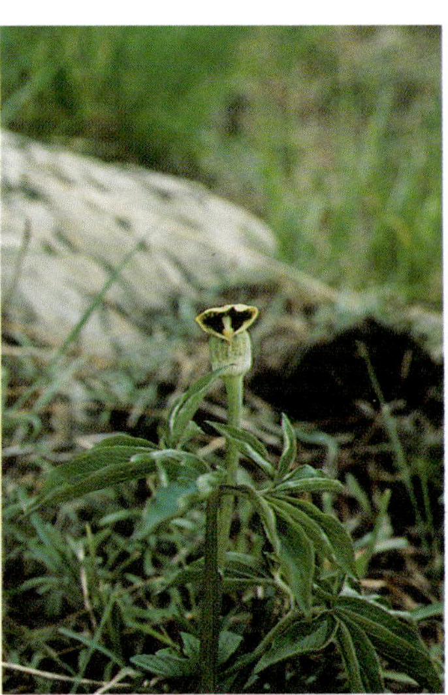

637. *Arisaema flavum* Changla Gali, Y. J. Nasir

638. *Eichornia crassipes* Lahore, Rubina A. Rafiq

640. *Nannorhops ritchieana* Awaran, Makran, Y. J. Nasir

647. *Pinus gerardiana* Chaprot, Hunza, Y. J. Nasir

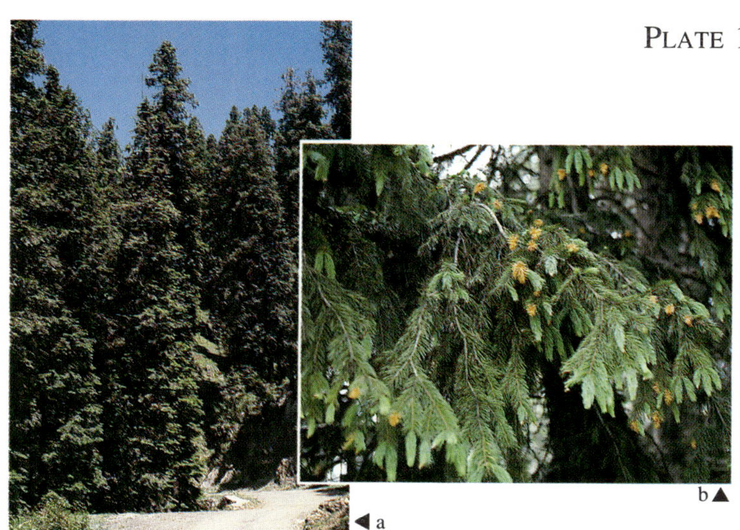

643. *Abies pindrow* a, b: Thandiani, Y. J. Nasir

645. *Cedrus deodara* Kalam, Fritz Berger

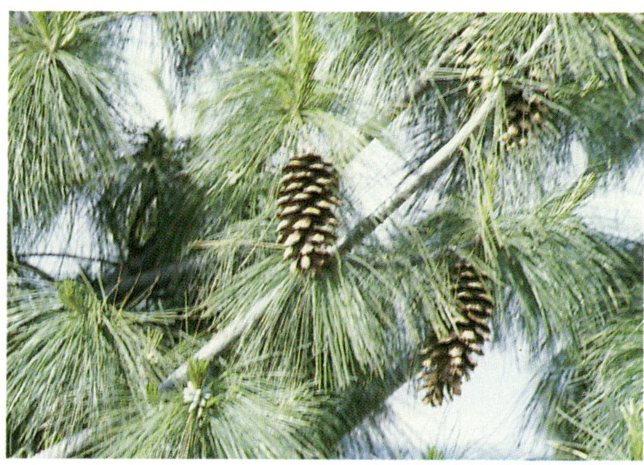

648. *Pinus roxburghii* Shogran, Kaghan, Y. J. Nasir

644. *Picea smithiana* Kalam, Y. J. Nasir

646. *Cedrus deodara & Pinus wallichiana* Kalam, Fritz Berger

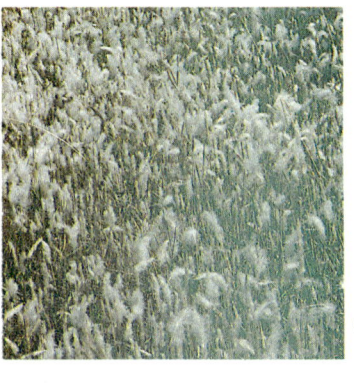

655. *Imperata cylindrica* Islamabad, Rubina A. Rafiq

651. *Ephedra gerardiana* Mahodand, N. Swat, Fritz Berger

652. *Arundo donax* Taxila, Y. J. Nasir

654. *Saccharum spontaneum* Islamabad, Y. J. Nasir

653. *Phragmites karka* Islamabad, Y. J. Nasir

649. *Juniperus excelsa* Kalam, Fritz Berger

650. *Juniperus communis var. saxatilis* Ushu, N. Swat, Fritz Berger

Uncommon in the Margalla hills and Lehtrar area east of Islamabad. Forests from 900-2000 m in Himalaya. June-July.

637. **A. flavum** (Forssk.) Schott Pl. 102

Monoecious. Bulb subglobose, 20-35 mm broad. Leaves 1 or 2. Leaflets 7-11. Spathe present below the foliage, yellowish-green, tube ovoid-globose, 10-12 mm long; limb longer than or equalling the tube. Spadix included in spathe.

Distribution: Yemen, Afghanistan, Pakistan, Kashmir, south Tibet, west China.

A more xeric form, preferring open sunny places in Waziristan, Dir, Swat and Hazara from 1700-3000 m. June-July.

PONTEDERIACEAE *Water Hyacinth Family*

A tropical family of fresh water or marshy perennial herbs with submerged or floating leaves. Flowers bisexual, in spikes or racemes, subtended by a spathe-like sheath. Perianth segments 6, free or united. Stamens 6. Fruit capsular.

Eichornia 1 sp.

Aquatic free floating herbs. Leaves mostly basal. Perianth of 6 united segments. Stamens 6. Ovary 3 locular. Capsule ovoid to linear.

638. **E. crassipes** (Mart.) Schlecht *Water Hyacinth* Pl. 102

Mat forming herbs up to 40 cm tall. Roots many, fibrous. Leaves rhomboid to broadly ovate. Petiole 6-30 cm long, inflated in the middle, sac-like. Flowers very showy, 6-15, mauve to lilac. Lower segments of perianth with a yellow spot.

Distribution: Native of Brazil. Introduced and naturalized in many tropical and subtropical countries.

Troublesome weed that chokes ponds and water channels. Source of potash. April-July.

JUNCACEAE *Rush Family*

Perennial tufted herbs. Leaves basal, flat or cylindrical, narrow, sheathing. Flowers small, regular, bisexual. Perianth of 6 sepals, in two series. Stamens 6. Ovary superior, 3 locular. Stigmas 3. Capsule 3-angled, 3-valved.

Juncus 16 spp.

Rhizomatous herbs. Flowers brownish or greenish-white. Stems free. Capsules often beaked, yellowish to brownish.

639. **J. membranaceus** Royle ex D. Don Pl. 102

Plants 15-40 cm tall, slender. Leaves 1-3, 4-14 cm long, linear. Flowers 8-20 per head, whitish, 6-7 mm long. Perianth segments oblanceolate. Stamens included. Capsule ovoid, 5-8 mm long, brownish-black.

Distribution: Afghanistan, Pakistan, Kashmir to Sikkim.

Common in wet places from Chitral eastwards to Baltistan from 1800-4000 m in mountainous areas. May-October.

PALMAE *Palm Family*

Monoecious trees or shrubs. Leaves pinnately or palmately divided, often very large (2.5 m or more), plicate. Flowers small, regular, in paniculate clusters. Bracts large, often sheathing the inflorescence. Perianth of 6 segments, in two series. Stamens 6. Fruit a berry or a drupe, sometimes very large as in coconut.

+	Leaves palmate or fan-like. Shrub or small tree up to 2.5 m tall.	**1. Nannorrhops**
–	Leaves pinnate or feather-like. Trees up to 16 m tall, with a well developed trunk.	**2. Phoenix**

1. Nannorrhops 1 sp.

Gregarious, tufted leafy shrubs with extensive branching rhizomes. Stem unarmed. Leaves fan-like, up to 1 mm broad. Flowers bisexual, paniculate. Bracts many, sheathing.

640. **N. ritchieanna** (Griff.) Aitchison *Mazri or Dwarf Palm* Pl. 102

Leaves 40-130 cm long, glabrous. Petiole 15-30 cm long, base sheathing, fibrous. Panicles up to 1 m broad. Perianth segments 4-5 mm long. Seeds 10-12 mm long.

Distribution: South Iran, Afghanistan, Pakistan.

Hot dry mountainous places to 1800 m from Sindh Kohistan, lower Balochistan up to Kurram valley. Leaves valuable for making mats, sandals, baskets etc. July-November.

2. **Phoenix** c.4 spp. *Palm*

Dioecious Palms with large pinnate leaves. Inflorescence branched, axillary. Flowers unisexual. Fruit a 1-seeded drupe, fleshy.

641. **P. sylvestris** (L.) Roxb. *Khajoor, Wild Date Palm* Pl. 102

Tree up to 10 m tall. Stem clothed with remains of the leaf bases. Leaves 80-100 cm long. Leaflets numerous. Drupe 20-25 mm long, orange-yellow.

Distribution: Native to Pakistan. Also cultivated.

Plains to 1000 m. Self sown. Quite conspicuous in Punjab Salt Range and around Taxila. The leaves are used for weaving mats and bags. The fruit is edible. March-April, fruiting September-October.

ALISMATACEAE *Water-Plantain Family*

Annual or perennial marshy herbs. Rhizome stout. Leaves radical. Flowers scapose, small, bisexual, regular, 3-merous, disposed in racemes or panicles. Sepals 3. Petals 3, free. Fruit a collection of achenes.

Sagittaria 1 sp. *Arrowhead*

Erect monoecious or dioecious plants. Leaves dimorphic, aerial ones arrow-like, submerged ones ribbon-like. Flowers racemose, white. Male flowers on the upper part of raceme. Achenes compressed laterally.

642. **S. guayanensis** H. B. & K. Pl. 102
subsp. **lappula** (D. Don) Bogm.

Aquatic herb with emergent leaves. Leaves 5-8 cm long. Stalk up to 40 cm. Racemes 30-45 cm long. Flowers 12-15 mm broad, the lower ones bisexual. Petals obovate-suborbicular, white. Head of achenes 10-12 mm across.

Distribution: Tropical and subtropical Africa and Asia.

Near streams etc. From plains to 1000 m, mainly in rice growing tracts of northern Punjab, N.W.F.P., lower Swat. August-September.

PINACEAE *Pine Family*

Monoecious trees with needle-like leaves. Flowers arranged in cone-like structure. Female cones enlarged and woody, the male small. Fruit a woody cone. Seeds winged.

1.	+	Leaves solitary on the main stems. Shoots absent.	2
	–	Leaves all or mainly on short lateral shoots.	3
2.	+	Leaves flat in cross section. Cones erect.	**1. Abies**
	–	Leaves 4-angled in cross section. Cones pendulous.	**2. Picea**
3.	+	Leaves short, needle-like, spiral on long shoots, in clusters on short ones. Male cones solitary on short branches.	**3. Cedrus**
	–	Leaves scale-like on long shoots. Long needle-like leaves on short ones. Male cones in fascicles on short shoots.	**4. Pinus**

1. Abies 2 spp. *Fir*

Evergreen trees to 30 m or more tall. Leaves borne directly on stems, linear, flat. Male cones cylindric-oblong, flowers in leaf axils on undersurface of shoots. Female cones with seed scales falling away at maturity, leaving the cone axis naked.

643. **A. pindrow** Royle *Palundar, Silver Fir* Pl. 103

Trees pyramidal shaped with drooping branches, 20 m or more tall. Bark light grey to brownish. Leaves 15-40 mm long, dark green, glaucous. Male cones 1-2 cm long, axillary, reddish-green. Female cones up to 12 cm long, cylindric-oblong, violet-purple. Seed scale obovate, 15-20 mm long.

Distribution: Afghanistan, Pakistan, Kashmir to Nepal.

Common and gregarious from 2100-3500 m, preferring northern aspects. Wood used for construction purposes. Cones ripen in September and disintegrate upon ripening.

2. Picea 1 sp. *Spruce*

Branches whorled, with peg-like projection (remains of leaf bases). Leaves needle-like, 4-angled. Female cones pendulous.

644. **P. smithiana** (Wall.) Boiss. *Kachal, Himalayan Spruce* Pl. 103

Trees up to 30 m or more tall with drooping branchlets. Leaves 20-40 mm long. Male cones axillary, solitary, pendulous, oblong, 20-25 mm long.

Distribution: Afghanistan, Pakistan, Kashmir to central Nepal.

Common from 2500-3400 m. Often in association with fir and pine. Generally considered hardier than *Abies,* favouring more exposed slopes. Cones ripen October-November.

3. Cedrus 1 sp. *Cedar*

Monoecious or dioecious trees up to 20 m or more tall. Both long shoots (with scattered leaves) and short shoots (with whorled leaves) present. Seed scales of female cones deciduous.

645. **C. deodara** (Roxb. ex D. Don) G. Don *Deodar, Himalayan Cedar* Pl. 103

Branches horizontal. Leaves needle-like, 20-25 mm long, angled. Male cones solitary, erect, cylindrical, 25-60 mm long. Mature female cones barrel-shaped, 7-11 cm long, brown.

Distribution: Afghanistan, Pakistan, Kashmir to Nepal.

Gregarious at altitudes of 2100-3000 m. Wood straight grained and used for door and window frames for construction purposes. Cones ripen October-November.

4. Pinus 3 spp. *Pine*

Evergreen tree with flaking bark. Wood resinous. Short shoot with fascicles of 2,3 or 5 long needle-like leaves. Male cones in dense clusters at base of short shoots. Female cones becoming large and woody at the end of second year. Seeds winged.

1.	+	Leaves in fascicles of 5.	646. **P. wallichiana**
	–	Leaves in fascicles of 3.	2
2.	+	Bark silvery. Leaves 6-12 cm long, stiff. Seed winged. Wing less than seed length.	647. **P. gerardiana**
	–	Bark not silvery. Leaves 15-22 cm long, flexuous. Seed winged. Wing 2-3 times seed length.	648. **P. roxburghii**

646. **P. wallichiana** A.B. Jackson *Kail, Biar, Blue Pine* Pl. 103

Trees up to 30 m tall with greyish scaly bark. Leaves bluish to grey-green. Male cones 10-15 mm long, in clusters. Female cones 2-4, at the tips of branches, 16-28 cm long, cylindrical, drooping.

Distribution: Afghanistan, Pakistan, Kashmir to Nepal.

Common and gregarious from 1800-3500 m. Often favouring southern aspects too dry for *Abies.* Forming mixed stands with *P. roxburghii* between 1500-1800 m and with fir and spruce at higher levels. Wood used for construction purposes. Cones ripen. October-November.

647. **P. gerardiana** Wall. ex Lamb. *Chilghoza, Pinenut Tree* Pl. 103

Trees up to 30 m tall with beautifully patterned silvery bark and irregular spreading branches. Leaves short, stiff, in clusters of 3. Seeds 15-20 mm long, cylindric, with a short wing.

Distribution: Afghanistan, Pakistan, Kashmir.

Inner dry valleys from 2100-3000 m. Much threatened by felling both in Chilas, Gilgit and Balochistan. Seeds edible when roasted and of a good quality. Cones ripen August-September.

648. **P. roxburghii** Sargent *Chir, Chil* Pl. 103

Trees with a soft flaky bark and resinous wood. Branches spreading, irregular. Leaves needle-like, 15-22 cm long. Male cones 12-15 mm long, yellowish-brown, in dense terminal clusters. Female cones solitary or 2-3, subglobose, up to 12 cm broad woody at maturity. Wing of seed 2-3 times seed length.

Distribution: Afghanistan, Pakistan, Kashmir to Sikkim.

Common from 600-1800 m. Forming pure stands or mixed with blue pine. The thick bark is highly resistant to forest fires. The wood is not durable. The resin is extracted for making turpentine. Male cones ripen February-March.

CUPRESSACEAE *Cypress Family*

Monoecious or dioecious trees or shrubs. Leaves needle-like closely appressed to the stem. Male cones small. Female fruit woody or berry-like. Seeds winged or not.

Juniperus 4 spp. *Juniper*

Trees or shrubs with scale-like or needle-like opposite leaves. Male cones terminal axillary. Female cones of 3-8 scales, becoming fleshy when mature. Seeds not winged.

+	Trees. Leaves dimorphic, with both scale-like and linear-lanceolate leaves.	649. **J. excelsa**
–	Prostrate or low growing shrubs. Leaves linear-lanceolate or broadly so.	650. **J. communis**

649. **J. excelsa** M. Bieb. *Sanober, Apurs* Pl. 104

Medium sized trees. Leaves on upper branches scale-like, opposite, 1.5-2 mm long, broadly ovate; on lower branches subulate, 5-8 mm long, tip spinose. Berries blue-black, c. 10 mm broad, subglobose.

Distribution: South west Europe, Caucasus, Arabia, Iran, Afghanistan, Pakistan, Kashmir to Nepal.

Common from 2000-4000 m, forming open forests especially in higher ranges of Balochistan. One of the longest lived tree species, some specimens reputed to be well over thousand years old. Much lopped for fuel and in pencil making. October.

650. **J. communis** var. **saxatilis** Pallas Pl. 104

Dioecious shrub with a carpet-like habit. Leaves 3 per node, 8-10 x 2-4 mm, needle-like, recurved. Male cone 6-8 mm long, axillary. Female cones solitary, 18-20 mm long. Fruit berry-like, 8-12 mm broad, bluish-black.

Distribution: Alpine and arctic Europe and Asia, north Africa, America.

Alpine and near the tree line from 2200-4500 m throughout the northern areas. August-September.

EPHEDRACEAE *Ephedra Family*

Dioecious shrubs with opposite leaves. Male flowers in ovoid spikes. Female flowers 2, enclosed in a membranous sheath, and surrounded basally by many overlapping bracts. Fruit berry-like, red.

Ephedra 8-9 spp.

Erect to decumbent or scandent plants. Branches glabrous or scabrid, dark green. Leaves scale-like or not. Male cones 2-4, clustered at the nodes. Female strobili 2-4, each with several pairs of bracts.

651. **E. gerardiana** Wall. ex Stapf Pl. 104

Densely tufted shrub up to 120 cm tall, with green ascending branchlets, that are striate and smooth. Male flowers 4-8. Bracts obtuse, connate. Female cones solitary with 2-3 pairs of bracts. Berry 6-8 mm long, ovoid, red.

Distribution: Afghanistan, Pakistan, Kashmir to Bhutan, south west China.

Dry rocky places from 2300-4100 m. Exploited medicinally. A source of the alkaloid ephedrine. August-September.

POACEAE

Grass Family

A large and economically important family of annual or perennial herbs, often with rhizomes or stolons. Leaves alternate, narrow, base sheathing the stem, venation parallel. Inflorescence of spikelets disposed in panicles, spikes or racemes. Flowers usually bisexual, small. Perianth reduced to 2 scales (lodicules). Stamens 1-6. Styles 2, with feathery stigma. Fruit a caryopsis.

1. + Tall plants 3-6 m with broad leaves. Ligule short and membranous. **1. Arundo**

   – Smaller plants up to 2.5-4 m tall, with narrow leaves. Ligule a ring of long hairs or membranous. 2

2. + Plants low-growing up to 60 cm tall. Inflorescence spike-like, 3-18 x 2-3 cm. **4. Imperata**

   – Plants more than 1.5 m tall. Inflorescence 20-60 x 6-20 cm. 3

3. + Spikelets 1-many flowered, at maturity generally not falling off as a whole. **2. Phragmites**

   – Spikelets 2-flowered, falling off as a whole at maturity. **3. Saccharum**

1. Arundo 1 sp.

Perennials characterised by a reed-like habit. Inflorescence paniculate, large, plumose. Ligule of leaf membranous.

652. **A. donax** L. *Nara* Pl. 104

Rhizome creeping, well developed. Plants up to 5 m tall. Leaves broad, 30-50 x 3-5 cm. Panicle up to 65 cm long, bushy, 5-10 cm broad.

Distribution: Mediterranean eastwards to Burma. Introduced in many places.

Damp and marshy places. Common up to 1400 m almost throughout Pakistan. The leaves are used for making mats etc. Often planted as a screen or hedge in the plains. June-December.

2. Phragmites 2 spp. *Nal, Reed*

Perennial plants of wet or semi-wet areas. Leaf blades lanceolate or narrowly so. Ligule a ring of long or short hairs. Panicles large, dense, silky. Spike 3-11-flowered.

653. **P. karka** (Retz.) Trin. ex Steud. Pl. 104

Tall reed with leaves 30-80 x 1.2-4.5 cm. The blade scabrid on the undersurface. Apex narrow and stiff. Panicle up to 55 x 10-23 cm.

Distribution: Tropical Africa and Asia.

Swamps, lake margins and streams from sea level to 2400 m throughout the country. April-November.

3. Saccharum 8 spp.

Tall perennials with narrow leaves. Ligule membranous or with a fringe of hairs. Panicles large, plumose. Spikelets with long silky hairs.

654. **S. spontaneum** L. *Kahi* Pl. 104

Plants up to 4 m tall. Leaves 0.3-1.8 m long, up to 8 mm broad. Panicle 25-50 cm long. Spikelets silky hairy.

Distribution: Widely distributed in tropical regions of the Old World.

Open places or sandy areas where periodic flooding occurs. Flowering stems are widely used to make cane furniture, verandah blinds etc. Good sand binder. Plains to 1300 m. July-September.

4. Imperata 1 sp.

Perennial with well developed rhizome and narrow leaves. Inflorescence a spike-like silvery panicle.

655. **I. cylindrica** (L.) Raeuschel *Sword or Cotton Grass* Pl. 104

Tuft forming grass generally up to 30 cm tall. Leaf blades flat, more or less erect. Panicles cylindrical, silky white at maturity.

Distribution: Old World tropics. Mediterranean and Middle East, Chile.

Open places, poorly drained soil from plains to 1800 m. April-May.

A SELECT BIBLIOGRAPHY

The reader will find the following works on the Flora of Pakistan useful :

1. Ali, S. I. & Qaiser, M, (1986), *A Phytogeographical Analysis of the Phanerogams of Pakistan and Kashmir*, proceedings of the Royal Society of Edinburgh, 89B, 89-101.

2. Biswas, A. K. (1987), *Environmental Concerns in Pakistan with Special Reference to Water and Forests*, Environmental Conservation 14, No. 4.

3. Blatter, E. (1928-29), *Beautiful Flowers of Kashmir,* 2 vols, John Bale, Danielsson, London.

4. Boissier, E. (1867), *Flora Orientalis,* Vol. 1, Basel-Geneva-Lyon, H. George.

5. Brandis, D. (1906), *Indian Trees,* Constable, London.

6. Coventry, B. O. (1923-30), *Wild Flowers of Kashmir,* 3 vols. Raithbay, Lawrence, London.

7. Eig, A. (1931), *Les Elements at Les Groups Phytogeographiques Auxiliariaires dans la Flore Palestinienne,* 2 pts. Feddes Repertorium, Beihefte 63, 1-201.

8. Hara, H. (1966), *The Flora of Eastern Himalaya,* University of Tokyo Press, Tokyo.

9. Jafri, S. M. H. (1966), *Flora of Karachi,* p. 375, Book Corp Karachi.

10. Leonard, J. A. K. (1987), *Contribution a l'etude de la Flora et de la Vegetation des Deserts d'Iran Fasc.* 9., Jardin Botanique Nation de Belgique.

11. Nasir, E. & Ali, S. I, eds. (1970-). *Flora of Pakistan,* National Herbarium, NARC, Islamabad & Department of Botany, University of Karachi, Karachi.

12. Nasir, E. & Y. & Akhter, R. (1987), *Wild Flowers of Rawalpindi & Islamabad Districts,* Text 82, pp + 52 Plates, National Herbarium, NARC, Islamabad.

13. Parker, R. N. (1918), *A Forest Flora for the Punjab with Hazara & Delhi,* Supdt. Govt. Printing, Lahore.

14. Polunin, O. & Stainton, A. (1984, 1986), *Flowers of the Himalaya,* 2 vols. Oxford University Press, New Delhi.

15. Stewart, R. R. (1982), *History & Exploration of Plants in Pakistan & Adjoining Areas,* p. 186, National Herbarium, NARC, Islamabad.

16. Zohary, M. 1950. *The Flora of Iraq and its Phytogeographical Division,* Bulletin Directorate General of Agricultural Iraq, No. 31. Baghdad, Government Press, 1973.

17. *Geobotanical Foundations of the Middle East,* vols. 1 & 2. Stuttgart, Gustav Fischer Verlag.

GLOSSARY OF BOTANICAL TERMS

A

Achene — A small dry 1-seeded indehiscent fruit, as in *Ranunculus*.

Actinomorphic — Flowers radially symmetrical.

Adpressed — Usually pertaining to hairs or leaves that are pressed to the surface.

Adventitious — Roots that do not arise from the seed, but from any other part of the plant.

Alternate — Leaves arranged singly at different heights on the stem axis.

Amplexicaul — Base of leaves that clasp the stem horizontally.

Androecium — The male part of flower; the stamens as a unit of flower.

Annual — Plants that flower and fruit in one year.

Anterior — That portion of an axillary or lateral flower that faces outwards from the axis of the inflorescence.

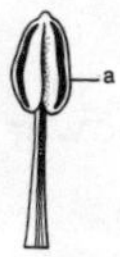

Anther — Part of the stamen that bears pollen (a).

Apiculate — Apex with a short pointed tip.

Apomict — Plants that do not reproduce sexually for a period of their life-span.

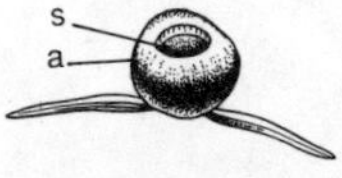

Aril — A fleshy out-growth of seed coat (a), that envelopes the seed(s).

Ascending	Generally pertaining to plant habit where the stem and branches are produced upward obliquely.
Attenuate	Pertaining to a leaf base or apex which tapers gradually.
Auriculate	A leaf base with small rounded lobes.

B

Baccate	Berry-like.
Berry	A juicy, indehiscent fruit (b).
Biennial	Plants of 2 seasons duration, from seed to maturity and death.
Bifid	Apices of leaves or petals that are 2-cleft.
Bilabiate	Generally applied to parts of a flower (i.e. corolla, calyx), which are 2-lipped, an upper lip (u) and lower lip (l).
Bipinnate	The leaflets of a pinnate leaf again pinnate.
Bisexual	Having both sexes.
Biternate	Twice ternate. See under ternate.
Bract	Usually a small structure present at the base of a flower stalk (b).
Bracteole	A secondary bract, generally present on the flower stalk (br).
Bulb	An underground reduced stem, in which the inner leaves (i.l.) are fleshy and outer ones (o.l.) scaly, as in onion.
Bulbil	A small bulb, usually present in the leaf axil, that serves as an asexual reproductive body.

C

Caducous Falling off soon.

Caespitose Growing in tufts.

Calyx The outermost whorl of a flower, made up of individual parts, the sepals.

Campanulate Pertaining to a bell shape. Such as the corolla in *Campanula*.

Capitate Forming a head or head-like cluster.

Capsule A dry fruit, splitting by valves or shedding seeds by pores as in poppy, lady fingers.

Carpel A simple pistil or one of the component parts of a compound pistil. See under pistil.

Caryopsis Nut-like fruit of the grass.

Catkin A pendulous spike-like cluster of flowers, as in mulberry.

Cauline Pertaining to a stem. Such as leaves inserted on the stem.

Ciliate Bearing hairs on the margin.

Cladode A leaf-like stem (c), as in *Asparagus*.

Clavate Club shaped.

Claw Stalk-like portion of a petal.

Cocci 1-seeded parts of a lobed fruit that splits, as in *Euphorbia*.

Compound A structure made up of several similar parts.

Cone A fruiting body comprising of bracts arranged on an elongated axis, in the axil of which lie the male or female flowers as in pine, fir.

Cordate Heart shaped. Pertaining to leaf shape.

Coriaceous Leathery texure. As surface of some leaves.

Corm An underground stem, modified and compact.

Corolla The second whorl of a flower. Individual parts are the petals (p).

Corona A cup-like or crown-like process from the throat of the corolla.

Corymb A flowering axis in which the flowers all appear to be at the same level. The younger ones placed towards the center, as in mustard.

Crenate Margin of leaf that is lobed. The lobes being rounded.

Crenulate As above, but very small lobes.

Cuneate Margin of leaf or petal which is narrow or tapering towards the base or apex.

Cyathium The specialized inflorescence of *Euphorbia,* in which the unisexual flowers are included in a cup-like structure.

Cyme A flower cluster in which the central flowers open first.

D

Deciduous Falling off. Such as leaves of non-evergreen trees as in poplar.

Decumbent Trailing along the ground with the end ascending.

Dentate Generally pertaining to leaf margin in which teeth are at more or less right angles to the mid-vein of the leaf.

Denticulate Finely dentate.

Diadelphous Stamens in two sets, such as in sweet pea where the 10 stamens(s) are in two groups of 9 and 1, respectively.

Dichotomous Branching of stem forked once or twice.

Digitate Palm or hand-like. Such as the leaf of castor oil.

Dimorphous Generally a flower with two forms.

Dioecious In which the sexes of plants (i.e. male and female flowers) are on different plants.

Divaricate Spreading or widely apart.

Dorsal The back or outer surface of an organ such as the lower side of a leaf.

Dorsifixed Pertaining to the stamens in which the stalk is attached to the back of the anther.

Drupe A 1-seeded fleshy indehiscent fruit, in which the seed is enclosed in a hard covering.

E

Ebracteate Without bracts

Elliptic Pertaining to an oval leaf or petal in which both ends taper.

Emarginate Apex of leaf or petal with a shallow notched end.

Endemic Generally plants of restricted or local distribution.

Epicalyx A whorl of bract-like structures below the calyx.

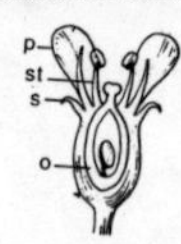

Epigynous Flowers in which the ovary(o) is present below sepals (s), petals(p) and stamens(st).

Epipetalous Attached to the petal of corollas, such as stamens.

Evergreen Plants that remain green throughout the year, due to persistence of the leaves. Such as fir, spruce etc.

Extrorse Anthers in which dehiscence is by the outer face. i.e. towards the petals.

F

Farina A mealy covering generally on the leaf, as in *Primula* sp.

Fid Suffix for cleft or incision, like 2-fid or bi-fid.

Filament The stalk of stamen.

Fimbriate Fringed.

Fistular Usually said of a stem that is hollow within.

Floccose Woolly with tufts of soft hair.

Floriferous Flower bearing.

Foliaceous Leaf-like, usually pertaining to bract.

Follicle Dry dehiscent fruit opening by a single suture.

Funnelform Shaped like a funnel. i.e. corolla shape where the tube gradually widens upwards, as in *Convolvulus* spp.

G

Gamopetalous A flower in which the petals are united together.

Gamosepalous A flower in which the sepals are united together.

Gibbous Swollen on one side.

Glabrescent Becoming glabrous.

Glabrous A surface which is not hairy.

Glandular Bearing secreting glands or hairs.

Gynoecium The female part of a flower. Individual parts are the pistils.

H

Hastate Shape of an arrow head. Pertaining to a leaf.

Head A short dense aggregate, as a collection of flowers.

Herb A plant that is not woody.

Heterogamous A flower head (capitulum) bearing male, female or neuter flowers or when male and female flowers are borne on the same plant.

Heterostylous Styles of different lengths i.e. long-styled or short-styled flowers.

Hirsute With rough or coarse hairs.

Hispid Having stiff or bristly hairs.

Homo- Prefix meaning the same or alike.

Hypanthium An enlargement or development of the receptacle.

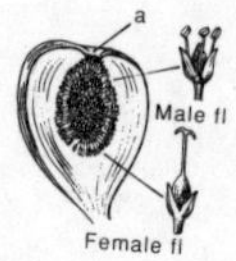

Hypanthodium A condensed inflorescence in which the fleshy receptacle encloses the male and female, with an opening(a) at the top of the flowers. As in fig.

Hypogynous Sepals, petals and stamens attached below the ovary.

I

Imparipinnate A compound leaf in which the number of leaflets are odd.

Indehiscent Generally said of fruits or anthers that do not split.

Indumentum Pertaining to hairs.

Inferior An ovary which lies below the sepals, petals and stamens. See also under epigynous.

Inflorescence Arrangement of flowers on the flowering axis.

Infundibuliform Funnel shaped.

Internode The interval between two nodes i.e. that part of a stem from where the leaf arises.

Involucre A collection of bracts(br) usually present near the flower base.

Irregular A flower that has two similar halves, if cut along one plane only or incapable of being dvided into two equal halves.

K

Keel Said of a flower in which the 2 front petals are united as in a pea flower (k).

L

Labellum The lower petal of an orchid flower(1).

Lamina Expanded or green portion of a leaf.

Lanceolate A leaf which widens above the base and tapers towards the apex.

Latex The milky juice of such plants as the dandelion or spurge.

Lax Loosely arranged.

Leaflet A part of a compound leaf.

Ligulate A strap-shaped leaf or petal.

Limb The expanded portion of a gamopetalous corolla.

Linear Long and narrow as in a grass leaf.

Lip One of the two parts of a gamopetalous corolla or calyx. Usually termed as upper or lower lip. Labellum.

Lomentum A constricted pod which breaks up at maturity into indehiscent 1-seeded parts.

Lyrate A shallow incised leaf with a terminal large lobe and smaller lower lobes.

M

Mericarp A fruit with dry 1-seeded carpels(m).

-merous Used as a suffix to denote parts or numbers, as 4-merous.

Mesophyte Plants that have intermediate water requirements.

Midrib Main rib of a leaf.

Monadelphous Stamens united into one set.

Moniliform Beaded.

Monoecious With male and female flowers on the same plant.

Monotypic Having only one exponent, as a genus with but one species.

Mucronate A short and sharp abrupt point.

Muricate Rough with short hard points.

N

Napiform Turnip shaped.

Node See under Internode.

Non Not; generally used immediately before word or words to imply negative sense.

Nut A hard indehiscent 1-seeded fruit, resulting from a compound ovary.

O

Ob- (Prefix), inversed, reversed, the other way round.

Obcordate Deeply lobed at apex; reverse cordate.

Oblanceolate Tapering towards both ends but broader towards the base.

Oblique Unequal sided, as the divisions of a leaf on either side of the midrib unequal.

Oblong Two or three times as long as broad.

Obovate Egg shaped in outline with the broader end towards apex.

Obpyramidal	Inversely pyramidal shaped.
Obtuse	Blunt or rounded at the apex.
Ochreate	A membranous, tube-like stipule(o) through which the stem passes.
Opposite	Leaf pair at a node, one on either side of the stem axis.
Ovate	Egg shaped outline.

P

Palmate	Lobed or divided in a palm-like manner, i.e. like the extended fingers of an open hand.
Panicle	A loose irregularly branched inflorescence.
Papillose	Small nipple-like projections(p).
Pappus	Generally a ring of hairs of some fruits as in sunflower.
Parallel	Veins in a leaf that do not unite or anastomose.
Parietal	Borne on the walls within an ovary. Usually referring to the location and arrangement of ovules.
Paripinnate	Term generally used for a pinnate leaf (see under pinnate) in which the number of leaflets(l) is even.
Pedate	With the leaflets arranged in a palm-like manner and the side ones divided.
Pedicel	The stalk of a flower.
Peduncle	The stalk of a fruit or fruit cluster.

Pentamerous	A flower in which calyx, sepals, petals, etc are 5 in number (5-merous).
Pepo	The fruit of the melon family, which is many-seeded and formed from an inferior ovary.
Perennial	A plant with 3 or more seasons duration.
Perfect	Said of a flower that has stamens and pistil.
Perianth	A term used for both sepals or petals collectively.
Petal	The individual part(s) of a corolla.
Petiole	Stalk of a leaf.
Pinnae	The primary units of a pinnately divided compound leaf.
Pinnate	Said of a compound leaf with leaflets arranged on either side of the midrib. Feather-like.
Pinnatisect	Cut down to the midrib in a pinnate way.
Pistil	The female part of a flower consisting of ovary, style and stigma.
Placentation	The arrangement of ovules within the ovary.
Plicate	Folded lengthwise, like a closed paper fan.
Pod	A dry many-seeded fruit, opening at maturity, as in a pea pod.
Polypetalous	Having separate petals.
Polysepalous	Having separate sepals.
Pome	A fruit formed from an inferior flower, as in apple.

Porose	Capsule dehiscing and shedding seeds by pores, as in the poppy.
Posterior	Said of that portion (i.e. sepal or petal) of an axillary or lateral flower that faces towards the axis of inflorescence.
Precocious	Flowers that bloom before the leaves come out.
Procumbent	Lying for the whole or greater part of its length close to the ground. Trailing. Opposite of decumbent.
Pruinose	Having a bloom of waxy secretion on the surface.
Pubescent	Hairs on the surface which are short, soft and straight.
Pyrene	The seed or stone of a drupe.

R

Raceme	An elongated inflorescence in which the flowers are stalked and the younger ones are present towards the top.
Rachis	The midrib of a leaf.
Radical	Leaves that arise from the base of the stem.
Receptacle	That portion of a flower that bears the sepals, petals, stamens and pistil. The thalamus.
Regular	A flower with petals or lobes alike in size and shape.
Reniform	Kidney-shaped.
Repand	With a wavy margin.
Reticulate	Veins of a leaf that are netted i.e. anastomose.
Retuse	Apex with a small, shallow notch.

Rhizome An underground root-like stem.

Ringent A corolla with a wide mouth.

Rotate A regular gamopetalous corolla with a short tube and flat spreading limb.

Rugose Wrinkled.

Runcinate Margin of a leaf toothed, in which the teeth point backwards.

Runner An enlongated lateral shoot, rooting at regular intervals, as in grass.

S

Saccate Bag-like or pouched

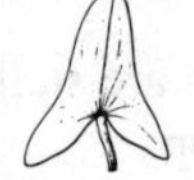

Sagittate An arrow-shaped leaf.

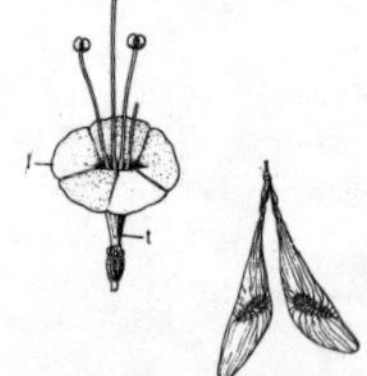

Salverform A corolla with a slender cylindric tube(t) and flat expanded upper portion, the limb(l).

Samara An indehiscent dry fruit that is winged.

Scabrid Rough to the touch.

Scandent Climbing.

Scape A leafless flowering stem, as in onion.

Scarious Dry, thin and membranous, not green.

Schizocarp A general name for 2 or more united carpels splitting at maturity into as many 1-seeded units.

Secund Said of an inflorescence in which the flowers are all on one side of the axis.

Sepal Individual part(s) of a calyx.

Serrate Leaf margin in which the teeth are directed upwards.

Sessile Without a stalk.

Setose Bristle-like.

Silicula Fruit of the *Brassicaceae* (Mustard Family) which is not much longer than broad.

Siliqua An elongated fruit of the *Brassicaceae* in which the two halves fall away from the frame.

Simple Undivided.

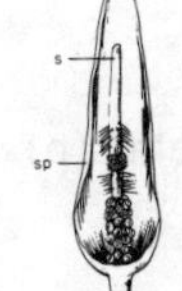

Spadix An inflorescence in which sessile flowers are arranged around a fleshy axis(s).

Spathe A large sheath-like bract (sp) enclosing a spadix.

Spathulate Spoon shaped.

Spicate An inflorescence which is spike-like.

Spur An extension(s) of some part of a flower which is hollow and has nectar. As in violet or larkspur.

Stamen Individual part of the androecium or male part of the flower.

Staminode A non-functional stamen.

Standard The large posterior petal of a pea or similar flower.

Stellate Star-like.

Stigma The part of the pistil that receives the pollen.

Stipule An appendage present at the base of a leaf(s).

Stolon A prostrate branch rooting and producing shoot(s) at regular intervals.

Striate Surface with longitudinal furrows.

Style The portion of the pistil which lies above the ovary and bears the stigma.

Subspecies A subunit of a species.

Subulate An organ which is narrow, stiff and tapering.

Suffruticose Said of low, woody branched plants.

Superior An ovary which lies above the sepals(s), petals(p) and stamens(st). See also under hypogynous.

Syncarpous With united carpels.

T

Tendril A slender process(t) serving as a holdfast or for climbing.

Tepals Used for sepals and petals of similar form and not readily differentiated, as in tulip and onion.

Ternate Three in a whorl or cluster.

Thalamus See receptacle.

Tomentose Dense hairs which are short, soft and tangled.

Torulose See moniliform.

Trifoliolate With 3 leaflets.

Truncate Usually said of a leaf base that ends abruptly.

Tuber A swollen underground stem as in potato.

U

Umbel An inflorescence in which the pedicels radiate from the top of a common peduncle and are of nearly the same length.

Unarmed Without spines.

Undulate A wavy margin. Repand.

Unifoliolate Compound leaf with 1 leaflet.

Urceolate Urn or pitcher-shaped.

V

Valvate The margins of the members of a whorl (like sepals or petals) meet but do not overlap.

Variety A subunit of a species below the rank of a subspecies.

Venose Having veins.

Versatile An anther attached at a point to the filament, so that it can swing to and fro, such as the anthers in grasses.

Verticillaster A false whorl formed by a pair of cymes which proceed from the axils of opposite leaves and appear to form a whorl of flowers around the stem, as in many members of the Mint family.

Villous Covered with long soft hairs.

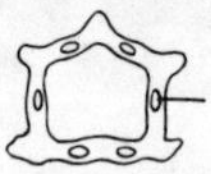

Vittae Oil bearing canals(v) in the fruit of the *Umbelliferae* (carrot family).

Vix Scarcely or barely.

Z

Zygomorphic A flower which has similar halves if cut in on plane only. Opposite of actinomorphic.

INDEX OF LOCAL NAMES

INDEX OF ENGLISH NAMES

INDEX OF SCIENTIFIC NAMES